Current Topics in Behavioral Neurosciences

Volume 39

About this Series

Current Topics in Behavioral Neurosciences provides critical and comprehensive discussions of the most significant areas of behavioral neuroscience research, written by leading international authorities. Each volume offers an informative and contemporary account of its subject, making it an unrivalled reference source. Titles in this series are available in both print and electronic formats.

With the development of new methodologies for brain imaging, genetic and genomic analyses, molecular engineering of mutant animals, novel routes for drug delivery, and sophisticated cross-species behavioral assessments, it is now possible to study behavior relevant to psychiatric and neurological diseases and disorders on the physiological level. The *Behavioral Neurosciences* series focuses on "translational medicine" and cutting-edge technologies. Preclinical and clinical trials for the development of new diagnostics and therapeutics as well as prevention efforts are covered whenever possible.

More information about this series at http://www.springer.com/series/7854

Joseph H. Porter • Adam J. Prus
Editors

The Behavioral Neuroscience of Drug Discrimination

 Springer

Editors
Joseph H. Porter
Department of Psychology
Virginia Commonwealth University
Richmond, VA
USA

Adam J. Prus
Department of Psychological Science
Northern Michigan University
Marquette, MI
USA

ISSN 1866-3370   ISSN 1866-3389   (electronic)
Current Topics in Behavioral Neurosciences
ISBN 978-3-030-07507-1   ISBN 978-3-319-98561-9   (eBook)
https://doi.org/10.1007/978-3-319-98561-9

This Springer imprint is published by the registered company Springer Nature Switzerland AG
The registered company address is: Gewerbestrasse 11, 6330 Cham, Switzerland

Preface

The year 2018 marks the 40th anniversary of the founding of the Society for the Stimulus Properties of Drugs (SSPD). The events that led to the founding of SSPD with its first official meeting on June 3, 1978, in Baltimore, MD, USA, have been described by the society's first three presidents Donald A. Overton, John A. Rosecrans, and Herbert Barry III in a special issue on drug discrimination published in *Pharmacology Biochemistry and Behavior* (Overton et al. 1999). Even prior to that first meeting of SSPD in 1978, books were beginning to appear about this new, exciting area of research that allowed behavioral pharmacologists to measure the "subjective effects" of drugs and, perhaps even more importantly, to demonstrate that the discriminative stimulus properties were related to specific pharmacological activity at the receptors of brain neurotransmitters in the central nervous system. There have been at least six different books written specially about drug discrimination and the discriminative stimulus properties of drugs (Colpaert and Balster 1988; Colpaert and Rosecrans 1978; Colpaert and Slangen 1982; Glennon and Young 2011; Ho et al. 1978; Thompson and Pickens 1971), and there also have been several special issues of journals that focused on drug discrimination studies published over the years [*Drug Development Research*, Vol. 16, 1989; *NIDA Research Monograph* 116 (DHHS pub. # ADM 92-1878), 1991; *Behavioural Pharmacology*, Vol. 2, 1991; *Pharmacology Biochemistry and Behavior*, Vol. 64 (2), 1999; *Psychopharmacology*, Vol., 2009]. Thus, there is a rich history of researchers in this field periodically coming together to present an update on the most current information about the discriminative stimulus properties of drugs. This book continues this tradition and is published as part of the *Current Topics in Behavioral Neurosciences (CTBN)* series published by Springer and is titled *The Behavioural Neuroscience of Drug Discrimination*. The goal of this volume is to provide up-to-date summaries on a number of diverse topics that encompass the current research literature for the stimulus properties of drugs.

As with any writing project like this, there are many people to thank. First and foremost, we would like to thank Bart Ellenbroek who first approached me (JHP) several years ago to see if I would be interested in editing a book on drug discrimination. I of course said yes and then didn't think anything else about it until Bart contacted me some time later and said the project had been approved. I immediately asked Adam Prus to join me as a co-editor as he was a logical choice for a co-editor, plus I knew that this project would require a tremendous amount of time—recruiting potential authors for the various chapters and, of course, the actual editing of each chapter as they were completed. As with any writing project of this scope, there were many delays along the way and the editors and staff at Springer (K. Adeitia, Alameluh Damodharan, Susanne Dathe, Wilma McHugh, Sujitha Shiney, and Nayak SulataKumari) have displayed an amazing level of patience dealing with a large group of authors (including the co-editors!) who failed to meet deadlines much too often. Finally though, we have a finished project with 14 chapters that cover a wide diversity of topics in the field of drug discrimination. We want to thank all the contributors to each chapter as this book would not have been possible without them. Also, there are several individuals (Scott Bowen, Herbert Covington, and Richard Young) who graciously gave of their time to help out with the review process for one or more chapters. Their input was extremely valuable and helped to improve the quality and clarity of the individual chapters and of course the entire book itself. At the end of the chapter "Drug Discrimination: Historical Origins, Important Concepts, and Principles," we discuss the individual chapters and authors, and in the chapter "A Prospective Evaluation of Drug Discrimination in Pharmacology," Ellen Walker helps to put all of these diverse chapter topics into perspective.

Finally, as we noted in the first paragraph, completion of this book marks the 40th anniversary of the founding Society for the Stimulus Properties of Drugs, but it is also a bit bittersweet, as it also marks the end of this society. As often happens with small research societies, they have a natural life span—the birth of the society, the growth and development of that research field into adulthood, the maturing of that field into old age, and of course the natural ending of its existence. While we are a bit sad about this, we realize that drug discrimination has become an extremely valuable behavioral assay in the field of behavioral pharmacology and that it is utilized in many research labs around the world. It still remains the best (and perhaps the only) approach for studying the subjective effects of drugs that possess psychostimulant properties. For that reason, we are confident that drug discrimination will remain an extremely valuable research assay, with no demise in sight!

Richmond, VA, USA   Joseph H. Porter
Marquette, MI, USA   Adam J. Prus

References

Colpaert FC, Balster RL (eds) (1988) Transduction mechanisms of drug stimuli. Springer, Berlin
Colpaert FC, Rosecrans JA (eds) (1978) Stimulus properties of drugs: ten years of progress. Elsevier/North-Holland Biomedical Press, Amsterdam

Colpaert FC, Slangen JL (eds) (1982) Drug discrimination: applications in CNS pharmacology. Elsevier Biomedical Press, Amsterdam

Glennon RA, Young R (eds) (2011) Drug discrimination: applications to medicinal chemistry and drug studies. Wiley, Hoboken

Ho BT, Richards DW III, Chute DL (eds) (1978) Drug discrimination and state dependent learning. Academic, New York

Overton DA, Rosecrans JA, Barry H III (1999) Creation and first 20 years of the society for the stimulus properties of drugs (SSPD). Pharmacol Biochem Behav 64(2):347–352

Thompson T, Pickens R (eds) (1971) Stimulus properties of drugs. Appleton Century Crofts, New York

Contents

Part I
Overview

Drug Discrimination: Historical Origins, Important Concepts, and Principles

Joseph H. Porter, Adam J. Prus, and Donald A. Overton

Abstract Research on the stimulus properties of drugs began with studies on state dependent learning during the first half of the twentieth century. From that research, an entirely new approach evolved called drug discrimination. Animals (including humans) could discriminate the presence or absence of a drug; once learned, the drug could serve as a discriminative stimulus, signaling the availability or nonavailability of reinforcement. Early drug discrimination research involved the use of a T-maze task, which evolved in the 1970s into a two-lever operant drug discrimination task that is still used today. A number of important concepts and principles of drug discrimination are discussed. (1) The discriminative stimulus properties of drugs are believed in large part to reflect the subjective effects of drugs. While it has been impossible to directly measure subjective effects in nonhuman animals, drug discrimination studies in human subjects have generally supported the belief that discriminative stimulus properties of drugs in nonhuman animals correlate highly with subjective effects of drugs in humans. In addition to the ability of the drug discrimination procedure to measure the subjective effects of drugs, it has a number of other strengths that help make it a valuable preclinical assay. (2) Drug discrimination can be used for classification of drugs based on shared discriminative stimulus properties. (3) The phenomena of tolerance and cross-tolerance can be studied with drug discrimination. (4) Discriminative stimulus properties of drugs typically have been found to be stereospecific, if a drug is comprised of enantiomers. (5) Discriminative stimulus properties of drugs reflect specific CNS activity at neurotransmitter receptors. (6) Both human and nonhuman subjects display individual differences in their sensitivity to discriminative stimuli and drugs. (7) The drug discrimination procedure has been used extensively as a preclinical assay in drug development. This

J. H. Porter (✉)
Department of Psychology, Virginia Commonwealth University, Richmond, VA, USA
e-mail: jporter@vcu.edu

A. J. Prus
Northern Michigan University, Marquette, MI, USA

D. A. Overton
Temple University, Philadelphia, PA, USA

© Springer International Publishing AG 2018
Curr Topics Behav Neurosci (2018) 39: 3–26
DOI 10.1007/7854_2018_40
Published Online: 31 March 2018

chapter is the first in the volume The Behavioural Neuroscience of Drug Discrimination, which includes chapters concerning the discriminative stimulus properties of various classes of psychoactive drugs as well as sections on the applications and approaches for using this procedure.

Keywords Cross-tolerance · Discriminative stimulus · Drug development · Drug discrimination · Individual differences · State dependent learning · Stereospecific · Stimulus properties · Subjective effects · Tolerance

Contents

1   Introduction

Drug discrimination is a paradigm in which an organism learns to discriminate the pharmacological effects of a drug from the absence of drug or from the noticeably different pharmacologically effects produced by other drugs. The procedure as established today primarily relies on operant responding procedures (however, see Riley et al. 2016, this volume) and has been used in a wide variety of species, most commonly including rats, mice, and pigeons, and also in nonhuman primates and humans. Operant drug discrimination procedures require extensive training in order for an organism to accurately learn to identify the effects produced by a drug (or a combination of drugs) and the dose of that drug. The drug is referred to as a *training drug*. An appeal of this procedure is that discriminative stimulus properties of a drug can consist of those identified as *subjective*, rather than *objective*, and that the drug is a stimulus (see Catania 1971). As Catania (1971) emphasizes, discriminative control by a drug represents a *behavioral* relationship between environmental events (a drug in this case) and responses. Also, it is not necessary to understand the underlying receptor mechanisms responsible for this stimulus control, in order to understand the relationship between the interoceptive event and the response. Regardless, drug discrimination has been used extensively to study recreational and abused substances in order to identify underlying pharmacological actions and mechanisms responsible for their subjective effects. Drug discrimination also has been utilized for studying therapeutic psychoactive drugs, such as antidepressants, anxiolytics, and antipsychotics. For a number of years, the drug discrimination paradigm has been used in both academia and industry to help elucidate the pharmacological basis of psychoactive substances. This volume, titled *"The Behavioural Neuroscience of Drug Discrimination"* as part of the *Current Topics in Behavioral Neurosciences* (CTBN) series provides reviews of the current literature for a number of either specific drugs or categories of drugs. This introductory chapter provides an overview of the historical origins of the drug discrimination procedure and discusses some important concepts and principles regarding the drug discrimination procedure.

The individual chapters in this book review the current state of the art regarding the discriminative stimulus effects of the primary classes of psychoactive drugs. The chapters highlight seminal and key findings in these areas sufficient to cover general scope of knowledge from these fields and focus on the utility of these procedures for CNS pharmacology research. Whenever possible, chapters connect the stimulus properties of drugs to mediating neuropharmacological actions (i.e., effects on specific receptor mechanisms). Moreover, the chapters in this book all document how the discriminative stimulus effects shown in animals translate to humans. Finally, by featuring leading experts in their respective areas, the chapters update and provide insight into future avenues of study with the drug discrimination paradigm.

2 Historical Origins

2.1 State Dependent Learning

As a number of excellent reviews have been written documenting the early history and concepts of the control of behavior by drugs as stimuli (e.g., Overton 1971, 1982, 1991; Schuster and Balster 1977), we will only briefly describe the historical antecedents to drug discrimination before focusing primarily on the transition from the T-maze drug discrimination developed by Overton (see Overton 1991), to the currently used operant drug discrimination procedure.

The first report of *dissociated* learning produced by drugs, later called *state dependent learning*, was by Combe (1835) who published a report of a delivery man who left a package at an incorrect address while drunk and then could not remember where he had left it until he was again intoxicated. The idea that memories might be linked to a drug state was later popularized in Wilkie Collins' classic novel "The Moonstone" (1868), which cited Combe's report as proof of the possibility. In both of these sources, the amnesic effect was apparently asymmetrical in that memories formed while drugged were unavailable without drug; however, memories formed while sober would generalize into the drug state. Later at the end of the nineteenth century, Théodule-Armand Ribot, a famous French psychologist, developed a theory for memory retrieval in which interoceptive stimulus cues played an important role (Ribot 1882, 1891). His model predicted symmetrical amnesias with retrieval impairments after either normal to abnormal or abnormal to normal changes in body physiology. Ribot presented no new data and his theory apparently was an integral part of the intense interest in dissociation that existed in Europe throughout the last half of the nineteenth century.

The next real data about drug state dependent learning was not published until 1937 when Girden and Culler (1937) reported impaired retrieval of conditioned leg flexion responses in dogs after drug (D) to no drug (N) transitions. The effects of N to D transitions were not tested. These findings made their way into contemporary textbooks (Morgan and Stellar 1950, p 449) but seem not to have been very well integrated into the neuroscience of the time and led to only one replication attempt Gardner and McCullough (1962). It would be hard to argue that the scattered reports just described were part of a program of research about drug effects on memory retrieval. Instead, it appears that they were put in the "scientific curiosities" category and received little attention. However, the beginnings of the drug discrimination procedure can be traced back to this early work on state dependent learning conducted during the 19th and the first half of the twentieth century (Overton 1991).

2.2 Drug Discrimination

A major advancement in understanding state dependent learning came from the theories of Neal Miller, a widely respected psychologist, who argued that drug effects should act as memory retrieval cues and that laboratory experiments using a 2 × 2 design could show these effects (Grossman and Miller 1961; Miller 1957; Miller and Barry 1960). Incidentally, the 2 × 2 design employs four groups of subjects that are trained and later tested for retrieval using the drug conditions N–N, N–D, D–N, and D–D. Studies by one of the present authors (Overton 1961, 1964, 1966) also played an important role during this transition period, as it obtained convincing results and was widely read and cited. It used escape training in a T-maze drug discrimination paradigm and showed that the frank dissociative amnesias produced by high dosages were replaced by gradually acquired discriminative control at lower dosages – hence linking state dependent learning and drug discrimination phenomena.

As a better understanding of the ability of drugs to serve as stimuli (in a manner analogous to sensory stimuli) was obtained, the state dependent learning procedure evolved to produce a drug discrimination procedure. This allowed researchers to demonstrate for the first time that animals could reliably distinguish a drug state from a nondrug state and that the effects of drugs could be established as discriminative stimuli. The first drug discrimination study actually had been conducted several years earlier by Conger (1951) who used an approach/avoidance task in which rats were trained to approach when "inebriated" and to avoid when "sober," or vice versa. Thus, Conger was able to demonstrate that ethanol exerted discriminative control over the behavior of the rats and, like other stimuli, drugs could set the occasion for responding – i.e., drugs could serve as *discriminative stimuli*. Overton (1961, 1964) further refined the drug discrimination procedure, introducing the two-choice T-maze (escape from shock), which was a symmetrical procedure in that the discriminative cue properties of the drug and nondrug conditions were demonstrated by a response selection rather than by response occurrence. Another early study reported that rats could discriminate the typical antipsychotic drug chlorpromazine from saline (Stewart 1962). Using a three-compartment test chamber (somewhat similar to a T-maze), rats were successfully trained to discriminate 4.0 mg/kg (i.p.) chlorpromazine from saline in a shock-avoidance task, and tests showed that several phenothiazines fully substituted for chlorpromazine, while the tricyclic antidepressant imipramine did not. A definitive review of this early research was published by Overton (1968). Over the next 20 years, a large number of drug discrimination studies were conducted with the T-maze procedure (see Overton 1982), but a major change in the drug discrimination procedure took place in the 1970s with the introduction of an operant task requiring rats to press response levers instead of running in a T-maze.

2.3 Two-Lever Operant Drug Discrimination

In 1968, Harris and Balster trained three rats on a two-lever multiple fixed ratio 50/differential-reinforcement-of-low rate 20 s (MULT FR 50 DRL 20 s) to discriminate DL-amphetamine from saline. After completion of discrimination testing, the rats were tested under extinction conditions in the presence of the drug cue and nondrug cue (saline). All three rats successfully acquired the amphetamine discrimination and responded primarily on the condition-appropriate lever. Harris and Balster concluded that the internal state (i.e., subjective effects of drug or no drug) of the animal controlled this responding, and that "*a more complete understanding of drug behavior interactions can be achieved by considering the stimulus properties of drugs* [emphasis added] *in addition to their traditionally emphasized pharmacological effects.*" This last statement really was both insightful and predictive as the use of the drug discrimination paradigm exploded over the next 30–40 years and became one of the most important assays for understanding the behavioral effects of drugs in vivo (see McMahon 2015). Following the Harris and Balster (1968) publication, there were a number of studies that employed this new two-lever operant drug discrimination approach, but it took several years before the procedures became more standardized. Also, as noted by Overton (1991), it soon became that two-lever operant drug discrimination procedures were more sensitive to drug stimulus effects at doses much lower than those needed in the T-maze studies and many other behavioral tests (see Kubena and Barry 1969a, b) – clearly, an advantage.

In 1971, Harris and Balster subsequently published a chapter exploring multiple two-lever drug discrimination procedures (they also tested single-lever multiple schedules which we will not address) (Harris and Balster 1971). While they only tested a few rats under each schedule condition, they generally obtained comparable results for each drug tested in four different multiple schedules (MULT CRF CRF, MULT DRL DRL, MULT FR FR, and MULT FR EXT; CRF = continuous reinforcement, DRL = differential reinforcement of low rate 20 s, FR = fixed ratio 50, and EXT = extinction). One obvious advantage of the FR operant schedule was that it engendered higher response rates, which could be advantageous when testing was conducted under extinction conditions.

Kubena and Barry (1969a, b) subsequently demonstrated that the two-lever drug discrimination procedure could be used not only to train rats to discriminate subjective drug effects but also that novel drugs could be tested to determine if they shared discriminative stimulus properties with the training drug. Kubena and Berry (1969b) trained rats to discriminate either alcohol (1,200 mg/kg) or atropine sulfate (10 mg/kg) from saline in a two-lever drug discrimination procedure using a variable interval (VI) 1 min food reinforcement schedule. In the alcohol-trained rats, they found that pentobarbital, chlordiazepoxide, and chloral hydrate shared discriminative stimulus properties with alcohol producing almost complete alcohol-appropriate lever responding at the higher tested doses (i.e., full substitution). In the atropine-trained rats, scopolamine produced full substitution for atropine producing 100% atropine-appropriate lever responding at a dose of 1.0 mg/kg. This study demonstrated several useful properties of the two-lever operant drug discrimination. First, similar to Overton (1961) this study demonstrated that *the discriminative stimulus properties of drugs are mediated primarily by their central nervous system effects* as atropine's discriminative cue did not generalize to atropine methyl bromide, which does not cross the blood–brain barrier (i.e., it only has peripheral nervous system effects). Second, *the ED_{50} values for the dose–effect curves in the operant drug discrimination procedure were much lower than those seen in studies that used the T-maze drug discrimination procedure*. This suggested that the operant two-lever procedure was more sensitive to the behavioral effects of drugs than the T-maze procedure. Third, having response requirements for both drug- and saline-appropriate responding that are equivalent and physically adjacent is an advantage, as this makes it easier to measure the non-discriminative stimulus properties of a drug (e.g., decreasing response rates because of sedative effects).

A series of studies by Colpaert in the 1970s played a major role in helping to demonstrate the value of the two-lever operant drug discrimination procedure and to standardize the two-lever operant drug discrimination procedure. For example, Colpaert and Niemegeers (1975) trained rats to discriminate the narcotic (opioid) fentanyl (0.04 mg/kg, s.c.) from vehicle in a two-lever drug discrimination procedure using a fixed ratio (FR) food reinforcement schedule in which a food pellet was delivered after every tenth response on the condition-appropriate lever (responses on the incorrect lever had no consequence). Then, they did substitution tests with four opioids (dextromoramide, phenoperidine, piritramide, and morphine) and found that the fentanyl cue fully generalized to each of these drugs. In contrast, the neuroleptic haloperidol did not generate fentanyl-appropriate responding. Thus, this study was able to demonstrate that the "narcotic" cue produced by fentanyl generalized to other opioids, but not to a drug from another behavioral classification (i.e., the neuroleptic haloperidol) – this showed that the discriminative stimulus properties of a drug appeared to be specific to drugs in a single pharmacological category. This study also was important as it helped to make the use of FR schedules the standard approach in two-lever operant studies. Colpaert et al. emphasized that response rates could reveal drug-induced stimulatory or inhibitory effects, while the animal's lever selection indicated difference in the drug-induced stimulus. In a subsequent study, Colpaert et al. (1975) were able to further confirm the specificity of the "narcotic" cue in 0.04 mg/kg fentanyl-trained rats and demonstrated dose-dependent generalization curves for five narcotic drugs: dextromoramide, fentanyl (the training drug), fentatienil (Sufenta® is a synthetic opioid analgesic that is more potent than its parent drug, fentanyl), morphine, and piritramide, which fully substituted for the fentanyl cue and significant reductions in responding at the highest dose tested for each drug. In contrast, the nonnarcotic drugs amphetamine, atropine, caffeine, cocaine, chlordiazepoxide, chlorpromazine, desipramine, dexetimide, haloperidol, imipramine, isopropamide, LSD, mescaline, nicotine, pentobarbital, and spiperone did not engender fentanyl-appropriate responding.

An important paper by Shannon and Holtzman in 1976 helped to lay the groundwork for establishing many of the standard approaches for studying drug effects in the two-lever operant procedure and also demonstrated the utility of the two-lever drug discrimination procedure for understanding the pharmacology underlying the discriminative stimulus properties of a drug. They trained male rats to discriminate 3.0 mg/kg morphine (s.c.) from saline in a shock-avoidance procedure (thus, the rats were not food deprived) and between 50% and 60% of the rats acquired the discrimination with greater than 90% accuracy in 6–8 weeks of training. In this study, they demonstrated a number of important features about the discriminative stimulus properties of morphine: (1) Morphine's discriminative cue was *time-dependent* with full generalization being observed within 30 min after injection and lever choice returning to the saline lever by 3.5 h after injection, (2) morphine's discriminative stimulus was *dose dependent* as they found that 0.1 mg/kg produced saline-appropriate responding while 3.0 (training dose) and 10 mg/kg produced greater than 90% morphine-lever responding (a 100-fold range), (3) morphine's discriminative cue was *pharmacologically specific* as the narcotic antagonist naloxone tested with morphine produced a rightward shift in morphine's generalization curve, (4) morphine's discriminative cue was *stereoselective* as inactive isomers of morphine and levorphanol (thebaine and dextrorphan, respectively) did not produce morphine-appropriate responding, and (5) finally, they demonstrated *cross-tolerance* to morphine's discriminative cue with methadone and the lack of cross-tolerance to pentobarbital.

Thus, by the mid-1970s the drug discrimination procedure was being used by increasing numbers of behavioral pharmacologists. As described by Overton, Rosecrans, and Barry, this increased interest in the drug discrimination paradigm led to the creation of the Society for the Stimulus Properties of Drugs (SSPD) in 1978 with the first official meeting being held that year in Baltimore, Maryland in conjunction with the CPDD (College on Problems of Drug Dependence) meeting (Overton et al.

1999). SSPD continued to hold yearly meetings until its last official meeting in 2012 in New Orleans, Louisiana. There were an increasing number of studies utilizing the drug discrimination procedure that grew exponentially through the late 1980s and peaked in 1998 with a little over 200 publications that year, followed by a subsequent decline (Bolin et al. 2016a, this volume; McMahon 2015; Stolerman et al. 1989).

There have been a few theories regarding explanation for the recent decline in the number of drug discrimination studies. One explanation is offered by McMahon (2015). Much of the drug discrimination literature has focused on drugs of abuse and many studies have used the drug discrimination assay to help determine abuse liability of new compounds. The drug self-administration paradigm appears to be replacing its use to a certain extent. As McMahon describes, self-administration certainly has greater face validity than drug discrimination with regard to drug-taking behavior in that the animals have to "work" in order to obtain the drug. Furthermore, there has been a downsizing of preclinical neuropharmacological research by many pharmaceutical companies in recent years. This has further contributed to a decline in this line of research. Despite this, drug discrimination remains a valuable tool for preclinical behavioral research.

3 Important Concepts and Principles of the Drug Discrimination Paradigm

We will not try to provide a detailed methodology for how to conduct drug discrimination studies, as there are several excellent articles/book chapters, which have been written and provided comprehensive details of training and testing methods for both human (Bolin et al. 2016b) and nonhuman (Solinas et al. 2006; Young 2009; Glennon and Young 2011a) drug discrimination studies. Rather, this section focuses on a number of important concepts and principles inherent to the drug discrimination paradigm that make it such a valuable preclinical assay for studying in vivo behavioral effects of drugs and relating those effects to specific pharmacological mechanisms.

3.1 A Method to Measure Subjective Effects

One of the most important questions to ask about drug discrimination is *what does it measure?* One commonly held assumption has been that the "*discriminative stimulus effects of drugs may be based entirely or in part upon their **subjective effects***
[emphasis added]" (Balster 1988). Balster further argues that understanding the underlying neural (pharmacological) mechanisms of these discriminative stimulus effects should aid in the understanding of the neural mechanisms of subjective

experiences and mood states in humans. While there are procedures for assessing and quantifying verbal reports of drugs' subjective effects in humans, verbal reports obviously cannot be obtained from nonhuman animals. This is where the drug discrimination procedure has proven to be so valuable as it allows us *"to ask"* animals "how they perceive (*feel*)" the subjective effects of drug administration. Drug discrimination is the only procedure known to the current authors to allow this unique insight into the subjective effects of drugs in animals. For example, Schuster and Johanson (1988) provided a nice review of the relationship between discriminative stimulus properties and subjective effects of drugs in both human and nonhuman studies. In experienced drug users, the subjective effects of psychotropic drugs have been assessed to help evaluate their abuse potential by comparing these effects to the subjective effects produced by known drugs of abuse. Human drug discrimination studies (e.g., Chait et al. 1985, 1986; see Bolin et al. 2016a, this volume) have helped to demonstrate that the discriminative stimulus effects of drugs correlate highly with subjective effects as assessed by verbal reports. Although Preston and Bigelow (1991) caution that *"there is a relationship* [between subjective and discriminative drug effects in human subjects], *though not a simple one, and that the nature of the relationship is likely to be influenced by the procedural details of specific drug discrimination training and testing paradigms."* Schuster and Johanson (1988) conclude that it is very reasonable to assume that these two processes are similar in nonhuman animals. This is what makes the drug discrimination procedure so valuable for studying the subjective effects of both known and novel drugs.

Colpaert argues that since the morphine discriminative cue is due to its central narcotic action (i.e., CNS effects as opposed to peripheral effects), drug discrimination provides *"an original means by which to investigate subjectively experienced drug effects"* (Colpaert and Niemegeers 1975; see also Colpaert et al. 1975). Colpaert extended this idea to state *"that the discriminative stimulus properties of drugs, as assessed by this animal method* [i.e. drug discrimination], *may be relevant to subjectively experienced drug effects in humans"* (Colpaert et al. 1976). Other researchers in this newly emerging field of discrimination shared this viewpoint. Hirschhorn and Rosecrans (1976) stated that *"The observation that certain drugs can serve as discriminative stimuli for laboratory animals ... demonstrates that animals can distinguish the effects of these drugs from the non-drug condition and suggests a possible method by which* **subjective drug effects** [emphasis added] *can be studied in animals."* Shannon and Holtzman (1976) argued that the results of their two-lever morphine discrimination study with rats *"suggest that the component of action of morphine that enables it to function as a discriminative stimulus in the rat is analogous to the component of action of morphine responsible for producing subjective effects in man."* Thus, although definitive proofs may still be elusive, there has been widespread agreement that the drug discrimination procedure provides a unique opportunity to measure the *subjective effects* of drugs by studying their discriminative stimulus properties.

3.2   Classification of Drugs

In addition to the ability of the drug discrimination procedure to measure the subjective effects of drugs, it has a number of other strengths that help make it such a valuable preclinical assay. One of these is that drug discrimination can be used to create a *classification of drugs based on shared discriminative stimulus properties*. Herbert Barry, III was one of the first to stress the utility of the drug discrimination procedure for classification of drugs according to their discriminable effects (Barry 1974). Reviewing findings from the drug discrimination literature that included both T-maze and early operant procedures, Barry summarized that the study of the discriminative stimulus properties of a large number of drugs has identified several categories including: (1) central sedatives (e.g., barbiturates and minor tranquilizers like chlordiazepoxide), (2) central anticholinergics (specifically antimuscarinic drugs), (3) nicotine, (4) marihuana (Δ^9-THC), and (5) hallucinogens (e.g., mescaline and LSD). Not surprisingly, much of the focus in the drug discrimination field has been on drugs of abuse as it was hoped that the drug discrimination paradigm would provide a unique insight into subjective effects of drugs that could relate to the abuse potential of drugs in humans. While this has been realized to a great extent, Barry stressed in his early paper the need to develop uniform procedures in the drug discrimination field. As described above (Sect. 2), the introduction of the two-lever operant drug discrimination procedure (primarily with FR schedules of reinforcement) answered this need for the most part, although as Barry (1974) pointed out there is a *"special need for the development of techniques for more rapid training of drug discrimination in rats and other laboratory animals."* This objective still has not been realized with operant drug discrimination procedures; although, a more rapid approach has been developed utilizing the *"conditioned taste aversion discrimination procedure"* (see Riley et al. 2016, this volume).

Classification of drugs with the drug discrimination procedure has been a major use of this procedure over the years. In the 1970s, as discussed above in Sect. 2, Colpaert and Niemegeers (1975) and Colpaert et al. (1975) utilized drug discrimination to identify the specificity of the stimulus properties of narcotic drugs (fentanyl was the training drug); however, the drug discrimination procedure can also be used to classify drugs for other behavioral classifications. For example, Porter et al. (2000) trained rats to discriminate a low dose of the atypical antipsychotic clozapine (1.25 mg/kg, i.p.). As shown in Table 1, all but two of the atypical antipsychotic drugs tested fully substituted for clozapine (i.e., they generated >80% clozapine-appropriate responding) and one of those produced partial substitution (>60% and <80% clozapine-appropriate responding). In contrast, none of the four typical antipsychotics fully substituted for clozapine, although thioridazine did produce partial substitution. These studies demonstrate the usefulness of the drug discrimination procedure for assigning drugs to different categories.

Table 1 Results of generalization testing in rats trained to discriminate a low dose (1.25 mg/kg) of the atypical antipsychotic clozapine form vehicle (adapted from results in Porter et al. 2000)

Test drug	Maximum percentage of clozapine-lever responding	Level of substitution[a]
Atypical antipsychotics		
Clozapine (training drug)	96.7% at 5.0 mg/kg	Full
Olanzapine	90.3% at 1.0 mg/kg	Full
Sertindole	99.8% at 5.0 mg/kg	Full
Risperidone	87.1% at 0.5 mg/kg	Full
Quetiapine	66.4% at 10.0 mg/kg	Partial
Remoxipride[b]	23.1% at 4.0 mg/kg	No
Typical antipsychotics		
Chlorpromazine	27.9% at 1.0 mg/kg	No
Fluphenazine	29.5% at 0.25 mg/kg	No
Thioridazine	74.3% at 5.0 mg/kg	Partial

[a]Level of substitution: Full = >80% drug lever responding (DRL); Partial = >60–<80% DLR; No = <60% DLR

[b]Although remoxipride is typically classified as an atypical antipsychotic, it is sometimes considered to be a typical antipsychotic (see Nadal 2001); lack of full or partial substitution for clozapine supports this conclusion

3.3 Tolerance and Cross-Tolerance

The phenomenon of *tolerance* to effects of drugs after repeated (chronic) administration has been common knowledge for a long time. As defined on the National Institute on Drug Abuse (NIDA) website (https://www.drugabuse.gov/publications/teaching-packets/neurobiology-drug-addiction/section-iii-action-heroin-morphine/6-definition-tolerance), "*When drugs such as heroin are used repeatedly over time, tolerance may develop. Tolerance occurs when the person no longer responds to the drug in the way that person initially responded. Stated another way, it takes a higher dose of the drug to achieve the same level of response achieved initially. For example, in the case of heroin or morphine, tolerance develops rapidly to the analgesic effects of the drug.*" They also point out that tolerance is not the same thing as addiction, although addiction may occur to drugs that produce tolerance.

The drug discrimination procedure requires repeated administration of drugs over long periods of time (usually months of training and testing), yet the discriminative stimulus properties of drugs typically remain very stable and no evidence of tolerance or sensitivity is usually seen. For example, Colpaert et al. (1976) trained rats to discriminate 0.04 mg/kg fentanyl (i.p.) from saline and then over a period of 17 weeks fentanyl or morphine generalization curves were obtained. During each week, the rats received either two or three doses of fentanyl and/or saline (five injections each week) as part of the training regimen. The ED_{50} values for these generalization curves did not change over the 4-month period. However, the same rats used in the drug discrimination experiments did develop a marked tolerance to the analgesic effects of fentanyl and morphine. Based on these results, the authors concluded that tolerance did not develop to the discriminative stimulus properties of narcotic analgesics.

However, under the right testing conditions, it is possible to demonstrate tolerance to drugs in the drug discrimination paradigm. Young (1991) has provided an excellent review of the conditions required to demonstrate tolerance in the drug discrimination procedure. After establishing morphine (3.2 mg/kg) as a discriminative stimulus in rats, training and testing are suspended and then the subjects were treated daily with various doses of morphine for approximately 2 weeks (varied across the studies she reviewed). Tolerance to morphine was dose dependent as low doses (3.2 or 10 mg/kg) produced little or no tolerance (i.e., the generalization curve did not change from baseline); however, the generalization curve displayed increasingly greater rightward shifts (increased tolerance) as the morphine dose was increased (maximum of 17.8 mg/kg, 2×/day). This tolerance disappeared after 3–5 days of suspending morphine treatments. Other studies she reviewed found that tolerance increased as a function of the length of morphine treatment (up to 2 weeks) and that *cross-tolerance to methadone also was evident*. (Cross-tolerance occurs when tolerance to a drug's effects produces tolerance to another drug's effects. These drugs typically belong to the same classification group and often affect the same receptor mechanisms.)

While most of the studies examining the phenomena of tolerance and cross-tolerance to the discriminative stimulus properties of drugs have focused on drugs of abuse, a series of studies also have shown tolerance and cross-tolerance with antipsychotic drugs in several drug discrimination studies. Goudie et al. (2007a) first established clozapine (5.0 mg/kg, i.p.) as a training drug in rats and determined a dose–effect curve (DEC1). Then, training and testing were suspended for 10 days and then a second DEC2 was determined. Finally, a third DEC3 was determined after a 10-day "wash-out" period during which no drug was administered and testing and training were suspended. Results revealed a significant rightward shift after the 10 days of repeated clozapine dosing (5.0 mg/kg, 2×/day) – i.e., *tolerance* to clozapine's discriminative cue was obtained. Following the 10 days of no drug treatment, the tolerance to clozapine's cue was lost and DEC3 was similar to DEC1. Using the same procedures, *cross-tolerance* was obtained with cyproheptadine (an anti-allergy/appetite stimulant), which has a binding profile very similar to clozapine. A second study by Goudie et al. (2007b) reported similar findings (cross-tolerance) with the atypical antipsychotic olanzapine and the compound JL13 (a clozapine congener). Goudie et al. concluded that the tolerance between these compounds provides a further demonstration of shared mechanisms of action. Wiebelhaus et al. (2011) used a similar procedure and demonstrated that repeated dosing with N-desmethylclozapine (major active metabolite of clozapine) and N-desmethylolanzapine (major active metabolite of olanzapine) produced cross-tolerance to clozapine (2.5 mg/kg, s.c.) in C57BL/6 mice. Cross-tolerance between these two metabolites and the atypical antipsychotic clozapine was interpreted as evidence that the discriminative stimulus properties of all three compounds shared common underlying pharmacological mechanisms.

It should be noted that Colpaert (1995) has argued that studies reporting tolerance to the discriminative stimulus of opiate drugs have in fact *not* demonstrated tolerance, although we feel that the articles discussed above (and others not cited in this review) have demonstrated tolerance. It should be noted, however, that it does appear to require specific testing conditions in order to demonstrate tolerance to the discriminative stimulus properties of drugs. While it is beyond the scope of this chapter to explore these issues thoroughly, the interested reader is encouraged to read Colpaert's (1995) review article.

3.4 Stereospecificity of Discriminative Stimulus Effects

Another important aspect of drug effects that is often not addressed in drug discrimination studies is the *stereospecificity* of the training drug. Glennon and Young (2011b) devoted an entire chapter to this topic and provided many examples of this. As they discuss, many drugs are composed of enantiomers (isomers) in a 50–50 composition, and, unless otherwise stated in a study, it should be assumed that the racemic (+) form of the drug is being used. As Glennon and Young state, "*Structural isomers are chemical entities with identical empirical formulas that differ in the nature or sequence of their atoms.*" Importantly, these isomers can differ in terms of their pharmacological effects or to the extent that they are responsible for the discriminative stimulus properties of the racemic drug (see Glennon and Young 2011b for full discussion on this topic). An example of this is shown in Fig. 2. Donahue et al. (2014) trained mice to discriminate the (*S*)-isomer (10 mg/kg, s.c.) of the atypical antipsychotic drug amisulpride from vehicle. In substitution tests with *rac*-amisulpride and the (*R*)-isomer, they found that rac-amisulpride was about 3 times less potent than (*S*)-amisulpride and that (*R*)-amisulpride was about 10 times less potent than (*S*)-amisulpride in producing (S)-amisulpride-like responding. Figure 1 shows significant rightward shifts in the dose–response curves for *rac*-amisulpride and (*R*)-amisulpride relative to (*S*)-amisulpride. This demonstrated that the discriminative stimulus effects of amisulpride are *stereoselective* and that the (*S*)-isomer contributes more to the stimulus properties of *rac*-amisulpride than does the (*R*)-isomer (see Donahue et al. 2017 for additional confirmation of this finding). Interestingly, the potency relationships between (*S*)-, (*R*)-, and *rac*-amisulpride suggested that the stimulus effects of amisulpride could be mediated, at least in part, by activity at dopamine receptors as these potency relationships were somewhat similar to those reported in binding studies. These studies found that (*S*)-amisulpride is approximately 2 times and 20–50 times more potent than *rac*-amisulpride and (*R*)-amisulpride, respectively, with regard to binding affinity to dopamine $D_{2/3}$ receptors (Castelli et al. 2001; Marchese et al. 2002a, b). Thus, stereochemistry of a drug can be an important aspect of understanding the discriminative stimulus properties of a drug. The isomers of a drug may both contribute to the discriminative stimulus properties of the drug with one isomer being more potent than the other (i.e., stereoselectivity), or one of the isomers may have similar properties to the parent drug and contribute to its discriminative stimulus, while

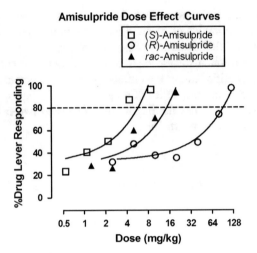

Fig. 1 This figure presents a direct comparison of the dose–response curves of %drug lever responding (DLR) for rac-amisulpride and its (S)- and (R)-enantiomers with regression lines. Doses for (S)-amisulpride were adjusted to the base form for direct comparison (ED50¼ 1.57 mg/kg [95% C.I.¼1.14–2.15 mg/kg]). From Donahue et al. (2014) – reproduced with permission

the other does not (i.e., stereospecificity) (for more complete discussion of this, see Chapter 4 in Glennon and Young 2011b).

3.5 Receptor Mechanisms and Discriminative Stimulus Effects

In 1988, an entire book was devoted to *transduction mechanisms of drug stimuli* (Colpaert and Balster 1988). A major theme of this book was that discriminative stimulus properties of drugs reflect specific CNS effects at *neurotransmitter receptors*. As Balster (1988) states *"If discriminative effects are related to subjective effects, then it seems reasonable to hope that studies of the neural mechanisms for these effects may lead us toward an understanding of the neural mechanisms of some of the subjective experiences and mood states that are the basis of our perception of drug effects."* Balster concludes with *"Studies of discriminative stimulus properties of drugs and their mechanisms of transduction can provide us important insights into basic brain-behavior relationships."* An early example of this was a study by Rosecrans and Glennon (1979) in which they used drug cues to study psychoactive mechanisms by comparison to other drugs (i.e., substitution tests) and by determining if the drug cue could be antagonized (i.e., blocked) with specific receptor antagonists. For example, in morphine-trained rats they demonstrated that methadone was equally potent in producing a similar dose-dependent generalization curve;

whereas, meperidine was significantly less potent as indicated by a rightward shift in the generalization curve. In antagonism studies, they found that the discriminative stimulus of LSD could be antagonized in a dose-dependent manner (i.e., greater antagonism with increasing doses) by the serotonin antagonist (BC105). Perhaps more interestingly, they presented findings comparing serotonin binding affinities (data presented as pA2 values from rat fundus assays) in a series of tryptamine analogs to the ability of these compounds to substitute for 5-OMeDMT (1.5 mg/kg training dose). As shown in Fig. 2 from that study, there was a strong correlation between the pA2 values and the equivalent dose at which each compound substituted for 5-OMeDMT, which is a hallucinogen with strong affinity for serotonin 5-HT2 and 5-HT1A receptors. These data clearly demonstrated that the discriminative stimulus properties of 5-OMeDMT were related to its binding affinity to serotonin receptors. In a study in which two separate groups of Sprague-Dawley rats were trained to discriminate either the atypical antipsychotic clozapine (5.0 mg/kg, i.p.) or the muscarinic antagonist scopolamine (0.5 mg/kg, i.p.) from vehicle, it was found that complete *cross-generalization* occurred between clozapine and scopolamine, indicating a shared underlying mechanism for their respective discriminative stimuli. In addition, only drugs that display high binding affinities for muscarinic cholinergic

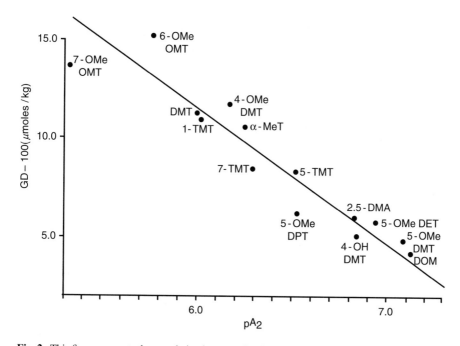

Fig. 2 This figure presents the correlation between the discriminative stimulus (DS) properties and pA2 values of a series of tryptamine and phenylisopropylamine analogs. The GD-100 value represents the equivalent dose at which each compound generalized with the training dose of 5-OMeDMT (1.5 mg/kg) when it was used as the DS. Each drug was administered at various doses 15 min prior to being placed in the operant chamber for a 15-min test session. From Rosecrans and Glennon (1979) – reproduced with permission

receptors substituted for these two training drugs. Based on these results, the authors concluded that antagonism of muscarinic receptors (especially M_1) plays an important role in the discriminative stimulus properties of clozapine in rats (Kelley and Porter 1997). In contrast, in C57BL/6 mice, clozapine's discriminative cue appears to be mediated by antagonism of serotonergic $5\text{-}HT_2$ receptors and α_1 adrenoreceptors (Philibin et al. 2005, 2009). Thus, it is often possible to ascertain the underlying receptor mechanisms that mediate the discriminative stimulus properties of a drug. However, as these studies show, these mechanisms may differ across species. Therefore, some caution must be exercised when making inferences across different species, including humans.

3.6   Individual Differences Between Subjects

An important, but understudied topic in drug discrimination research concerns *individual differences* between subjects in their sensitivity to the discriminative stimulus properties of drugs. These differences are often reflected in the number of training sessions required for individual subjects to meet the discrimination criterion – i.e., some subjects will acquire the discrimination in fewer training sessions than other subjects. The first operant study to systematically explore the importance of the speed of acquisition of the discrimination was by Martin Schechter in 1983. Twelve male Sprague-Dawley rats were trained to discriminate 0.16 mg/kg apomorphine from saline with responding reinforced according to an FR 10 reinforcement schedule. Half of the rats acquired the discrimination in a mean of 22.5 sessions (*early learners*) and the other half in a mean of 44.2 sessions (*late learners*) (significantly different, $p < 0.001$). When apomorphine generalization curves were established, the early learners had an $ED_{50} = 0.01$ mg/kg; whereas, the late learners had an $ED_{50} = 0.07$ mg/kg, which represented a 3.9-fold rightward shift in the generalization curve (the dose–response curves were parallel). Thus, the early learning group was more sensitive to apomorphine's discriminative stimulus than was the late learning group. While it has been well established that higher training doses result in higher ED_{50} values for the dose–response curves (see review by Stolerman et al. 2011), Schechter's (1983) study was the first to demonstrate that sensitivity to the training drug's discriminative cue also could affect the ED_{50}.

A second study by O'Neal et al. (1988) examined how the rate of acquisition of Δ^9-THC (delta-9-tetrahydrocannabinol) discrimination reflected sensitivity to Δ^9-THC's discriminative stimulus. Male Sprague-Dawley rats were trained to discriminate 3.0 mg/kg Δ^9-THC from saline in a two-lever operant discrimination task (FR 10) and after acquisition of the Δ^9-THC discriminative cue, the rats were divided into two groups using a median split – slow learners and fast learners. For the slow learners, the mean number of sessions to criterion (STC) = 50.0; for the fast learners, the STC = 27.3 (significantly different, $p < 0.001$). Similar to results found in the Schechter's (1983) study, the slow learners displayed a rightward shift in the Δ^9-THC generalization curve with an $ED_{50} = 1.63$ mg/kg; whereas, the ED_{50} for the fast learners = 0.77 mg/kg. Thus, the fast learners displayed a greater sensitivity to Δ^9-THC, replicating the greater sensitivity to apomorphine shown by the fast learners in the Schechter's (1983) study. We have not been able to find additional studies that have examined the relationship between speed of acquisition and the subsequent sensitivity of individual subjects to a drug's discriminative cue. Nonetheless, both of these studies suggest that reporting the number of sessions required to reach the training criteria should be information routinely provided in publications.

Bevins et al. (1997) reported that individual differences in rats in the sensitivity to amphetamine in several behavioral assays (novelty-induced activity, novelty-induced place preference, novel-object interaction, and amphetamine-induced activity) were related to differences in amphetamine discrimination. For example, rats more sensitive to the activating effects of amphetamine were also more sensitive to amphetamine in the drug discrimination assay. Individual differences in *human subjects* also have been shown in nicotine drug discrimination studies. Perkins (2011) summarizes some of the factors that contribute to individual differences in human nicotine drug discrimination studies. For example, these studies find that women, generally, are less sensitive to nicotine's discriminative stimulus properties as reflected in more difficulty in acquiring the cue or showed flattened generalization curves. Individual differences in animal studies with nicotine drug discrimination also have been shown that may be related to genetic differences (e.g., Quarta et al. 2009). Finally, Morgan and Picker (1996) reported three- to tenfold differences in the lowest doses of several opiates that would substitute for morphine in rats trained to discriminate morphine (3.0 mg/kg) from vehicle in a two-lever drug discrimination study. Individual differences were also observed in the antinociceptive effects of these opiates in a hot water tail-withdrawal procedure. These authors concluded that these individual differences between subjects are probably determined in large part by the relative efficacy of these drugs at the *mu* opioid receptor.

Finally, it is also possible that differential sensitivity among subjects to the discriminative stimulus properties of drugs may reflect the fact that different subjects may "tune" into different components of a cue. It has been well established that "compound" discriminative stimuli can be demonstrated with drug mixtures as the cue (see review by Stolerman et al. 1999). However, it is also possible for a single drug to have a *compound* discriminative cue. For example, in rats trained to discriminate ethanol from water, asymmetrical generalization of ethanol to gamma-Aminobutyric acid (GABA) enhancers (e.g., chlordiazepoxide), to N-methyl-D-aspartate (NMDA) antagonists (e.g., dizocilpine [MK-801]), and to serotonin (5-HT) agonists (e.g., trifluoromethylphenylpiperazine) was found. Stolerman et al. (1999) concluded from these studies that their finding supported the concepts of ethanol having a compound stimulus (see also Grant 1999), since ethanol generalized to drugs of more than one pharmacological classification. It certainly seems reasonable to assume that subjects might attend to one or more components of a drug's pharmacological actions, which make up its compound cue and might explain individual differences in acquisition to a drug's discriminative cue.

3.7 Drug Development

As a preclinical behavioral assay, drug discrimination has proven to be a useful tool. For example, clozapine is an atypical antipsychotic drug that is considered to be the "gold standard," prototypical of the second generation of antipsychotic drugs and it remains the standard by which other atypical antipsychotic drugs are compared (Hippius 1991; Porter and Prus 2009). When the antipsychotic olanzapine was being developed by Eli Lilly and Company, Moore et al. (1992) published an article on the behavioral pharmacology of olanzapine. One of the behavioral assays employed in that study was two-lever drug discrimination in which clozapine 5.0 mg/kg, i.p. was trained as a discriminative stimulus. Olanzapine fully substituted for clozapine's cue, indicating that olanzapine's discriminative stimulus properties were similar to those of clozapine. Based on these results, and results from a number of other behavioral assays used in this study, the authors concluded that olanzapine would have the profile of an atypical antipsychotic drug (like clozapine). Olanzapine was later approved by the FDA in 1996 for treatment of schizophrenia.

Drug discrimination has been frequently used by pharmaceutical companies (e.g., Millan et al. 1999) and in academia (e.g., Burgdorf et al. 2013) to help characterize the behavioral pharmacology of novel compounds and by government agencies like the Drug Enforcement Agency (DEA) to aid in scheduling the abuse liability of drugs (see Ator and Griffiths 2003; Balster and Bigelow 2003). In addition to the atypical antipsychotic drug olanzapine (see above), another good example is the atypical antipsychotic risperidone. Colpaert (2003) has written an excellent review of the discovery process for risperidone and how important the study of subjective effects in laboratory animals was to this process. He concluded that *"the pathway to risperidone chiefly cut across the field of in vivo pharmacology, and in particular behavioral pharmacology, underscoring the unique contribution of the field to drug discovery."* In 2002, the In Vivo Pharmacology Training Group published a commentary on *"The rise and fall of in vivo pharmacology."* In this article, they stated *"Pharmacology is, by definition, the study of the mechanism of action of drugs, and requires a knowledge and understanding of responses to drugs induced both in vitro and in vivo. Such analysis of drug action is needed to transform molecular or cellular discoveries into clinical practice and, equally, to identify the molecular questions that arise from clinical observations. These studies are essential because responses observed in vitro can be magnified, diminished or totally different in the more complex integrated system. . . . . . .This article . . . . . . outlines why in vivo work is vital for the analysis of drug action and for the discovery and development of new therapeutic agents."* (In Vivo Pharmacology Training Group 2002). We concur with these conclusions and recognize the utility and value of preclinical behavioral assays in the drug development process. Behavioral (in vivo) assays (like drug discrimination) are just as important as in vitro assays for this process and the two approaches go hand-in-hand in the discovery and development of new therapeutic drugs.

4 Summary and Overview of This Volume

The current chapter (Chapter 1 in Part 1, this volume) provided a brief overview of the historical origins of the drug discrimination procedure and described how its beginnings can be traced to state dependent learning, and then how it transitioned from the first drug discrimination studies in a T-maze task in the 1960s to a two-lever operant procedure in the 1970s. Then, we discussed how the discriminative stimulus properties of drugs are believed in large part to reflect the subjective effects of drugs and that drug discrimination studies in human subjects have generally supported the belief that discriminative stimulus properties of drugs in nonhuman animals correlate highly with subjective effects of drugs in humans. Finally, we discussed a number of other concepts and principles that help make drug discrimination a valuable preclinical assay.

The chapters in Part 2 of this volume review the current state of the art regarding the discriminative stimulus effects of the primary classes of psychoactive drugs. In Chapter 2, William Fantegrossi provides an overview of early drug discrimination work on psychostimulant drugs but also includes coverage of recent findings on the discriminative stimulus effects of bath salts. Chapter 3 provides a thorough summary about the discriminative stimulus effects of nicotine and recognizes the influential work of John Rosecrans, who is posthumous co-author of this chapter with Richard Young. The discriminative stimulus effects of ethanol are addressed extensively by Kathleen Grant's group in Chapter 4, and the chapter pays particular attention to the stimulus effects of ethanol across different species, including humans. Of particular relevance for interpreting the subjective effects of ethanol from these studies, this chapter points out the qualitatively different stimulus properties of low versus higher training doses of ethanol. In Chapter 5, Keith Shelton takes us through studies designed to evaluate the subjective effects of inhalants and devotes some emphasis to the unique methodological challenges involved in this work. In Chapter 6, Tsutomu Suzuki, with Tomohisa Mori, reviews an extensive literature on the discriminative stimulus effects of hallucinogens and dissociative anesthetic drugs, e.g., ketamine, and gives a glimpse of future directions in drug discrimination research as he associates intracellular signaling processes to the mediation of certain stimulus effects. In Chapter 7, two of the leading experts on the behavioral pharmacology of cannabinoids, Jenny Wiley and Aron Lichtman, contribute to a review on the discriminative stimulus effects of cannabinoids, which includes stimulus effects of endocannabinoids as well as synthetic cannabinoid compounds. Eduardo Butelman and Mary Jeanne Kreek, in Chapter 8, gave an up-to-date account on drug discrimination for opioid compounds and provided novel thoughts on future directions in this area. Chapter 9 is the first of two chapters that focused on the discriminative stimulus effects of drugs for mental illness. Chapter 9, written by the co-editors of this volume, along with the assistance of Kevin Webster, reviews studies evaluating the stimulus effects of antipsychotic drugs, with an emphasis on the utility of this procedure for identifying effective antipsychotic drugs for schizophrenia. The chapter also connects reported subjective effects of antipsychotic drugs in human patients to certain receptors known to mediate stimulus effects of antipsychotics in animals. Chapter 10, also co-authored by the editors of this volume, uses the same approach to evaluate the discriminative stimulus effects of antidepressants and anxiolytics.

Part 3 of this volume, called "Approaches to Drug Discrimination," provides a variety of perspectives on ways to understand the drug discrimination procedures along with some of its applications. In Chapter 11, Steve Negus and Matthew Banks discuss analyzing drug discrimination data using pharmacokinetic–pharmacodynamic analyses. In Chapter 12, Craig Rush reviews drug discrimination studies conducted in humans and some of the methodological advantages and challenges. In Chapter 13, Anthony Riley and others from his group demonstrate how the stimulus properties of drugs can be studied using conditioned taste aversion procedures. In this volume's final chapter (Chapter 14), Ellen Walker provides commentary on the chapters in this volume and discusses new directions for the use of drug discrimination in pharmacology research. Overall, this volume on the drug discrimination provides an insightful evaluation of a wide array of critical topics in this field written by leading experts on this procedure. The editors of this volume are grateful to all of the authors who have made this a notable addition to the literature in behavioral neuroscience.

Finally, we would like to dedicate this volume to the memory of two pioneer researchers in the field of drug discrimination. Dr. John A. Rosecrans and Dr. Torbjörn U.C. Järbe were two of the early scientists in drug discrimination research who did so much to help shape this newly emerging area of research back in the 1970s and whose influence continued into this century. Their legacy and influence in this field lives on and will be remembered always. We will miss both of them.

John A. Rosecrans   Torbjörn U.C. Järbe
(1935–2015)   (1946–2017)

References

Ator NA, Griffiths RR (2003) Principles of drug abuse liability assessment in laboratory animals. Drug Alcohol Depend 70:S55–S72

Balster RL (1988) Drugs as chemical stimuli. In: Colpaert FA, Balster RL (eds) Transduction mechanisms of drug stimuli, Psychopharmacology series, vol 4. Springer, Berlin, pp 3–11

Balster RL, Bigelow GE (2003) Guidelines and methodological reviews concerning drug abuse liability assessment. Drug Alcohol Depend 70:S13–S40

Barry H III (1974) Classification of drugs according to their discriminable effects in rats. Fed Proc 33(7):1814–1824

Bevins RA, Klebaur JE, Bardo MT (1997) Individual differences in response to novelty, amphetamine-induced activity and drug discrimination in rats. Behav Pharmacol 8(2–3): 113–123

Bolin BL, Alcorn JL III, Reynolds AR, Lile JA, Rush CR (2016a) (Chapter 12, this volume) human drug discrimination: elucidating the neuropharmacology of commonly abused illicit drugs. In: Porter JH, Prus AJ (eds) The behavioural neuroscience of drug discrimination. Springer, New York, NY

Bolin BL, Alcorn JL III, Reynolds AR, Lile JA, Rush CR (2016b) Human drug discrimination: a primer and methodological review. Exp Clin Psychopharmacol 24(4):214–228

Burgdorf J, Zhang XL, Nicholson KL, Balster RL, Leander JD, Stanton PK, Gross AL, Kroes RA, Moskal JR (2013) GLYX-13, a NMDA receptor glycine-site functional partial agonist, induces antidepressant-like effects without ketamine-like side effects. Neuropsychopharmacology 38(5):729–742

Castelli MP, Mocci I, Sanna AM, Gessa GL, Pani L (2001) (S)-amisulpride binds with high affinity to cloned dopamine D3 and D2 receptors. Eur J Pharmacol 432:143–147

Catania AC (1971) Discriminative stimulus functions of drugs: interpretations. I. In: Thompson T, Pickens R (eds) Stimulus properties of drugs. Appleton-Century-Crofts, New York, NY, pp 87–110

Chait LD, Uhlenhuth EH, Johanson CE (1985) The discriminative stimulus and subjective effects of d-amphetamine in humans. Psychopharmacology 86:301–312

Chait LD, Uhlenhuth EH, Johanson CE (1986) The discriminative stimulus and subjective effects of phenylpropanolamine, mazindol and d-amphetamine in humans. Pharmacol Biochem Behav 24: 1665–1672

Collins W (1868) The moonstone. Tinsley Brothers, London

Colpaert FC (1995) Drug discrimination: no evidence for tolerance to opiates. Pharmacol Rev 47(4):605–629

Colpaert FC (2003) Discovering risperidone: the LSD model of psychopathology. Nat Rev Drug Discov 2(4):315–320

Colpaert FC, Balster RL (eds) (1988) Transduction mechanisms of drug stimuli, Psychopharmacology series, vol 4. Springer, New York, NY

Colpaert FC, Niemegeers CJ (1975) On the narcotic cuing action of fentanyl and other narcotic analgesic drugs. Arch Int Pharmacodyn Ther 217(1):170–172

Colpaert FC, Niemegeers CJ, Janssen PA (1975) The narcotic cue: evidence for the specificity of the stimulus properties of narcotic drugs. Arch Int Pharmacodyn Ther 218(2):268–276

Colpaert FC, Kuyps JJ, Niemegeers CJ, Janssen PA (1976) Discriminative stimulus properties of fentanyl and morphine: tolerance and dependence. Pharmacol Biochem Behav 5(4):401–408

Combe G (1835) A system of phrenology, 3rd edn. Marsh, Capen and Lyon, Boston, MA

Conger JJ (1951) The effects of alcohol on conflict behavior in the albino rat. Q J Stud Alcohol 12:1–29

Donahue TJ, Hillhouse TM, Webster KA, Young R, De Oliveira EO, Porter JH (2014) (S)-amisulpride as a discriminative stimulus in C57BL/6 mice and its comparison to the stimulus effects of typical and atypical antipsychotics. Eur J Pharmacol 734:15–22

Donahue TJ, Hillhouse TM, Webster KA, Young R, De Oliveira EO, Porter JH (2017) Discriminative stimulus properties of the atypical antipsychotic amisulpride: comparison to its isomers and to other benzamide derivatives, antipsychotic, antidepressant, and antianxiety drugs in C57BL/6 mice. Psychopharmacology 234(23–24):3507–3520

Gardner LA, McCullough C (1962) A reinvestigation of the dissociative effect of curareform drugs. (abstract). Am Psychol 17:398

Girden E, Culler EA (1937) Conditioned responses in curarized striate muscle in dogs. J Comp Psychol 23:261–274

Glennon RA, Young R (2011a) Chapter 2. Methodological considerations. In: Glennon RA, Young R (eds) Drug discrimination: applications to medicinal chemistry and drug studies. Wiley, Hoboken, NJ, pp 19–40

Glennon RA, Young R (2011b) Chapter 4. Role of stereochemistry in drug discrimination studies. In: Glennon RA, Young R (eds) Drug discrimination: applications to medicinal chemistry and drug studies. Wiley, Hoboken, NJ, pp 129–161

Goudie AJ, Cole JC, Sumnall HR (2007a) Olanzapine and JL13 induce cross-tolerance to the clozapine discriminative stimulus in rats. Behav Pharmacol 18:9–17

Goudie AJ, Cooper GD, Cole JC, Sumnall HR (2007b) Cyproheptadine resembles clozapine in vivo following both acute and chronic administration in rats. J Psychopharmacol 21:179–190

Grant KA (1999) Strategies for understanding the pharmacological effects of ethanol with discrimination procedures. Pharmacol Biochem Behav 64(2):261–267

Grossman SP, Miller NE (1961) Control for stimulus-change in the evaluation of alcohol and chlorpromazine as fear-reducing drugs. Psychopharmacology 2:342–351

Harris RT, Balster BL (1968) Discriminative control by d1-amphetamine and saline of lever choice and response patterning. Psychon Sci 10(3):105–106

Harris RT, Balster BL (1971) An analysis of the function of drugs in the stimulus control of operant behavior. In: Thompson T, Pickens R (eds) Stimulus properties of drugs. Appleton Century Crofts, New York, NY, pp 111–132

Hippius H (1991) A historical perspective of clozapine. J Clin Psychiatry 60(suppl 12):22–23

Hirschhorn ID, Rosecrans JA (1976) Generalization of morphine and lysergic acid diethylamide (LSD) stimulus properties to narcotic analgesics. Psychopharmacology (Berl) 47(1):65–69

In Vivo Pharmacology Training Group (2002) The fall and rise of in vivo pharmacology. Trends Pharmacol Sci 23(1):13–18

Kelley BM, Porter JH (1997) The role of muscarinic cholinergic receptors in the discriminative stimulus properties of clozapine in rats. Pharmacol Biochem Behav 57(4):707–719

Kubena RK, Barry H III (1969a) Two procedures for training differential responses in alcohol and nondrug conditions. J Pharm Sci 58(1):99–101

Kubena RK, Barry H III (1969b) Generalization by rats of alcohol and atropine stimulus characteristics to other drugs. Psychopharmacologia (Berl) 15:196–206

Marchese G, Bartholini F, Ruiu S, Casti P, Saba P, Gessa G, Pan L (2002a) Effect of the amisulpride isomers on rat catalepsy. Eur J Pharmacol 444:69–74

Marchese G, Ruiu S, Casti P, Saba P, Gessa GL, Pani L (2002b) Effect of the amisulpride isomers on rat prolactinemia. Eur J Pharmacol 448:263–266

McMahon LR (2015) The rise (and fall?) of drug discrimination research. Drug Alcohol Depend 151:284–288

Millan MJ, Schreiber R, Monneyron S, Denorme B, Melon C, Queriaux S, Dekeyne A (1999) S-16924, a novel, potential antipsychotic with marked serotonin1A agonist properties. IV. A drug discrimination comparison with clozapine. Journal of Pharmacology & Experimental Therapeutics 289(1):427–436

Miller NE (1957) Objective techniques for studying motivational effects on animals. In: Garattini S, Ghetti V (eds) Psychotropic drugs, proceedings of the international symposium on psychotropic drugs. Elsevier, North-Holland, Amsterdam, pp 83–103

Miller NE, Barry H III (1960) Motivational effects of drugs: methods which illustrate some general problems in psychopharmacology. Psychopharmacologia 1:169–199

Moore NA, Tye NC, Axton MS, Risius FC (1992) The behavioral pharmacology of olanzapine, a novel "atypical" antipsychotic agent. J Pharmacol Exp Ther 262:545–551

Morgan D, Picker MJ (1996) Contribution of individual differences to discriminative stimulus, antinociceptive and rate-decreasing effects of opioids: importance of the drug's relative intrinsic efficacy at the mu receptor. Behav Pharmacol 7(3):261–284

Morgan CT, Stellar E (1950) Physiological psychology, 2nd edn. McGraw-Hill, New York, NY

Nadal R (2001) Pharmacology of the atypical antipsychotic remoxipride, a dopamine D2 receptor antagonist. CNS Drug Rev 7(3):26–282

O'Neal MF, Means LW, Porter JH, Rosecrans JA, Mokler DJ (1988) Rats that acquire a THC discrimination more rapidly are more sensitive to THC and faster in reaching operant criteria. Pharmacol Biochem Behav 29:67–71

Overton DA (1961) Discriminative behavior based on the presence or absence of drug effects (abstract). Am Psychol 16:453–454

Overton DA (1964) State-dependent or "dissociated" learning produced with pentobarbital. J Comp Physiol Psychol 57:3–12

Overton DA (1966) State-dependent learning produced by depressant and atropine-like drugs. Psychopharmacologia 10:6–31

Overton DA (1968) Dissociated learning in drug states (state-dependent learning). In: Efron DH, Colle JO, Levine J, Wittenborn R (eds) Psychopharmacology, a review of progress, 1957–1967. PHS Pub No 1836. Sept. of Docs., US Govt. Print. Office, Washington, DC, pp 918–930

Overton DA (1971) Discriminative control of behavior by drug states. In: Thompson T, Pickens R (eds) Stimulus properties of drugs. Appleton-Century-Crofts, New York, NY, pp 87–110

Overton DA (1982) Comparison of the degree of discriminability of various drugs using the T-maze drug discrimination paradigm. Psychopharmacology 76:385–395

Overton DA (1991) Historical context of state dependent learning and discriminative drug effects. Behav Pharmacol 2:253–264

Overton DA, Rosecrans JA, Barry H III (1999) Creation and first 20 years of the society for the stimulus properties of drugs (SSPD). Pharmacol Biochem Behav 64(2):347–352

Perkins KA (2011) Nicotine discrimination in humans. In: Glennon RA, Young R (eds) Drug discrimination: applications to medicinal chemistry and drug studies. Wiley, Hoboken, NJ, pp 129–161

Philibin SD, Prus AJ, Pehrson AL, Porter JH (2005) Serotonin receptor mechanisms mediate the discriminative stimulus properties of the atypical antipsychotic clozapine in C57BL/6 mice. Psychopharmacology 180:49–56

Philibin SD, Walentiny DM, Vunck SA, Prus AJ, Meltzer HY, Porter JH (2009) Further characterization of the discriminative stimulus properties of the atypical antipsychotic drug clozapine in C57BL/6 mice and a comparison to clozapine's major metabolite N-desmethylclozapine. Psychopharmacology 203:303–315

Porter JH, Prus AJ (2009) Discriminative stimulus properties of atypical and typical antipsychotic drugs: a review of preclinical studies. Psychopharmacology 203:279–294

Porter JH, Varvel SA, Vann RE, Philibin SD, Wise LE (2000) Clozapine discrimination with a low training dose distinguishes atypical from typical antipsychotic drugs in rats. Psychopharmacology 149:189–193

Preston KL, Bigelow GE (1991) Subjective and discriminative effects of drugs. Behav Pharmacol 2:293–313

Quarta D, Naylor CG, Barik J, Fernandes C, Wonnacott S, Stolerman IP (2009) Drug discrimination and neurochemical studies in alpha7 null mutant mice: tests for the role of nicotinic alpha7 receptors in dopamine release. Psychopharmacology (Berl) 203(2):399–410

Ribot T (1882) Diseases of memory. Kegan, Paul Trench and Co., London

Ribot T (1891) The diseases of personality, 4th edn. Open Court, Chicago, IL

Riley AL, Clasen MM, Friar MA (2016) (Chapter 13, this volume) conditioned taste avoidance drug discrimination procedure: assessments and applications. In: Porter JH, Prus AJ (eds) The behavioural neuroscience of drug discrimination. Springer, New York, NY

Rosecrans JA, Glennon RA (1979) Drug-induced cues in studying mechanisms of drug action. Neuropsychopharmacology 18:981–989

Schechter MD (1983) Drug sensitivity of individual rats determines degree of drug discrimination. Pharmacol Biochem Behav 19:1–4

Schuster CR, Balster RL (1977) The discriminative stimulus properties of drugs. In: Thompson T, Dews PB (eds) Advances in behavioral pharmacology, vol 1. Academic Press, New York, NY, pp 86–138

Schuster CR, Johanson CE (1988) Relationship between the discriminative stimulus properties and subjective effects of drugs. Psychopharmacol Ser 4:161–175

Shannon HE, Holtzman SG (1976) Evaluation of the discriminative effects of morphine in the rats. Evaluation of the discriminative effects of morphine in the rat. J Pharmacol Exp Ther 198:54–65

Solinas M, Panlilio LV, Justinova Z, Yasar S, Goldberg SR (2006) Using drug-discrimination techniques to study the abuse-related effects of psychoactive drugs in rats. Nat Protoc 1(3): 1194–1206

Stewart J (1962) Differential responses based on the physiological consequences of pharmacological agents. Psychopharmacologia 3:132–138

Stolerman IP, Rasul F, Shine PJ (1989) Trends in drug discrimination research analysed with a cross-indexed bibliography, 1984–1987. Psychopharmacology (Berl) 98(1):1–19

Stolerman IP, Mariathasan EA, White J-AW, Olufsen KS (1999) Drug mixtures and ethanol as compound internal stimuli. Pharmacol Biochem Behav 64(2):221–228

Stolerman IP, Childs E, Matthew M, Ford MM, Grant KA (2011) Role of training dose in drug discrimination: a review. Behav Pharmacol 22:415–429

Wiebelhaus JM, Webster KA, Meltzer HY, Porter JH (2011) The metabolites N-desmethylclozapine and N-desmethylolanzapine produce cross-tolerance to the discriminative stimulus of the atypical antipsychotic clozapine in C57BL/6 mice. Behav Pharmacol 22:458–467

Young A (1991) Tolerance to drugs acting as discriminative stimuli. In: Glennon R, Jarbe T, Frankenheim J (eds) Drug discrimination: applications to drug abuse research, NIDA research monograph, vol 116. National Institute of Drug Abuse, Rockville, MD, pp 197–212

Young R (2009) Chapter 3. Drug discrimination. In: Buccafusco JL (ed) Methods of behavioral analysis in neuroscience, 2nd edn. CRC Press, Boca Raton, FL, pp 39–58

Part II
The Discriminative Stimulus
Properties of Drugs

Discriminative Stimulus Effects of Psychostimulants

Michael D. Berquist and William E. Fantegrossi

Abstract Numerous drugs elicit locomotor stimulant effects at appropriate doses; however, we typically reserve the term psychostimulant to refer to drugs with affinity for monoamine reuptake transporters. This chapter comprises select experiments that have characterized the discriminative stimulus effects of psychostimulants using drug discrimination procedures. The substitution profiles of psychostimulants in laboratory rodents are generally consistent with those observed in human and nonhuman primate drug discrimination experiments. Notably, two major classes of psychostimulants can be distinguished as those that function as passive monoamine reuptake inhibitors (such as cocaine) and those that function as substrates for monoamine transporters and stimulate monoamine release (such as the amphetamines). Nevertheless, the discriminative stimulus effects of both classes of psychostimulant are quite similar, and drugs from different classes will substitute for one another. Most importantly, for both the cocaine-like and amphetamine-like psychostimulants, dopaminergic mechanisms most saliently determine discriminative stimulus effects, but these effects can be modulated by alterations in noradrenergic and serotonergic neurotransmission as well. Thusly, the drug discrimination assay is useful for characterizing the interoceptive effects of psychostimulants and determining the mechanisms that contribute to their subjective effects in humans.

Keywords Discriminative stimulus • Drug discrimination • Monoamine transporter blocker • Monoamine transporter releaser/substrate • Preclinical model • Psychostimulants • Stimulus properties

M.D. Berquist and W.E. Fantegrossi (✉)
Department of Pharmacology and Toxicology, College of Medicine, University of Arkansas for Medical Sciences, 4301 West Markham Street, Slot 638, Little Rock, AR 72205, USA
e-mail: WEFantegrossi@uams.edu

© Springer International Publishing AG 2017
Curr Topics Behav Neurosci (2018) 39: 29–50
DOI 10.1007/7854_2017_5
Published Online: 24 March 2017

Contents

1 Introduction

This chapter reviews select experiments that have characterized the discriminative stimulus effects of psychostimulants using drug discrimination procedures. We limit our discussion to psychostimulants that primarily serve as reuptake inhibitors or substrates for release at monoamine transporters with a special emphasis on drugs that directly modulate dopamine (DA) neurotransmission (e.g., through pharmacological effects at the dopamine transporter [DAT]). Drugs that directly modulate serotonergic activity at the serotonin transporter (SERT), noradrenergic activity at the norepinephrine transporter (NET), intracellular mechanisms (e.g., intraterminal vesicular releasers), and neuronal activity via stimulation at postsynaptic dopaminergic cell-surface receptors will also be discussed. In addition, given the current interest in an emerging class of psychostimulants, the synthetic cathinones (casually referred to as "bath salts"), we will briefly discuss their discriminative stimulus effects where appropriate. Drugs that produce acute psychostimulant-like effects through other, non-monoaminergic pharmacological mechanisms (e.g., phencyclidine, caffeine, and nicotine) will not be included and we refer readers to chapters found elsewhere in this book or in other available resources.

2 Discriminative Stimulus Effects of Psychostimulant Drugs

As discussed within other chapters of this book, drug discrimination procedures are valuable for characterizing the discriminative stimulus effects of psychoactive substances. Although most psychostimulant drug discrimination studies are conducted in rodents (e.g., [1]), the substitution profiles of psychostimulants are generally consistent with those observed in human and nonhuman primate drug discrimination experiments. As such, where available, the similarities in drug substitution profiles observed across human and nonhuman drug discrimination findings will be reported in the present chapter to highlight the predictive utility of the drug discrimination assay in nonhuman subjects. The sections that follow are categorized by pharmacological effects at specific protein targets in the central nervous system. Each of the following sections will include a representative drug from the associated pharmacological class to demonstrate the utility and translatability of drug discrimination procedures. Furthermore, an overview of the experimental parameters in drug discrimination preparations (e.g., difference between partial substitution and full substitution) will not be discussed here, but drug discrimination terms and concepts will be mentioned throughout this chapter. Readers may refer to [2] within this book to glean basic concepts of drug discrimination methodology, if needed.

3 A Note on the Sex of Experimental Subjects

Most preclinical research has used males as experimental subjects and considerably fewer drug discrimination experiments have included females. Despite historical precedents for preferentially concentrating on male subjects instead of females (e.g., convenience, literature precedence), the consideration of sex as a biological variable is currently being promoted at all levels of NIH-funded research. Indeed, the NIH has recently added a policy requiring some discussion of experimental designs to study male and female animals in preclinical studies, unless sufficient justification can be given that such sex-specific inclusion would be unwarranted. In light of this new policy, we include below several drug discrimination experiments that used psychostimulants as training or test compounds and made direct comparisons of males to females in discriminative performance, but note that this area is considerably understudied in comparison to other common in vivo assays of psychostimulant effects which show sex differences, including locomotor activity [3, 4] and intravenous self-administration [5, 6].

4　Monoamine Transporter Blocker: Cocaine

4.1　Dopamine

Cocaine is a nonselective, passive reuptake inhibitor at DAT, NET, and SERT [7–9], and possesses pharmacologically-relevant binding affinities at serotonin (5-HT)$_3$ receptors [10], muscarinic M$_1$ and M$_2$ receptors [11], and σ-receptors [12]. In an early drug discrimination experiment that included cocaine as the training drug, Colpaert et al. [13] reported that 1.25 mg/kg racemic amphetamine (a relatively selective DA releaser, see below) and 0.31 mg/kg apomorphine (a direct agonist at DA receptors) produced partial substitution (intermediate percent drug-lever responses) in rats trained to discriminate 10 mg/kg cocaine (a common training dose of cocaine) from saline; however, the antipsychotics haloperidol (0.08–0.16 mg/kg) and pimozide (1.25–2.5 mg/kg), which are potent dopamine antagonists, failed to block cocaine's cue. In a similar study, McKenna and Ho [14] observed that d-amphetamine (0.25–0.5 mg/kg) (an isomer of amphetamine with greater in vivo potency, now referred to as $S(+)$-amphetamine to denote absolute configuration) and apomorphine (0.25–0.5 mg/kg) produced complete generalization in rats trained to discriminate 10 mg/kg cocaine from saline, whereas pretreatment with 0.5 mg/kg haloperidol attenuated cocaine's cue as evidenced by a downward shift in cocaine's dose–response curve. A downward shift of a training drug's dose–response curve in the presence of an antagonist generally indicates noncompetitive (or insurmountable) antagonism (assuming the dose of antagonist is held constant and the dose of the training drug varies), although it may also indicate a sedative/motoric effect or stimulus masking (i.e., the antagonist produces an interoceptive cue that reduces/competes with the saliency of the training drug cue). Unfortunately, McKenna and Ho [14] did not present response rate data to supplement the substitution test results. Disparate findings in drug discrimination research (e.g., haloperidol weakened cocaine's discriminative stimulus effects in one study [14], but not in another [13]) are common if one considers the context under which discrimination training and testing occur; indeed, a drug's discriminative stimulus effects are not an immutable property of the drug, but are rather the result of a complex interaction of variables related to the drug's pharmacological effects, the experimental environment, and the experimental subject's learning history (e.g., contingency of reinforcement).

As mentioned previously, relatively few drug discrimination experiments have included females as experimental subjects. Nevertheless, previous research has made direct comparisons between males and females in discrimination performance of cocaine and two of these studies will now be discussed. Craft and Stratmann [15] trained male and female Sprague–Dawley rats to discriminate 5.6 mg/kg cocaine (IP) from saline using a two-lever food-maintained drug discrimination procedure. There were no significant differences in acquisition of drug stimulus control, estimates of median effective dose (ED$_{50}$) values of cocaine (1.0–10 mg/kg) or d-amphetamine (0.1–0.56 mg/kg) following substitution tests, or in the acquisition and substitution profile when the training dose increased to 10 mg/kg cocaine. Moreover, these authors demonstrated that cocaine more potently stimulated locomotor activity in females than in males, but discrimination performance was virtually identical between sexes, indicating that a drug's locomotor effects must be considered separately from its discriminative stimulus effects.

In a later study, Anderson and van Haaren [16] trained male and female Wistar rats to discriminate 10 mg/kg from cocaine. Similar to the findings reported by Craft and Stratmann [15], there were no significant differences between males and females in acquisition of drug stimulus control (i.e., learning to discriminate the training drug from vehicle), substitution tests with cocaine (1.0–10 mg/kg), and the blocking effects of the D_1 receptor antagonist SCH-23390 (0.01–0.10 mg/kg) or the D_2 receptor antagonist raclopride (0.1–1.6 mg/kg) when each antagonist was injected prior to 10 mg/kg cocaine [16]. Based on these findings, it appears that males and females display comparable performance in discriminating cocaine's cue, and cocaine's interoceptive effects are mediated by common dopaminergic mechanisms.

The foregoing drug discrimination experiments and others (e.g., [17, 18]) have demonstrated that DA has an important role in cocaine's discriminative stimulus effects. Later research has confirmed these preliminary reports and provided further evidence that increased DA neurotransmission through DAT blockade is considered primarily responsible for mediating cocaine-like stimulus effects. For example, the selective DAT inhibitor GBR 12909 is 17-fold more potent at inhibiting DAT than SERT as measured in vitro [8]. GBR 12909 (2–16 mg/kg) produces complete generalization to cocaine in rats trained to discriminate 10 mg/kg cocaine from saline, and pretreatment with 2 mg/kg GBR 12909 potentiates cocaine's discriminative stimulus effects, as evidenced by a leftward shift in the cocaine dose–response curve [19]. In rats trained to discriminate 2.5 mg/kg cocaine from 10 mg/kg cocaine, Kleven and Koek [20] reported that pretreatment with the structurally related DAT-selective GBR 12935 potentiated the discriminative stimulus effects of 2.5 mg/kg cocaine; that is, as the test dose of GBR 12935 increased, rats shifted responding from the "low dose" 2.5 mg/kg cocaine lever to the "high dose" 10 mg/kg cocaine lever. In addition, GBR 12935 failed to fully substitute for 10 mg/kg cocaine in >50% of rats when injected prior to saline administration. It is noteworthy that GBR 12935 is approximately 78-fold more potent at inhibiting DAT than SERT as measured in vitro [8]. It is possible that the failure to fully substitute for cocaine is due to GBR 12935 possessing binding affinities to other protein receptors or membrane protein transporters [8], or perhaps low in vivo potency compared to its in vitro binding profile (see [20]). Nevertheless, rats can be successfully trained to discriminate GBR 12909 from cocaine [21], indicating that these drugs produce similar, but not identical, discriminative stimulus effects. As a final point, it is noteworthy that rats can be trained to discriminate 2.5 mg/kg from 10 mg/kg cocaine (a "dose–dose discrimination") ([20]; also see [22]). Indeed, the selected training dose of a drug is a primary determinant in its subsequent discriminative stimulus effects profile, which is unsurprising given that the magnitude of a drug's behavioral effects occur in a dose-dependent fashion. Indeed, there are oftentimes qualitative differences (e.g., dissimilar generalization gradients) between different doses of the same drug. In any event, Kleven and Koek [20] utilized this dose–dose discrimination preparation to detect cocaine-like effects that may not occur at the common 10 mg/kg cocaine training dose. Regardless of the drug discrimination methodology used, various DA uptake inhibitors (e.g., methylphenidate, WIN 35428, indatraline) and DA releasers (e.g., cathinone, fencamfamine, and methamphetamine) all produced full substitution in rats trained to discriminate 10 mg/kg cocaine from saline [23], further indicating that altered DA neurotransmission is generally considered to be the primary mediator of cocaine's discriminative stimulus effects. In a human drug discrimination experiment, Rush and Baker [24] observed that oral methylphenidate (a fairly selective DAT inhibitor, with at least eightfold selectivity for DAT over NET, and negligible SERT affinity) produced full substitution in humans trained to discriminate an oral dose of 200 mg cocaine, while the benzodiazepine triazolam produced an expected very low cocaine-appropriate responding. At this point it is worth noting that findings from positron emission tomography (PET) experiments in humans support the foregoing role of increased dopaminergic tone due to DAT blockade in mediating cocaine's discriminative stimulus effects as measured in rodents. For example, Volkow et al. [25] observed that the self-reported high following cocaine administration (IV) was significantly correlated to DAT occupancy. Indeed, blockade of $\geq 47\%$ of DAT was required to elicit cocaine's subjective effects in the volunteers [25].

It is therefore reasonable to speculate that increased dopaminergic tone (e.g., via inhibition of reuptake through the DAT) would necessarily lead to increases in postsynaptic receptor stimulation in the absence of factors that would limit dopaminergic neurotransmission (e.g., overexpression of catalytic enzymes, such as monoamine oxidase). In congruence with this notion, in addition to changes in overall dopaminergic tone, previous studies have demonstrated that stimulation of postsynaptic DA receptors are involved in mediating cocaine's discriminative stimulus effects. For example, Callahan et al. [26] observed that the D_2-like receptor agonist quinpirole (0.0313–0.125 mg/kg) produced full substitution for cocaine in rats trained to discriminate 10 mg/kg cocaine from saline, but the D_1-like receptor agonist SKF 38393 (5–20 mg/kg) produced only partial substitution before eliciting behavioral disruption in these subjects. In that same study, the D_2-like receptor antagonist haloperidol and the D_1-like receptor antagonist SCH 23390 (0.0063–0.25 mg/kg) reduced cocaine-lever responding when administered prior to the cocaine training dose. This study, and others not discussed here (e.g., [27, 28]), specify receptors that are involved in mediating cocaine's discriminative stimulus effects downstream from its direct pharmacological effects at DAT. For a review of DA's involvement in mediating cocaine's discriminative stimulus effects, see Callahan et al. [29].

As mentioned previously, synthetic cathinones are an emerging class of psychostimulants that increased in popularity in the early 2000s (for review, see [30]). Most of these compounds produce increases in extracellular monoamines through intraterminal release through a monoamine transporter, or via reuptake inhibition at monoamine transporters (for review, see [31]). In a recent study, Gatch et al. [32] trained groups of rats to discriminate 10 mg/kg cocaine or 1 mg/kg methamphetamine from saline. The synthetic cathinones 3,4-methylenedioxypyrovalerone (MDPV) (0.05–2.5 mg/kg), 4-methylmethcathinone (mephedrone) (0.5–5 mg/kg), methylone (0.5–5 mg/kg), napyrhone (0.5–5 mg/kg), flephedrone (0.5–10 mg/kg), and butylone (0.5–10 mg/kg) were tested for stimulus substitution. The results revealed that all synthetic cathinones produced full substitution in both the 10 mg/kg cocaine and 1 mg/kg methamphetamine training groups, indicating that these compounds produce interoceptive effects that are comparable to prototypical psychostimulants. Similarly, mice trained to discriminate 10 mg/kg cocaine from saline fully generalized responding to racemic MDPV and its enantiomers, with S(+)-MDPV being more potent than the racemate and R(−)-MDPV being dramatically (~30-fold) less potent than the racemate [33]. Importantly, mephedrone and naphyrone also fully substituted for cocaine in those same mice [34].

It should be noted that MDPV is a potent reuptake inhibitor at DAT and NET (and is approximately sixfold more selective for DAT over NET), with negligible affinity for SERT [35], whereas cocaine is a nonselective reuptake inhibitor at DAT, NET, and SERT (e.g., [35]). As such, substitution tests with drugs that increase extracellular DA and NE content, or stimulate postsynaptic dopamine and noradrenergic receptors may produce interoceptive effects that are similar to MDPV. In a recent report by Fantegrossi et al. [36] male NIH Swiss mice were trained to discriminate 0.3 mg/kg MDPV from saline. Substitution tests were performed with MDPV (0.01–0.3 mg/kg), MDMA (0.01–0.3 mg/kg), methamphetamine (0.01–0.3 mg/kg), morphine (1–30 mg/kg), and the synthetic cannabinoid JWH-018 (0.1–3 mg/kg). MDPV, MDMA, and methamphetamine engendered >75% MDPV-appropriate responding, whereas morphine and JWH-018 produced <50% MDPV-appropriate responding. These results indicate that MDPV, a drug with pharmacological actions that are similar to cocaine, produces interoceptive effects similar to prototypical drugs of abuse.

4.2 Norepinephrine

In addition to DAT, cocaine is also a nonselective reuptake inhibitor at NET and SERT. Comparatively fewer studies have investigated the role of NE neurotransmission in mediating cocaine's discriminative stimulus effects. In an early report, Colpaert et al. [13] observed that two compounds with effects at adrenoceptors, dibenamine (an α-adrenoceptor antagonist) and propranolol (a β-adrenoceptor antagonist), failed to block cocaine's discriminative effects in rats trained to discriminate 10 mg/kg cocaine from saline, indicating that the NE-releasing effects of cocaine may not contribute to its discriminative stimulus effects, or at least that antagonism of the α- and β-adrenoceptors do not *attenuate* cocaine's discriminative stimulus effects. As mentioned above, the initial training dose and subsequent methods used to test for substitution in drug discrimination experiments critically determine a drug's discriminative stimulus effects. Indeed, Young and Glennon [37] observed cross-substitution of cocaine and propranolol in rats that were trained to discriminate doses of either drug. That is, the β-adrenoceptor antagonist *substituted* for rather than *attenuated* cocaine's discriminative stimulus effects. In an earlier report that differentiated the role of adrenoceptors in mediating cocaine's discriminative stimulus effects, Kleven and Koek [22] demonstrated that pretreatment with the β_1/β_2-adrenoceptor antagonists (−)-propranolol and tertatolol, as well as the β_2-adrenoceptor antagonist ICI 118,551, produced high-dose lever selection (i.e., 10 mg/kg cocaine-lever selection) when administered prior to 2.5 mg/kg cocaine injection in rats trained to discriminate 2.5 mg/kg cocaine from 10 mg/kg cocaine. That is, the noradrenergic compounds potentiated the discriminative stimulus effects of a relatively low dose of cocaine, indicating that NE receptor stimulation may have an augmenting role in cocaine's discriminative stimulus effects. Moreover, Kleven and Koek [22] also found that stimulation of the β_2-adrenoceptor, but not the β_1-, α_1- or α_2-adrenoceptors, enhanced the discriminative stimulus effects of 2.5 mg/kg cocaine.

4.3 Serotonin

Schama et al. [38] investigated the effects of altered serotonergic tone on cocaine's discriminative stimulus effects. In that study, groups of squirrel monkeys were trained to discriminate intramuscular injections of 0.3 or 1.0 mg/kg cocaine from saline. Compared to monkeys trained to discriminate 1.0 mg/kg cocaine from saline, the nonselective serotonin receptor agonist quipazine produced a greater percent cocaine-lever selection in monkeys trained to discriminate the 0.3 mg/kg cocaine training dose. In addition, the selective serotonin reuptake inhibitor fluoxetine enhanced cocaine's discriminative stimulus effects in the low dose training group, but not in the high dose training group – findings that further demonstrate the importance of training dose in drug discrimination experiments. Last, Schama et al. [38] observed that pretreatment with ketanserin and ritanserin ($5\text{-}HT_2$ receptor antagonists) attenuated cocaine's discriminative stimulus effects in the low dose and high dose group, respectively, indicating a modulatory role of serotonin in mediating cocaine's discriminative stimulus effects. In a rodent drug discrimination experiment, Filip et al. [39] reported that pretreatment with the $5\text{-}HT_{2A}$ antagonist SR 46349B (0.5–1 mg/kg) produced a rightward shift in the dose–response curve for cocaine in rats trained to discriminate 10 mg/kg cocaine from saline, indicating that stimulation of $5\text{-}HT_{2A}$ receptors modulates the discriminative stimulus effects of cocaine. In that study, pretreatment with a $5\text{-}HT_{2B}$ (SB 204741; 1–3 mg/kg) or a $5\text{-}HT_{2C}$ antagonist (SDZ SER-082; 0.5–1 mg/kg) produced no change or a leftward shift (i.e., enhanced the discriminative stimulus effects) in the cocaine dose–response curve, respectively. Finally, as previously mentioned, cocaine does possess low binding affinity for serotonin $5\text{-}HT_3$ receptors as measured in vitro [10]. As such, Paris and Cunningham [40] investigated the role of 5-HT$_3$ receptors in mediating cocaine's discriminative stimulus effects. In that study, 5-HT$_3$ antagonists ICS 205930 (2–24 mg/kg) and MDL 72222 (2–16 mg/kg) both failed to substitute for cocaine's discriminative stimulus effects in rats trained to discriminate 10 mg/kg from saline. In addition, pretreatment with the 5-HT$_3$ antagonists failed to block cocaine's (5 mg/kg) discriminative stimulus effects. Thus, the results reported by Paris and Cunningham [40] indicate that altered activity of the 5-HT$_3$ receptor does not affect cocaine's discriminative stimulus effects, despite cocaine possessing pharmacologically relevant binding affinity at this receptor. In sum, these findings would seem to indicate subtle regulatory effects of altered serotonergic tone [38] and serotonin receptor stimulation [39] on cocaine's discriminative stimulus effects (for review, [41]), but certainly do not challenge the primacy of dopaminergic mechanisms in mediating cocaine-like interoceptive effects.

4.4   Non-monoaminergic Receptors

As mentioned previously, cocaine also possesses pharmacologically relevant binding affinity at muscarinic M$_1$ and M$_2$ receptors [11] and σ-receptors [12]. Tanda and Katz [42] investigated the effects of muscarinic M$_1$ receptor blockade on cocaine's discriminative stimulus effects. In that study, the M$_1$ antagonists telenzepine and trihexyphenidyl failed to substitute for cocaine's discriminative stimulus effects in rats trained to discriminate 10 mg/kg cocaine from saline; however, when the drugs were injected prior to cocaine administration, the antagonists enhanced cocaine's discriminative stimulus effects (i.e., produced a leftward shift in the cocaine dose–response curve) demonstrating that M$_1$ antagonism can increase the saliency of cocaine's interoceptive cue. In a recent study, Hiranita et al. [43] observed that the σ-receptor agonists PRE-084 and DTG (delivered at different pretreatment times via intraperitoneal, subcutaneous, or intravenous administration routes) both produced low percent cocaine-lever selection in rats trained to discriminate 10 mg/kg cocaine from saline. As above, these findings would seem to imply regulatory roles for some non-dopaminergic receptors in the discriminative stimulus effects of cocaine.

5   Monoamine Transporter Substrate/Releaser: MDMA

5.1   Stimulant: Hallucinogen Continuum

3,4-Methylenedioxymethamphetamine (MDMA) is a ring-substituted phenethylamine containing a chiral center which allows for stereoisomerism (see discussion on stereochemistry below). Although MDMA is abused (as a primary psychoactive ingredient in ecstasy or Molly) in its racemic form, it is informative to also consider the discriminative stimulus effects of its component enantiomers. Like cocaine, MDMA is a nonselective ligand at DAT, NET, and SERT, but unlike cocaine it functions as a substrate/releaser at these transporters (e.g., [44]). Although MDMA shares structural similarities to other phenethylamine derivatives (e.g., amphetamine; 2,4-dimethoxy-4-methylamphetamine [DOM] and bupropion [Wellbutrin®]), its discriminative stimulus effects are complex given its stereochemistry. Before discussing MDMA's stereochemical profile, it should be noted that MDMA produces multiple interoceptive effects that can be broadly categorized as stimulant- and hallucinogen-like. For example, in rats trained to discriminate 1.75 mg/kg MDMA from saline, Oberlander and Nichols [45] reported that $S(+)$-amphetamine produced full substitution for MDMA's discriminative stimulus effects in less than half of the subjects, while the hallucinogen lysergic acid diethylamide (LSD) produced 78% MDMA-lever responding. In addition, the hallucinogen DOM produced 56% MDMA-lever responding. These findings indicate that MDMA possesses a complex substitution profile that can be characterized as stimulant-like ($S(+)$-amphetamine, via DA-releasing effects) and hallucinogen-like (LSD and DOM, via agonist effects at serotonergic receptors). It is worth noting that rats can discriminate MDMA from d-amphetamine using a three-choice discrimination procedure (e.g., [46]), indicating that the discriminative stimulus effects of these compounds are similar, but nevertheless dissociable.

Broadbear et al. [47] trained female and male Sprague-Dawley rats to discriminate 1.5 mg/kg MDMA from 1.0 mg/mg dl-amphetamine and saline using a three-choice discrimination procedure. Any difference between males and females in acquisition of drug stimulus control was not reported. MDMA (0.38–1.5 mg/kg) and dl-amphetamine (0.25–1 mg/kg) equipotently substituted in males and females, although females were more sensitive to 0.75 mg/kg MDMA than males. Last, females were less sensitive than males to the rate-decreasing effects of dl-amphetamine. Overall, the report by Broadbear et al. [47] demonstrated that males and females do not display major differences in their ability to discriminate MDMA from dl-amphetamine and saline, as evidenced by comparable substitution profiles determined with these compounds.

In humans, MDMA produces increases in arousal, positive mood, vigor, and somaesthesia (user feels separate from their body) [48]. In addition, in humans trained to discriminate 20 mg d-amphetamine or $meta$-chlorophenylpiperazine (mCPP, a nonselective serotonergic agonist) from placebo, half of the subjects reported that MDMA felt similar to d-amphetamine and the other half reported effects similar to mCPP [48]. These findings are qualitatively similar to the previously described experiments conducted in rodents, and further buttress experimental observations that MDMA's discriminative stimulus effects comprises both dopaminergic and serotonergic components.

5.2   Stereoisomerism

As mentioned, MDMA contains a chiral center and thus exists as a pair of stereoisomers. Previous drug discrimination research has demonstrated that the discriminative stimulus effects of the stereoisomers (S(+)- and R(−)-isomers) produce different substitution profiles. In the first study to examine the stereoisomers of MDMA in mice using drug discrimination procedures, Murnane et al. [49] reported that the psychostimulants d-amphetamine and cocaine produced greater or more potent percent drug-appropriate responding in mice trained to discriminate 1.5 mg/kg of the S(+) isomer compared to mice trained to discriminate 1.5 mg/kg of the R(−)-isomer. In contrast, the hallucinogens 2,5-dimethoxy-4-(n)-propylthio-phenethylamine (2C-T-7) and N,N-dipropryltryptamine (DPT) produced greater or more potent percent drug-appropriate responding in the R(−)-isomer-trained mice than the S(+)-isomer-trained mice. These findings indicate that the discriminative stimulus effects of the S(+)-isomer of MDMA are more stimulant-like, while the R(−)-isomer of MDMA is more hallucinogenic-like [49].

5.3   Pharmacokinetic Considerations

An important determinant of a drug's discriminative stimulus effects is its pharmacokinetic and metabolic profile. Indeed, the onset of a drug's effects, its duration of action, and the potential influence of behaviorally active metabolites can affect a subject's performance in drug discrimination experiments. Fortunately, drug discrimination procedures permit time course analysis of a drug's discriminative stimulus effects, and if knowledge exists about a drug's metabolic disposition, the potential role of metabolites can be examined as well. Fantegrossi et al. [50] investigated the onset and duration of discriminative stimulus effects of MDMA and its enantiomers in mice. In this study, three different groups of mice were trained to discriminate 3.0 mg/kg racemic MDMA, 1.5 mg/kg S(+)-MDMA, or 1.5 mg/kg R(−)-MDMA from saline. Substitution tests were conducted with the three forms of MDMA in each training group. The results of the substitution tests revealed that racemic MDMA and the S(+) isomer produced full substitution in the 3.0 mg/kg racemic MDMA-trained mice, but the R(−) isomer failed to produce >20% racemic MDMA-appropriate responding. Results of the time course analysis revealed that 3.0 mg/kg racemic MDMA and 1.5 mg/kg S(+)-MDMA (IP) produced relatively rapid (<20 min) onset of discriminative stimulus effects and mice responded on the drug-paired lever for 60 min post-injection. In contrast, 1.5 mg/kg R(−)-MDMA reached peak discriminative stimulus effects at 20 min and responding on the drug-paired lever decreased at 40 min post-injection. These findings indicate that MDMA's discriminative stimulus effects may be primarily driven by the pharmacological effects of S(+)-MDMA, and to a lesser extent, R(−)-MDMA. Alternatively, the component enantiomers may determine different "phases" of the discriminative stimulus effects of racemic MDMA, with primary contributions of the $S(+)$-enantiomer at early time points, and a role for the $R(-)$-enantiomer emerging some time after administration. Furthermore, although the pharmacokinetic and metabolic profile of MDMA as observed in mice differs from observations conducted in humans [50, 51], the drug discrimination assay is useful for investigating the complex role of pharmacokinetic factors in determining a drug's discriminative stimulus effects.

5.4   Discriminative Stimulus Effects of MDMA

In addition to the discriminative stimulus effects of MDMA isomers, racemic MDMA produces an interesting, yet imperfectly understood, substitution profile. In rats trained to discriminate 0.5 mg/kg d-amphetamine from saline, Harper et al. [52] reported that MDMA produced approximately 50% drug-appropriate responding; however, in rats trained to discriminate 1.5 mg/kg MDMA from saline, d-amphetamine failed to produce >20% drug-appropriate responding. These findings indicate an asymmetrical substitution profile of MDMA (that is, MDMA may substitute for the discriminative stimulus effects of a different drug, but the other drug may not substitute for MDMA's discriminative stimulus effects, or vice versa). Other reports also support the asymmetric substitution profile observed with MDMA. Khorana et al. [53] trained groups of rats trained to discriminate 1.5 mg/kg MDMA or 8 mg/kg cocaine from saline. In the 1.5 mg/kg MDMA group, cocaine fully substituted for the MDMA cue. In contrast, MDMA failed to produce >36% cocaine-lever selection in the 8 mg/kg cocaine group. These findings indicate that the discriminative stimulus effects of MDMA depend on the training histories of the subjects (i.e., different training drugs in this example), further underscoring the important principle that discriminative stimulus effects of drugs are not immutable properties of those drugs, but are the result of a complex interaction of biological, environmental, and behavioral variables.

As a final point, it is noteworthy that any drug's discriminative stimulus effects is dose-dependent and, in the case of MDMA, possibly along a serotonergic-dopaminergic continuum as the dose increases. For example, Harper et al. [54] trained rats to discriminate 0.5 mg/kg d-amphetamine or 1.5 mg/kg MDMA from saline using a three-choice discrimination procedure. In that study, intermediate doses of MDMA (1.0 and 1.5 mg/kg) produced responding primarily on the MDMA-paired lever, but at larger doses of MDMA (3.0 and 4.5 mg/kg), subjects shifted responses to the 0.5 mg/kg d-amphetamine-paired lever. These results indicate that as the dose of MDMA increases, the discriminative stimulus effects of MDMA resemble that of a prototypical DA releaser. Using a three-choice discrimination procedure, Goodwin and Baker [46] trained rats to discriminate 1.5 mg/kg MDMA or 1 mg/kg d-amphetamine from saline. In that study, substitution tests with d-amphetamine produced equivalent percent responding on the d-amphetamine, MDMA, and saline levers at 0.25 mg/kg d-amphetamine; however, rats shifted responding to the *d*-amphetamine lever following 0.5 and 1 mg/kg *d*-amphetamine injections. Substitution tests with cocaine produced little responding on the MDMA lever, but instead elicited only dose-dependent responding on the *d*-amphetamine lever. The hallucinogens LSD and DOM produced responding on the saline lever at low doses, but rats shifted their responding to the MDMA lever following larger doses of LSD, and saline and MDMA lever selection became equivalent around 40–50% at larger doses of DOM. Finally, rats tested with the serotonin releaser fenfluramine shifted responding to the MDMA lever at larger doses, and pretreatment with the 5-HT$_2$ receptor antagonist pirenperone (0.16–0.64 mg/kg) (for more information on pirenperone, [55]) reduced MDMA-lever selection when administered in combination with 1.5 mg/kg MDMA. These findings indicate that drugs with agonist effects at serotonergic receptors (LSD and DOM) elicit interoceptive effects similar to those of MDMA, while drugs with antagonist effects at serotonergic receptors (pirenperone) attenuate the discriminative stimulus effects of MDMA.

To conclude this section, it should be mentioned that a previous report by Harvey and Baker [56] included substitution tests with two synthetic cathinones (MDPV [0.125–3 mg/kg], mephedrone [0.25–2 mg/kg]) in groups of rats trained to discriminate 1.5 mg/kg MDMA or 1.5 mg/kg MDMA +0.5 mg/kg *d*-amphetamine from saline. The mixture group was presumably included to add a more salient dopaminergic component to the 1.5 mg/kg MDMA cue. MDPV produced full substitution in the 1.5 mg/kg MDMA +0.5 *d*-amphetamine group, but only produced partial substitution in the 1.5 mg/kg MDMA group, indicating that the DA-releasing effects of *d*-amphetamine was necessary to generalize to MDPV's interoceptive effects. Mephedrone equipotently produced full substitution in both training groups, indicating that mephedrone's cue involves dopaminergic, serotonergic, and possibly noradrenergic components.

6 Monoamine Transporter Substrate/Releaser: *d*-Amphetamine and Methamphetamine

6.1 d-*Amphetamine: Prototypical DA Releaser*

The phenethylamine derivative *d*-amphetamine produces increases in cytoplasmic DA concentrations through pharmacological effects at intraterminal vesicles, DAT, and monoamine oxidase (a catalytic enzyme) (see [57]). In addition to increasing DA concentrations in extracellular space, *d*-amphetamine alters levels of other neurotransmitters, such as norepinephrine, serotonin, and acetylcholine [57]. Despite these other pharmacological effects, *d*-amphetamine is considered a prototypical DA releaser and is used extensively in drug discrimination research to evaluate the role of increased DA neurotransmission in a novel drug's discriminative stimulus effects.

Smith et al. [58] demonstrated that activation of dopamine D_1-like and D_2-like receptors mediates d-amphetamine's discriminative stimulus effects. In rats trained to discriminate 1 mg/kg d-amphetamine from saline, the D_2-like receptor agonist quinpirole produced complete generalization; however, the D_1-like receptor agonist SKF 38392 produced no substitution. Pretreatment with 0.2 or 0.5 mg/kg quinpirole, followed by an injection of 0.3 mg/kg d-amphetamine, produced complete generalization, whereas 0.3 mg/kg d-amphetamine delivered alone produced only partial substitution [58]. Although SKF 38392 produced no substitution when administered alone, pretreatment with doses of SKF 38392 followed by an injection of 0.3 mg/kg d-amphetamine produced complete generalization. These findings indicate that the dopamine D_1-like and D_2-like receptors are involved in mediating d-amphetamine's discriminative stimulus effects: D_2-like receptor stimulation produces discriminative stimulus effects that are similar to d-amphetamine's discriminative stimulus effects, and D_1-like receptor stimulation can potentiate d-amphetamine's discriminative stimulus effects. In human participants, Vansickel et al. [59] compiled the results of six studies performed in their laboratory to assess whether women discriminated 15 mg d-amphetamine differently than men. Substitution tests were performed with d-amphetamine (2.5–15 mg). There were no differences in acquisition of drug stimulus control or d-amphetamine substitution between males and females; however, male participants rated 10 and 15 mg d-amphetamine as producing a significantly greater high than in females, and males also reported significantly less nausea at 2.5 and 5 mg d-amphetamine compared to females. It should be emphasized that despite these qualitative differences between males and females after exposure to d-amphetamine, their discrimination performances were similar. As such, there may be qualitative differences between males and females in experiencing d-amphetamine's interoceptive cue (i.e., information that can be gathered by verbal reports), but no difference in males' and females' *ability to detect* said cues. Future drug discrimination experiments with human participants will be useful for further identifying the subjective qualities of drugs' interoceptive effects under experimental conditions.

The previous summary of cocaine's discriminative stimulus effects suggests that cocaine produces discriminative stimulus effects that are similar to d-amphetamine's discriminative stimulus effects (i.e., both drugs produce increases in extracellular DA concentrations), and vice versa. Indeed, both d-amphetamine [23, 60] and methamphetamine ([23]; see below for discussion of methamphetamine) produce complete generalization in rats trained to discriminate 10 mg/kg cocaine from saline.

6.2 Methamphetamine: Prototypical DA Releaser

Similar to d-amphetamine, methamphetamine disrupts vesicular dopamine storage (e.g., [61]) and produces regional increases in DA content (e.g., [62]), although methamphetamine also alters serotonergic neurotransmission to a greater extent than amphetamine [62]. In rats trained to discriminate 1 mg/kg methamphetamine, Munzar et al. [63] reported that phentermine (an amphetamine analog) produced complete generalization. In addition, phentermine administered in combination with 1 mg/kg fenfluramine produced a rightward shift in phentermine's dose–response curve, indicating that the 5-HT-releasing effects of fenfluramine decreased the potency of phentermine's discriminative stimulus effects. In a human drug discrimination experiment, Lamb and Henningfield [64] trained human volunteers to discriminate 30 mg d-amphetamine from placebo. Substitution tests were conducted with d-amphetamine, methamphetamine, and the μ-opioid agonist hydromorphone. Results revealed that d-amphetamine and methamphetamine produced full substitution for d-amphetamine's discriminative stimulus effects; however, hydromorphone failed to produce >30% drug-appropriate responding indicating that the interoceptive effects of hydromorphone were dissimilar from the effects produced by d-amphetamine. It should be noted that the drug discrimination assay is pharmacologically selective; that is, if participants are trained to discriminate a compound that increases intraterminal release of dopamine (e.g., d-amphetamine), then a compound with binding affinity to the μ-opioid receptor (e.g., hydromorphone) is unlikely to produce complete generalization. Indeed, many drug discrimination experiments include a substitution test compound with a different pharmacological mechanism of action to serve as a negative control (i.e., the researchers predict that the drug will engender low percent drug-lever selection).

Norepinephrine has also been studied as a potential modulator of methamphetamine's discriminative stimulus effects. In rats trained to discriminate 1 mg/kg methamphetamine from saline, the selective NET inhibitors desipramine and nisoxetine did not substitute for methamphetamine when administered alone, but each compound significantly shifted the methamphetamine dose–response curve to the left when administered as pretreatments [65] demonstrating that they potentiated methamphetamine's discriminative stimulus. Interestingly, and in apparent contrast with cocaine, neither the β-adrenoceptor agonist isoproterenol nor the antagonist propranolol generalized to methamphetamine when given alone nor altered the discriminative stimulus effects of methamphetamine when administered in combination [65]. No systematic or dose-related effects of the α-adrenoceptor agonists methoxamine (α_1) and clonidine (α_2) or the α-adrenoceptor antagonists prazosin (α_1) or yohimbine (α_2) were apparent when substituted for methamphetamine or when administered in combination with methamphetamine, although the α_2 ligands tended to produce larger magnitude effects in comparison with the α_1 ligands.

As previously noted with cocaine, 5-HT receptor subtypes appear to modulate the discriminative stimulus effects of methamphetamine in the rat. One notable study directly compared the effects of the hallucinogenic 5-HT$_2$ agonist DOI on cocaine-like interoceptive effects occasioned by either cocaine or methamphetamine in rats trained to discriminate 10 mg/kg cocaine from saline [66]. As expected, methamphetamine fully substituted for cocaine in these subjects, but while pretreatment with DOI did not alter the dose-effect curve for cocaine, it dramatically potentiated the cocaine-like discriminative stimulus effects of methamphetamine, as evidenced by a twofold shift in the methamphetamine dose-effect curve [66]. These data indicate that 5-HT$_2$ receptor activation is involved in psychostimulant discriminative stimulus effects, and may be particularly salient in the context of the interoceptive effects of methamphetamine.

7 Conclusion

The discriminative stimulus effects of psychostimulants have been well characterized using drug discrimination procedures. Regarding the psychostimulants presented in this chapter, monoaminergic neurotransmission largely mediates their discriminative stimulus effects, with dopamine being the most salient neurotransmitter in this regard (see Table 1). Moreover, comparable substitution profiles of psychostimulants are observed across species (e.g., mice, rats, nonhuman primates, and humans), indicating that drug discrimination procedures produce reliable observations of their discriminative stimulus effects. Although there is a paucity of drug discrimination experiments that directly compare the discrimination performance of females to males, there appears to be little difference in ability to discriminate interoceptive cues of psychostimulants as a function of sex in rodents. Instead, it is possible that the qualitative nature of a drug's interoceptive cue (i.e., the information that can be provided through verbal report in humans) differs between sexes. Future drug discrimination experiments in humans who possess a verbal repertoire are necessary to elucidate further any differences in qualitative aspects of a drug's discriminative cue.

In addition to providing translational value, drug discrimination procedures also permit analysis of drug stereochemistry, pharmacokinetics, and metabolic interactions. From this chapter, readers may glean the complexity that is involved in determining a drug's discriminative stimulus effects. Indeed, the discriminative stimulus effects of psychoactive substances vary with an innumerable number of biological, environmental, and behavioral factors, many of which were not addressed here. Nevertheless, the drug discrimination assay, in its most basic form, reveals pharmacological effects that occur within the central nervous system in species that display little to no verbal communication. We consider this an achievement in scientific research in general, and we submit that the drug discrimination approach is among the most useful in vivo analyses available to behavioral pharmacology.

Table 1 Protein mechanisms that produce full substitution, weaken, or enhance the interoceptive effects of psychostimulants

Psychostimulant training drug	Mechanisms producing full substitution	Mechanisms that weaken interoceptive effects	Mechanisms that enhance interoceptive effects	References/reviews
Cocaine	– Substrates/releasers at DAT – Reuptake inhibitors at DAT – D_2 receptor agonists – D_3 receptor agonists – β_2-adrenoceptor antagonist	– D_2 receptor antagonism – D_1 receptor antagonism – Nonselective 5-HT$_2$ receptor antagonists – 5-HT$_{2A}$ receptor antagonist	– Reuptake inhibitors at DAT – β_2-adrenoceptor antagonism – Selective reuptake inhibitors at SERT (SSRIs) – 5-HT$_{2C}$ antagonists – Muscarinic M_1 receptor antagonism	[13–20, 23, 24, 26–29, 37–39, 41, 42]
MDMA (racemic and stereoisomers included)	– Substrates/releasers at monoamine transporters – Reuptake inhibitors at monoamine transporters – 5-HT$_{1A/2A}$ agonists – 5-HT$_{1B/2C}$ agonists	– 5-HT$_2$ receptor antagonists – 5-HT$_3$ receptor antagonists – 5-HT depletion – D_1 receptor antagonists – D_2 receptor antagonists (time-dependent effect occurs 105 min post-injection of MDMA)	– 5-HT releaser (following neurotoxic regimen)	[45, 46, 49, 56, 67–73]
d-Amphetamine and Methamphetamine	– DA releasers – D_2 receptor agonists	– DA releasers +5-HT releasers (5-HT releasers have weakening effect)	– D_1 receptor agonists – D_2 receptor agonists – NET inhibitors – 5-HT$_2$ agonist (shifted methamphetamine curve to right in cocaine-trained rats)	[58, 63, 65, 66]

References

1. Glennon RA, Young R (2011) Methodological considerations. In: Glennon RA, Young R (eds) Drug discrimination: applications to medicinal chemistry and drug studies. John Wiley & Sons, Inc., Hoboken, pp. 19–40
2. Porter JH, Prus AJ (2017) Introduction and overview of drug discrimination. In Porter JH, Prus AJ (eds) The behavioral neuroscience of drug discrimination

3. Quiñones-Jenab V, Perrotti LI, Fabian SJ, Chin J, Russo SJ, Jenab S (2001) Endocrinological basis of sex differences in cocaine-induced behavioral responses. Ann N Y Acad Sci 937:140–171
4. Segarra AC, Agosto-Rivera JL, Febo M, Lugo-Escobar N, Menéndez-Delmestre R, Puig-Ramos A, Torres-Diaz YM (2010) Estradiol: a key biological substrate mediating the response to cocaine in female rats. Horm Behav 58:33–43
5. Becker JB, Koob GF (2016) Sex differences in animal models: focus on addiction. Pharmacol Rev 68:242–263
6. Lynch WJ (2006) Sex differences in vulnerability to drug self-administration. Exp Clin Psychopharmacol 14:34–41
7. Koe BK (1976) Molecular geometry of inhibitors of the uptake of catecholamines and serotonin in synaptosomal preparations of rat brain. J Pharmacol Exp Ther 199:649–661
8. Matecka D, Rothman RB, Radesca L, de Costa BR, Dersch CM, Partilla JS, Pert A, Glowa JR, Wojnicki HE, Rice KC (1996) Development of novel, potent, and selective dopamine reuptake inhibitors through alteration of the piperazine ring of 1-[2-(Diphenylmethoxy)ethyl]- and 1-[2-[Bis(4-flurophenyl)methoxy]ethyl]-4-(3-phenylpropyl)piperazines (GBR 12935 and GBR 12909). J Med Chem 39:4704–4716
9. Rothman RB, Baumann MH, Dersch CM, Romero DV, Rice KC, Carroll FI, Partilla JS (2001) Amphetamine-type central nervous system stimulants release norepinephrine more potently than they release dopamine and serotonin. Synapse 39:32–41
10. Kilpatrick GJ, Jones BJ, Tyers MB (1989) Binding of the 5-HT$_3$ ligand GR65630, to rat area postrema vagus nerve and the brain of several species. Eur J Pharmacol 159:157–164
11. Sharkey J, Ritz MC, Schenden JA, Hanson RC, Kuhar MJ (1988a) Cocaine inhibits muscarinic cholinergic receptors in heart and brain. J Pharmacol Exp Ther 246:1048–1052
12. Sharkey J, Glen KA, Wolfe S, Kuhar MJ (1988b) Cocaine binding at sigma receptors. Eur J Pharmacol 149:171–174
13. Colpaert FC, Niemegeers CJE, Janssen PAJ (1976) Cocaine cue in rats as it relates to subjective drug effects: a preliminary report. Eur J Pharmacol 40:195–199
14. McKenna ML, Ho BT (1980) The role of dopamine in the discriminative stimulus properties of cocaine. Neuropharmacology 19:297–303
15. Craft RM, Stratmann JA (1996) Discriminative stimulus effects of cocaine in female versus male rats. Drug Alcohol Depend 42:27–37
16. Anderson KG, van Haaren F (2000) Effects of SCH-23390 and raclopride on cocaine discrimination in male and female Wistar rats. Pharmacol Biochem Behav 65:671–675
17. Colpaert FC, Niemegeers CJE, Janssen PAJ (1978) Discriminative stimulus properties of cocaine and d-amphetamine, and antagonism by haloperidol: a comparative study. Neuropharmacology 17:937–942
18. D'Mello GD, Stolerman IP (1977) Comparison of the discriminative stimulus properties of cocaine in rats. Br J Pharmacol 61:415–422
19. Cunningham KA, Callahan PM (1991) Monoamine reuptake inhibitors enhance the discriminative state induced by cocaine in the rat. Psychopharmacology 104:117–180
20. Kleven MS, Koek W (1998) Discriminative stimulus properties of cocaine: enhancement by monoamine reuptake blockers. J Pharmacol Exp Ther 284:1015–1025
21. Tella SR, Goldberg SR (2001) Subtle differences in the discriminative stimulus effects of cocaine and GBR-12909. Prog Neuro-Psycopharmacol Biol Psychiatry 25:639–656
22. Kleven MS, Koek W (1997) Discriminative stimulus properties of cocaine: enhancement by β-adrenergic receptor antagonists. Psychopharmacology 131:307–312
23. Li SM, Campbell BL, Katz JL (2006) Interactions of cocaine with dopamine uptake inhibitors or dopamine releasers in rats discriminating cocaine. J Pharmacol Exp Ther 317:1088–1096
24. Rush CR, Baker RW (2001) Behavioral pharmacological similarities between methylphenidate and cocaine in cocaine abusers. Exp Clin Psychopharmacol 9:59–73

25. Volkow ND, Wang G-J, Fischman MW, Foltin RW, Fowler JS, Abumrad NN, Vitkun S, Logan J, Gatley SJ, Pappas N, Hitzemann R, Shea CE (1997) Relationship between subjective effects of cocaine and dopamine transporter occupancy. Nature 386:827–830

26. Callahan PM, Appel JB, Cunningham KA (1991) Dopamine D_1 and D_2 mediation of the discriminative stimulus properties of d-amphetamine and cocaine. Psychopharmacology 103:50–55

27. Acri JB, Carter SR, Alling K, Geter-Douglass B, Dijkstra D, Wikström H, Katz JL, Witkin JM (1995) Assessment of cocaine-like discriminative stimulus effects of dopamine D_3 receptor ligands. Eur J Pharmacol 281:R7–R9

28. Garner KJ, Baker LE (1999) Analysis of D_2 and D_3 receptor-selective ligands in rats trained to discriminate cocaine from saline. Pharmacol, Biochem Behav 64:373–378

29. Callahan PM, de la Garza IIR, Cunningham KA (1997) Mediation of the discriminative stimulus properties of cocaine by mesocorticolimbic dopamine systems. Pharmacol Biochem Behav 57:601–607

30. Valente MJ, De Pinho PG, de Lourdes BM, Carvalho F, Carvalho M (2014) Khat and synthetic cathinones: a review. Arch Toxicol 88:15–45

31. De Felice LJ, Glennon RA, Negus SS (2014) Synthetic cathinones: chemical phylogeny, physiology, and neuropharmacology. Life Sci 97:20–26

32. Gatch MB, Taylor CM, Forster MJ (2013) Locomotor stimulant and discriminative stimulus effects of 'bath salt' cathinones. Behav Pharmacol 24(5–6):437–447

33. Gannon BM, Williamson A, Suzuki M, Rice KC, Fantegrossi WE (2016) Stereoselective effects of abused "bath salt" constituent 3,4-methylenedioxypyrovalerone in mice: drug discrimination, locomotor activity, and thermoregulation. J Pharmacol Exp Ther 356 (3):615–623

34. Gannon BM, Fantegrossi WE (2016) Cocaine-like discriminative stimulus effects of mephedrone and naphyrone in mice. J Drug Alcohol Res 5:236009

35. Baumann MH, Partilla JS, Lehner KR, Thorndike EB, Hoffman AF, Holy M, Rothman RB, Goldberg SR, Lupica CR, Sitte HH, Brandt SD, Tella SR, Cozzi NV, Schindler CW (2013) Powerful cocaine-like actions of 3,4-methylenedioxypyrovalerone (MDPV), a principal consitutent of psychoactive 'bath salts' products. Neuropsychopharmacology 38:552–562

36. Fantegrossi WE, Gannon BM, Zimmerman SM, Rice KC (2013) In vivo effects of abused 'bath salt' constituent 3,4-methylenedioxypyrovalerone (MDPV) in mice: drug discrimination, thermoregulation, and locomotor activity. Neuropsychopharmacology 38:563–573

37. Young R, Glennon RA (2009) $S(-)$propranolol as a discriminative stimulus and its comparison to the stimulus effects of cocaine in rats. Psychopharmacology 203:369–382

38. Schama KF, Howell LL, Byrd LD (1997) Serotonergic modulation of the discriminative-stimulus effects of cocaine in squirrel monkeys. Psychopharmacology 132:27–34

39. Filip M, Bubar MJ, Cunningham KA (2006) Contribution of serotonin (5-HT) 5-HT$_2$ receptor subtypes to the discriminative stimulus effects of cocaine in rats. Psycopharmacology 183:482–489

40. Paris JM, Cunningham KA (1991) Serotonin 5-HT$_3$ antagonists do not alter the discriminative stimulus properties of cocaine. Psychopharmacology 104:475–478

41. Walsh SL, Cunningham KA (1997) Serotonergic mechanisms involved in the discriminative stimulus, reinforcing and subjective effects of cocaine. Psychopharmacology 130:41–58

42. Tanda G, Katz JL (2007) Muscarinic preferential M$_1$ receptor antagonists enhance the discriminative-stimulus effects of cocaine in rats. Pharmacol Biochem Behav 87:400–404

43. Hiranita T, Soto PL, Tanda G, Katz JL (2011) Lack of cocaine-like discriminative-stimulus effects of σ-receptor agonists in rats. Behav Pharmacol 22:525–530

44. Baumann MH, Ayestas Jr MA, Partilla JS, Sink JR, Shulgin AT, Daley PF, Brandt SD, Rothman RB, Ruoho AE, Cozzi NV (2012) The designer methcathinone analogs, mephedrone and methylone, are substrates for monoamine transporters in brain tissue. Neuropsychopharmacology 37:1192–1203

45. Oberlander R, Nichols DE (1988) Drug discrimination studies with MDMA and amphetamine. Psychopharmacology 95:71–76
46. Goodwin AK, Baker LE (2000) A three-choice discrimination procedure dissociates the discriminative stimulus effects of d-amphetamine and ($\pm$)-MDMA in rats. Exp Clin Psychopharmacol 8:415–423
47. Broadbear JH, Tunstall B, Beringer K (2011) Examining the role of oxytocin in the interoceptive effects of 3,4-methylenedioxymethamphetamine (MDMA, 'ecstasy') using a drug discrimination paradigm in the rat. Addict Biol 16:202–214
48. Johanson C-E, Kilbey M, Gatchalian K, Tancer M (2006) Discriminative stimulus effects of 3,4-methylenedioxymethamphetamine (MDMA) in humans trained to discriminate among d-amphetamine, $meta$-chlorophenylpiperazine and placebo. Drug Alcohol Depend 81:27–36
49. Murnane KS, Murai N, Howell LL, Fantegrossi WE (2009) Discriminative stimulus effects of psychostimulants and hallucinogens in S(+)-3,4methylenedioxymethamphetamine (MDMA) and R(−)-MDMA trained mice. J Pharmacol Exp Ther 331:717–723
50. Fantegrossi WE, Murai N, Mathúna BÓ, Pizarro N, de la Torre R (2009) Discriminative stimulus effects of 3,4-methylenedioxymethamphetamine and its enantiomers in mice: pharmacokinetic considerations. J Pharmacol Exp Ther 329:1006–1015
51. de la Torre R, Farré M, Ortuño J, Mas M, Brenneisen R, Roset PN, Segura J, Camí J (2000) Non-linear pharmacokinetics of MDMA ('ecstasy') in humans. Br J Clin Pharmacol 49:104–1098
52. Harper DN, Crowther A, Schenk S (2011) A comparison of MDMA and amphetamine in the drug discrimination paradigm. Open Addict J 4:22–23
53. Khorana N, Pullagurla MR, Young R, Glennon RA (2004) Comparison of the discriminative stimulus effects of 3,4-methylenedioxymethamphetamine (MDMA) and cocaine: asymmetric generalization. Drug Alcohol Depend 74:281–287
54. Harper DN, Langen A-L, Schenk S (2014) A 3-lever discrimination procedure reveals differences in the subjective effects of low and high doses of MDMA. Pharmacol Biochem Behav 116:9–15
55. Pawlowski L, Siwanowicz J, Bigajska K, Przegaliński E (1985) Central antiserotonergic and antidopaminergic action of pirenperone, a putative 5-HT2 receptor antagonist. Pol J Pharmacol Pharm 37:179–196
56. Harvey EL, Baker LE (2016) Differential effects of 3,4-methylenedioxypyrovalerone (MDPV) and 4-methylmethcathinone (mephedrone) in rats trained to discriminate MDMA or a d-amphetamine + MDMA mixture. Psychopharmacology 233:673–680
57. Kuczenski R, Segal DS (1994) Neurochemistry of amphetamine. In: Cho AK, Segal DS (eds) Amphetamine and its analogs: psychopharmacology, toxicology, and abuse. Academic Press, Inc., San Diego, pp. 81–114
58. Smith FL, St. John C, Yang TFT, Lyness WH (1989) Role of specific dopamine receptor subtypes in amphetamine discrimination. Psycopharmacology 97:501–506
59. Vansickel AR, Lile JA, Stoops WW, Rush CR (2007) Similar discriminative-stimulus effects of D-amphetamine in women and men. Pharmacol Biochem Behav 87:289–296
60. Kueh D, Baker LE (2007) Reinforcement schedule effects in rats trained to discriminate 3,4-methylenemethamphetamine (MDMA) or cocaine. Psychopharmacology 189:447–457
61. Brown JM, Hanson GR, Fleckenstein AE (2000) Methamphetamine rapidly decreases vesicular dopamine uptake. J Neurochem 74:2221–2223
62. Kuczenski R, Segal DS, Cho AK, Melega W (1995) Hippocampus norepinephrine, caudate dopamine and serotonin, and behavioral responses to stereoisomers of amphetamine and methamphetamine. J Neurosci 15:1308–1317
63. Munzar P, Baumann MH, Shoaib M, Goldberg SR (1999) Effects of dopamine and serotonin-releasing agents on methamphetamine discrimination and self-administration in rats. Psychopharmacology 141:287–296
64. Lamb RJ, Henningfield JE (1994) Human d-amphetamine drug discrimination: methamphetamine and hyrodmorphone. J Exp Anal Behav 61:169–180

65. Munzar P, Goldberg SR (1999) Noradrenergic modulation of the discriminative-stimulus effects of methamphetamine in rats. Psychopharmacology 143:293–301
66. Munzar P, Justinova Z, Kutkat SW, Goldberg SR (2002) Differential involvement of 5-HT (2A) receptors in the discriminative-stimulus effects of cocaine and methamphetamine. Eur J Pharmacol 436:75–82
67. Bubar MJ, Pack KM, Frankel PS, Cunningham KA (2004) Effects of dopamine D_1- and D_2-like receptor antagonists on the hypermotive and discriminative stimulus effects of (+)-MDMA. Psychopharmacology 173:326–336
68. Schechter MD (1988) Serotonergic-dopaminergic mediation of 3,4-methylenedioxymethamphetamine (MDMA, "ecstasy"). Pharmacol Biochem Behav 31:817–824
69. Glennon RA, Higgs R, Young R, Issa H (1992) Further studies on N-methyl-1 (3,4-methylenedioxyphenyl)-2-aminopropane as a discriminative stimulus: antagonism by 5-hydroxytryptamine3 antagonists. Pharmacol Biochem Behav 43:1099–1106
70. Yarosh HL, Katz EB, Coop A, Fantegrossi WE (2007) MDMA-like behavioral effects of N-substituted piperazines in the mouse. Pharmacol Biochem Behav 88:18–27
71. Schechter MD (1991) Effect of serotonin depletion by p-chlorophenylalanine upon discriminative behaviours. Gen Pharmacol 22:889–893
72. Baker LE, Makhay MM (1996) Effects of (+)-fenfluramine on 3,4-methylenedioxymethamphetamine (MDMA) discrimination in rats. Pharmacol Biochem Behav 53:455–461
73. Cole JC, Sumnall HR (2003) The pre-clinical behavioural pharmacology of 3,4-methylenedioxymethamphetamine (MDMA). Neurosci Biobehav Rev 27:199–217

Discriminative Stimulus Properties of S(−)-Nicotine: "A Drug for All Seasons"

John A. Rosecrans and Richard Young

Abstract S(−)-Nicotine is the major pharmacologically active substance in tobacco and can function as an effective discriminative stimulus in both experimental animals and humans. In this model, subjects must detect and communicate the nicotine drug state versus the non-drug state. This review describes the usefulness of the procedure to study nicotine, presents a general overview of the model, and provides some relevant methodological details for the establishment of this drug as a stimulus. Once established, the (−)-nicotine stimulus can be characterized for dose response and time course effects. Moreover, tests can be conducted to determine the similarity of effects produced by test drugs to those produced by the training dose of nicotine. Such tests have shown that the stimulus effects of nicotine are stereoselective [S(−)-nicotine >R(+)-nicotine] and that other "natural" tobacco alkaloids and (−)-nicotine metabolites can produce (−)-nicotine-like effects, but these drugs are much less potent than (−)-nicotine. Stimulus antagonism tests with mecamylamine and DHβE (dihydro-β-erythroidine) indicate that the (−)-nicotine stimulus is mediated via α4β2 nicotinic acetylcholine receptors (nAChRs) in brain; dopamine systems also are likely involved. Individuals who try to cease their use of

Sadly, **Dr. John A. Rosecrans** passed away during the writing of this chapter. John inspired both students and colleagues with his keen interest and enthusiasm in matters related to biomedical research and, especially, nicotine. Most of all, John will be remembered and missed for his friendship. The editors (J. H. Porter and A. J. Prus) of the book (*The Behavioural Neuroscience of Drug Discrimination*) this chapter is published in would also like to note that the book is dedicated to Dr. John A. Rosecrans.

J.A. Rosecrans
Department of Pharmacology and Toxicology, Virginia Commonwealth University, 410 North 12th Street, P.O. Box 980613, Richmond, VA 23298-0613, USA

R. Young (✉)
Department of Medicinal Chemistry, Virginia Commonwealth University, 800 East Leigh Street, P.O. Box 980540, Richmond, VA 23219-0540, USA
e-mail: ryoung@vcu.edu

© Springer International Publishing AG 2017
Curr Topics Behav Neurosci (2018) 39: 51–94
DOI 10.1007/7854_2017_3
Published Online: 4 July 2017

nicotine-based products are often unsuccessful. Bupropion (Zyban®) and varenicline (Chantix®) may be somewhat effective as anti-smoking medications because they probably produce stimulus effects that serve as suitable substitutes for (−)-nicotine in the individual who is motivated to quit smoking. Finally, it is proposed that future drug discrimination studies should apply the model to the issue of maintenance of abstinence from (−)-nicotine-based products.

Keywords Anabasine • Anatabine • Cotinine • Drug abuse • Drug discrimination • Lobeline • Methyllycaconitine • Nicotine • Nornicotine • Stereoisomers

Contents

1 Introduction

$S(−)$-Nicotine is one of the oldest and most widely used psychoactive drugs. Historically, (−)-nicotine ingestion, through tobacco smoking, has been traced to the Mayan civilization in Mexico (circa 600 A.D.). In the pre-Columbian Americas, it was smoked in pipes, chewed, and/or insufflated by itself or in combination with hallucinogenic snuffs (e.g., [1, 2]). Botanically, the tobacco plant belongs to the nightshade family *Solanaceae* and, therefore, is related to tomato and potato plants as well as to "deadly nightshade" (*Atropa belladonna*), from which belladonna (i.e., tropane alkaloids atropine, scopolamine, and hyoscyamine) is derived. Tobacco also belongs to the genus *Nicotiana*, named for Jean Nicot, French ambassador to

Portugal in the mid-sixteenth century. It was Nicot who first sent tobacco to the king of France. From France, its use spread throughout Europe. A South American species, *N. tabacum*, is the source for most of today's commercially marketed tobacco products (e.g., [3–5]).

Chemically, nicotine (1-methyl-2-(3-pyridyl)pyrrolidine; Fig. 1) is a tertiary amine composed of pyridine and pyrrolidine rings whose molecular structure was first proposed by Pinner [6] and confirmed by Pictet and Crepieux [7] and Spath and Bretschneider [8]. Moreover, nicotine has one chiral center (at carbon 2 of the pyrrolidine moiety) and natural nicotine, as constituted in tobacco, has a levorotatory [i.e., (−)] rotation (also called (−)-nicotine or *l*-nicotine). Most importantly, however, (−)-nicotine has the (*S*)-configuration, which provides information about the chemical structure of (−)-nicotine in three-dimensional space and how it may interact with receptors [9].

(−)-Nicotine-based products can generally be divided into two types: smoked tobacco (cigarette/cigar/pipe and hookah smoking of tobacco) and smokeless tobacco (chewing tobacco, snuff, and snus). Recently, however, electronic cigarettes (E-cigarette or E-cig) that produce (−)-nicotine in vaporized form have appeared in the marketplace. An E-cig is a battery-powered vaporizer that is thought to produce a similar sensation to tobacco smoking (a.k.a. "vaping"). The device employs a heating element that atomizes a liquid solution known as e-liquid, which usually contains a mixture of propylene glycol, glycerin, (−)-nicotine, and flavorings. All of these products contain (−)-nicotine but their use can vary significantly from person to person and product to product. For many tobacco users, continued nicotine consumption results in dependence (compulsive nicotine seeking and use), even at the risk of negative health consequences. Smokers, for example, may become physically addicted to (−)-nicotine and link smoking with many social- and work-related activities, which produce difficulties if the individual desires to cease smoking. Furthermore, if (−)-nicotine levels in the body are changed, smokers tend to compensate to reach their "comfort" level of drug by smoking more or less if the levels of nicotine are reduced [e.g., by administration of mecamylamine, a noncompetitive nicotinic receptor antagonist (see "mecamylamine" below)] or are increased [e.g., administration of exogenous (−)-nicotine], respectively. Moreover, smokers can "titrate" the level of (−)-nicotine in their system with adjustments in the number of puffs on a cigarette, duration of puffs, inter-puff intervals, and/or number of cigarettes smoked (e.g., [10]).

When a nicotine product is smoked, chewed, or inhaled, it is readily absorbed into the bloodstream and penetrates the blood–brain barrier to produce central effects. In addition, its peripheral actions include effects on the autonomic ganglia,

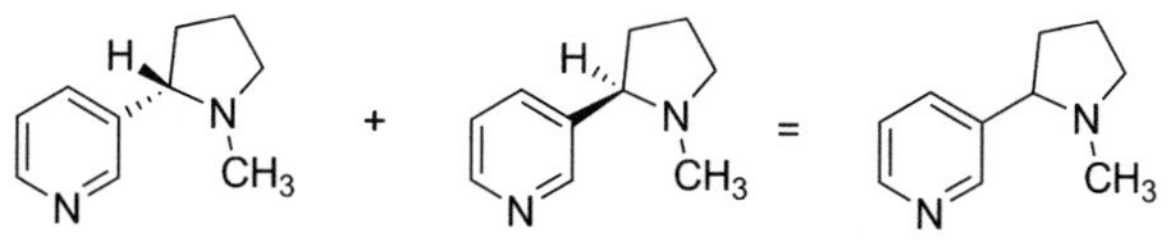

Fig. 1 Structures of S(−)-nicotine (*left*), R(+)-nicotine (*center*) and racemic nicotine (*right*)

adrenal medulla, and neuromuscular junction. It is important to note that (−)-nicotine acts as a stimulant at these sites only when administered at relatively low doses. When higher doses of the drug are administered, membranes are depolarized and maintained in the depolarized state for an extended period of time; i.e., blockade of nicotinic cholinergic (nACh) signals. This biphasic action of nicotine (stimulation followed by blockade of transmission due to a maintained depolarization) can complicate the formation of clear conclusions of its pharmacological actions. Consequently, (−)-nicotine often exhibits a "narrow-window" or steep dose-effect function between doses (or concentrations) that produce excitation and doses that exert blockade of biological actions. The interested reader is referred to Matta et al. [11], who have reviewed and compiled recommended doses of (−)-nicotine for in vivo research. In particular, these authors have noted that responses to (−)-nicotine often display a bell-shaped (inverted U-shaped) dose-response profile.

In humans, (−)-nicotine can function as both a "stimulant" and a "sedative." For example, immediately after exposure to nicotine, there is a "stimulant-kick" caused, in part, by its stimulation of the adrenal glands and resultant discharge of epinephrine (adrenaline). The release of epinephrine stimulates the body and causes a sudden release of glucose as well as an increase in blood pressure, respiration, and heart rate. Nicotine also suppresses insulin output from the pancreas, which indicates that smokers are usually hyperglycemic (higher blood sugar level). Centrally, (−)-nicotine has affinity for all brain nAChR subtypes, but binds preferentially and with high affinity to $\alpha4\beta2$ nAChRs (e.g., [12, 13]). Moreover, (−)-nicotine (indirectly) can produce a release of dopamine in brain regions that are thought to control pleasure and motivation; dopamine is thought to underlie the pleasurable sensations experienced by smokers (e.g., [14, 15] but see [16]). In addition, nicotine also can exert a sedative effect, depending on the smoker's level of arousal and administered dose of nicotine. Thus, (−)-nicotine seems to produce a unique combination of effects: when stimulation is needed, smokers may perceive the "smoke as a stimulant," and when they feel anxious and desire relief, they may perceive the "smoke as a tranquilizer." In this regard, the first author of this review has often referred to the human appeal for nicotine as "a drug for all seasons."

The dual effects of (−)-nicotine also can be seen in animal behavior. For example, in rodents, administration of low doses of nicotine produced increased motor activity whereas high doses produced decreased motor activity (e.g., [17, 18]). Moreover, the effects of nicotine on motor activity of animals can be dependent on pre-drug activity levels. That is, nicotine caused decreased activity of rodents that had a high pre-drug level of activity and produced increased activity of animals that had a low pre-drug level of activity. Also, pharmacological effects of nicotine have been observed to be contingent on whether subjects were pre-exposed to the behavioral paradigm under investigation. Lastly, different strains and gender of rodents have been shown to interact differentially in these aforementioned effects (e.g., [19–21]). Taken together, the effects of nicotine seem to be markedly dependent on the dose of (−)-nicotine as well as subjects' pre-drug level of activity, pre-exposure (i.e., level of tolerance) to nicotine and familiarity with the behavioral paradigm (e.g., [22]; reviewed in [23]). As such, these studies have provided important data as to how the acute effects of nicotine can affect the behavior of animals and, by extension, of humans. On the other hand, these assays of the acute effects of $(-)$-nicotine tend to engender much variability in results and this has led to searches of animal models in which dependent measures are more stable.

2 Drug Discrimination

A complete review of the drug discrimination literature of $(-)$-nicotine is beyond the scope of this review. Rather, the focus here is a description of the usefulness of the drug discrimination procedure to study $(-)$-nicotine, methodological issues and procedures, stereochemical aspects of nicotine, stimulus effects of other tobacco alkaloids and/or $(-)$-nicotine metabolites, nicotinic acetylcholine receptor (nAChR) mechanisms of action, and current pharmacotherapy for cessation of $(-)$-nicotine ingestion.

2.1 Why Use Drug Discrimination (DD) Procedures to Study S(−)-Nicotine?

An early study in humans by Johnston [24] demonstrated that the injection of $(-)$-nicotine was perceived as "pleasant" to smokers and "unpleasant" to nonsmokers. In fact, this study may have been the first scientific demonstration that nicotine has "appeal" to smokers that is not readily apparent to nonsmokers. A pivotal reason that nicotine has appeal to smokers, but not to nonsmokers, is that tolerance has developed to the unpleasant acute effects produced by nicotine or other tobacco constituents that are experienced by neophyte smokers; nausea and/or vomiting, dizziness, sweating, pallor, headache, and weakness (e.g., [25]). The inexperienced smoker cannot usually abide the amount of $(-)$-nicotine present in a single cigarette, but after sufficient experience with their consumption may be able to smoke many cigarettes over a relatively short period of time without the experiences of these adverse effects. Thus, the acute effects of $(-)$-nicotine may include more and/or different pharmacological actions than the chronic effects of $(-)$-nicotine. In fact, acquired tolerance to these adverse effects of nicotine probably exerts an important role in the acquisition and maintenance of dependence and consequent health problems that are linked to the use of tobacco products (e.g., [26]).

Basic research of the effects of $(-)$-nicotine on biological/behavioral variables has mainly employed acute nicotine treatment. However, human users of nicotine products are exposed to the substance chronically. Similarly, subjects in drug discrimination studies are exposed to training drug [e.g. $(-)$-nicotine] chronically. Thus, discrepancies in results between acute studies and chronic investigations of nicotine may be due to differences in responses from subjects who have different sensitivities to nicotine. Moreover, subjects' neural adaptations that result from repeated exposure to nicotine, such as with human smokers or participants in drug discrimination procedures, are very unlikely to be seen in subjects exposed to acute administration of nicotine (e.g., [27–29]). Taken together, drug discrimination procedures appear to simulate, to a reasonable degree, human involvement with (−)-nicotine over time (e.g., [30]). Moreover, the drug discrimination paradigm is one of only a few preclinical assays to have a counterpart procedure for humans (e.g., [31–33]).

Drug discrimination procedures are dependent on the ability of a subject to detect a specific drug state, which is similar to a human report of the subjective effects produced by a drug. This approach does not focus on the behavioral effects of a drug, but instead, is used to study subjects' internal reactions or "perceptions" of the drug effect(s). *In other words, the paradigm allows subjects to identify the effects of (−)-nicotine rather than being a procedure that studies the excitatory or disruptive effects of (−)-nicotine on behavior.* Thus, the DD procedure is not measuring the disruptive or other acute pharmacological effects (e.g., stimulation) of nicotine, but only the ability of an animal to detect the "state" produced by nicotine after chronic administration. As such, animals typically become behaviorally tolerant to the disruptive (acute) effects of (−)-nicotine given at the beginning of training so that experimental results are not encumbered by changes in rates of behavior. Importantly, however, tolerance to the stimulus effects of (−)-nicotine does not readily occur, which allows the experimenter to study the effects of nicotine in a repeated- or within-subjects experimental design over an extended period of time (often ≥2 years, see [30]). If tolerance did occur, then subjects would no longer be able to demonstrate that they recognize differences in effects between their training dose of nicotine and saline vehicle (control) states.

(−)-Nicotine, like many psychoactive drugs, can exert discriminative control over behavior (for review, see [34]). Historically, the first detailed publication on the stimulus properties of nicotine was reported by Morrison and Stephenson [35], who trained rats to discriminate the effects of 0.4 mg/kg (s.c.) of (−)-nicotine from saline in a two-lever operant conditioning task. Shortly thereafter, Rosecrans and colleagues published a series of studies in rats trained to discriminate (−)-nicotine from saline in both T-maze and two-lever operant tasks [36–41]. These studies firmly established that (−)-nicotine could serve as a centrally mediated discriminative stimulus in rats. Subsequently, other species have been used to establish nicotine as a discriminative stimulus; monkey, mouse, and human (e.g., [32, 33, 42–45]). The rat, however, is most commonly employed. Moreover, results of drug discrimination studies with non-human animal subjects and human research participants have shown a relatively high degree of concordance, which suggests that the DD model may reflect the internal or "subjective" effects of (−)-nicotine in humans (e.g., [32, 46]). The rationale and methods described in these early reports are still relevant today and are recapitulated below (also see [30, 47, 48]).

2.2   Rationale

The discriminative stimulus effects of $(-)$-nicotine involve procedures designed to assess the effects of this drug to exert control over behavior. In the paradigm, a subject is trained to make differential behavioral responses contingent upon administered treatments. An experimental participant (e.g., rodent) is trained to emit one response (such as pressing one lever in a two-lever operant chamber to obtain a reinforcer) following one treatment (e.g., dose of drug), and another response (that is, pressing the opposite-side lever) following a different treatment [e.g., saline vehicle (non-drug)]. These behavioral responses are highly dependent on the subject being able to detect a specific drug state and are similar to the requirement of a human to report the subjective effects of a given psychoactive drug. In these situations, behavioral responses performed by subjects are under the stimulus control of the administered dose of training drug. In other words, the animals' lever responses represent, or are reflective of, their subjective "experience" under a given treatment.

2.3   Methodology

Early drug discrimination studies of $(-)$-nicotine used the T-maze procedure in both positively reinforced and escape tasks, whereas later (and current) studies employed the use of two-lever operant chambers. T-maze tasks required subjects (usually rats) to choose between two alleys on each of several trials. In a typical maze experiment, a rat may have been trained to turn to the right-side alley (i.e., designated the drug-side for that rat) to obtain food reward or escape mild electric shock (i.e., consequences) after administration of its dose of training drug, and to turn to the left-side alley (i.e., designated the vehicle-side for that same rat) to receive consequences after injection of vehicle (usually saline). Experimenters considered the animals' first response during the first trial of sessions, before any consequence (e.g., reward or escape), as a reflection of the degree to which animals had learned to select the treatment-appropriate (i.e., correct) response. There were, however, a number of reasons for the decline in the use of the T-maze and the increased use of two-lever operant tasks. T-maze use declined because a consensus of thought among investigators was that (a) higher doses of drug [$(-)$-nicotine] were needed to train rats in T-maze procedures than in lever tasks and (b) data analysis was limited to the animals' choice on only the first trial within sessions of the T-maze versus the animals' many presses of the levers in the two-lever operant chamber. Thus, if only the first T-maze response was considered, the evaluation of stimulus control was based on a very small sample of responses. In comparison, the two-lever procedure allowed animals to respond at any rate on either lever, and the data could be expressed in terms of % drug [i.e., $(-)$-nicotine]-appropriate responding.

2.3.1 Initial Shaping of Behavior

Rats (usually) are food-restricted to 80–85% of their growing body-weight and shaped to lever-press for reinforcement (e.g., food pellet or sweet milk) with one lever in the operant chamber. The shaping procedure required subjects to be placed in the experimental chambers and taught to lever-press under a fixed ratio-one (FR-1) schedule of reinforcement, such that every lever-press was rewarded. During this initial exposure to training, rats were trained with only the right- or left-side lever in the operant chamber. Typically, half of the subjects were required to press the left-side lever and the other half the right-side lever to obtain reinforcement. The latter tactic was (is) important because of the finding that rodents may learn to use olfactory cues (hints) that remained on the levers by animals that preceded them [49]. After initial shaping, rats are exposed to at least four additional daily 15 min training sessions on the same lever (right or left), during which saline (1 mL/kg; s.c.) was administered 10 min prior to behavioral training; consequently, correct-lever responding in the non-drug (i.e., saline) state was established.

2.3.2 Training Under Both Drug and Non-drug Conditions

Once rats are shaped to lever-press under the saline condition, each subject was then trained to lever-press for food on the opposite lever (again with only one lever present in the chamber) 10 min after the administration of (−)-nicotine (typically, a training dose was chosen between 0.1 and 0.4 mg/kg, s.c.). As under the saline training condition, subjects are exposed to drug for, at least, four daily 15-min sessions. After subjects are trained to lever-press for food separately under both drug and saline conditions, with only one lever present in the chamber (approximately 8–10 training sessions), both levers are then introduced into the chamber and rats are trained daily for 15 min [10 min after either saline or (−)-nicotine administration]. (−)-Nicotine and saline treatment sessions are introduced with a double alternation design: i.e., 2 days with (−)-nicotine and 2 days with saline. This double alternation schedule is maintained throughout the study. During the training procedure, a specific schedule of reinforcement, typically a fixed ratio-10 (FR-10) or variable interval 15-s (VI-15 s) is introduced gradually in order to provide added control over behavior. In an FR schedule, the performer completes a fixed number of responses in order to obtain reinforcement; for example, on an FR10 schedule, every 10th response is reinforced. In VI schedules, the length of time that elapsed before reinforcement is delivered varies around the mean value specified by the schedule; for example, on a VI 15 s schedule, reinforcement is available, on average, after 15 s has elapsed since the last reinforcement, but may be available as shortly as 2 s later, or not until 60 s has elapsed. The first response after a time interval has elapsed produces reinforcement for the subject. Besides these operant schedules of reinforcement, subjects may learn other schedules of reinforcement or ways to discriminate a specific dose of drug from vehicle (for review, see [34]). In studies described below, however, rats learned to discriminate (−)-nicotine from saline on an FR or VI schedule of reinforcement in about 3 months, following 20–30 sessions under each treatment condition.

2.3.3 Testing Procedures

Behavioral data are collected during test sessions in which each animal is administered training dose of (−)-nicotine, other doses of (−)-nicotine, saline, or doses of other drugs in stimulus generalization or antagonism tests. Test sessions are conducted in which both levers are either reinforced (a technique used sometimes with FR schedules of reinforcement) or not reinforced (a tactic used with VI schedules of reinforcement). The degree of stimulus control exerted by (−)-nicotine is determined during these test sessions, which are interspersed between specific double alternation training sessions. The subjects' rate of learning to discriminate between (−)-nicotine and saline is easily monitored via these short sessions; sometimes, these sessions are followed by training under the treatment-correct condition. In addition, such tests are usually conducted on "crossover" days, in which the drug condition alternates from nicotine to saline or vice versa. The subjects' discrimination of (−)-nicotine from saline is considered optimal when they perform at least 90–95% of their lever-presses on the *nicotine-correct lever* following their training dose of (−)-nicotine, whereas they perform 0–5% of their lever-presses *on that same lever* after administration of saline. Experimental data are expressed as percent responses on the nicotine-appropriate lever. *Thus, all data are related to the nicotine-based discriminative stimulus.*

2.3.4 Challenge Experiments

The animals' discrimination of (−)-nicotine from saline is generally established within 2–4 months following initial shaping procedures, at which time a variety of experiments can be conducted. Moreover, the discriminative stimulus effects of nicotine are evaluated during both training and test sessions for up to 2 years in most experimental subjects [30, 47]. Consequently, many of the rats utilized in these studies were exposed to a minimum of 250 nicotine and saline training sessions. Once a group of test subjects had reached training criteria, drug challenge experiments termed (a) stimulus generalizations tests can be initiated to determine if other drugs produce the training drug-like response and (b) stimulus antagonism tests can be conducted to determine if substances (in combination with the training drug) can interfere with the animals' recognition of the training drug-like response (see mechanisms of action section below).

2.3.5 Stimulus Generalization

Stimulus generalization tests are used to determine if a training drug stimulus will generalize (i.e., substitute) to other drugs. The rationale of this approach is that an animal trained to discriminate a dose of training drug exhibits stimulus generalization only to drugs that exert a similar stimulus effect (though not necessarily through an identical mechanism of action). It is important to note that results of stimulus generalization tests are interpreted in relation to the dose of training drug-like effects. As such, and for example, a study of novel substances in $(-)$-nicotine-trained animals reflects the actions of the novel agents to produce "$(-)$-nicotine-like" stimulus effects. In most studies, stimulus generalization is said to have occurred when animals, after administration of a given dose of challenge drug, perform $\geq 80\%$ of their responses on the $(-)$-nicotine-appropriate lever. Where stimulus generalization occurred, an effective dose 50% (ED_{50}) value is calculated and reflects the dose at which animals would be expected to make 50% of their responses on the $(-)$-nicotine-appropriate lever. Besides complete stimulus generalization, two other types of results can occur: partial generalization and saline-like responding. Partial generalization occurs when animals, after being administered a thorough dose effect test, perform approximately ~40–70% of their responses on the nicotine-appropriate lever. Data of this type are very difficult to interpret. However, partial generalization may occur with a test drug because there are pharmacological effects that are common to both the training drug and the test drug; full generalization may not occur because the overlap of pharmacological effects to achieve full substitution is incomplete (for further discussion, see [34]). Lastly, administration of various doses of test drug may result in $\leq 20\%$ $(-)$-nicotine-appropriate responding. This type of result does not necessarily mean that a test drug is inert, but may indicate that the effect of the challenge drug is simply different from that produced by the dose of training drug. That is, the saline-designated lever also serves as a default response for a drug effect that is unlike that of the training drug and, hence, animals perform relatively few responses on the nicotine-appropriate lever. For example, Pratt et al. [50] trained rats to discriminate 0.4 mg/kg of $(-)$-nicotine from saline and reported that test doses between 0.25 and 4 mg/kg of fenfluramine, an appetite suppressant, produced saline-like responding; i.e. a maximum of ~20% $(-)$-nicotine-appropriate responding. Such doses of fenfluramine are not inert and indicate quite clearly that the stimulus effects produced by 0.4 mg/kg of $(-)$-nicotine are different from those produced by fenfluramine. Furthermore, some of the tested doses of fenfluramine have been shown to serve as discriminative stimuli (e.g., [51]; for review, see [34]).

3   Characterization of the Stimulus Effects of (−)-Nicotine

Drug discrimination studies of (−)-nicotine typically begin with an evaluation of the "strength" of the training stimulus and include both dose response and time course tests. Early studies indicated that both the rate of learning and sensitivity to the training drug were observed to be dose related; i.e., rats trained at relatively higher doses of (−)-nicotine learned the discrimination at a more rapid rate and appeared less sensitive to relatively lower doses of (−)-nicotine. After repeated training, however, rats exhibited fewer differences among training doses, but the dose response nature of the discrimination remained the same [30]. For example, drug discrimination learning curves of rats trained at three (−)-nicotine training dose levels (either 0.1, 0.2 or 0.4 mg/kg from saline, s.c.) under VI-15 s, FR-10 or differential reinforcement of low (DRL)-10 s rate of responding schedules of reinforcement were compared for nicotine sensitivity, as measured by ED_{50} doses (Table 1). As can be seen, ED_{50} values were proportional to training doses, but an asymptotic effect occurred above 0.2 mg/kg of (−)-nicotine. Thus, the stimulus effects of 0.2 and 0.4 mg/kg of (−)-nicotine were somewhat equipotent after the drug had exerted discriminative control over the animals' behavior. Table 1 also indicates that separate groups of rats trained to discriminate 0.4 mg/kg of (−)-nicotine from saline under the above FR, VI or DRL schedules of reinforcement displayed essentially equipotent ED_{50} doses of nicotine, which indicated that schedule of reinforcement did not markedly influence the "strength" of the stimulus effects of 0.4 mg/kg of (−)-nicotine [30, 48, 52].

Once a training dose of (−)-nicotine has been established as a discriminative stimulus, tests can be performed to determine its time course of action. Such tests investigate the effects of changing the pre-session injection interval of the training dose of drug and the beginning of a test session. For example, time course studies in the previously mentioned three groups of animals trained under the three doses of (−)-nicotine (under the VI-15 s schedule of reinforcement) were evaluated. The results revealed that percent (−)-nicotine-appropriate responding declined to 50% of its initial effect within 140–160 min after 0.4 mg/kg of nicotine, 100 min after 0.2 mg/kg of (−)-nicotine, and 70 min after 0.1 mg/kg of (−)-nicotine; thus, time course was proportional to dose. A further analysis of these time duration relationships also suggested a link between the appearance/disappearance of (−)-nicotine levels measured in brain areas (telencephalon, diencephalon, and brainstem) and

Table 1 $S(−)$-Nicotine dose response evaluations in rats trained under different schedules of reinforcement

Schedule and (−)-nicotine Training dose (s.c.)	ED_{50} dose mg/kg (95% C.L.)
VI-15 s; 0.1 mg/kg	0.026 (0.009–0.071)
VI-15 s; 0.2 mg/kg	0.079 (0.040–0.161)
VI-15 s; 0.4 mg/kg	0.086 (0.040–0.185)
FR-10; 0.4 mg/kg	0.098 (0.042–0.184)
DRL-10; 0.4 mg/kg	0.093 (0.040–0.215)

Data adapted from Chance et al. [52] and Rosecrans [30]

the time-related stimulus effects of (−)-nicotine [52, 53]. Other studies also have stressed the importance of training dose and pre-session injection intervals in the stimulus properties of (−)-nicotine (e.g., [54, 55]).

4 Nicotine Stereoisomers

Structurally, nicotine is a chiral substance that can exist as one of two stereoisomers: $S(−)$-nicotine or $R(+)$-nicotine (Fig. 1). Optical isomers of biologically active drugs typically display differences in potency, and, on occasion, also can display differences in effect. Comparative studies of the lethality and pharmacology of the optical isomers of nicotine are few, but the limited results are mostly consistent and indicate that the effects of the enantiomers are *stereoselective*: i.e. effects are qualitatively similar, but $S(−)$-nicotine is more potent than $R(+)$-nicotine (e.g., [56]). However, and unfortunately, a review of the literature failed to find even one study that directly compared the effects of racemic nicotine to effects of its stereoisomers in the same assay.

4.1 Lethality

$S(−)$- and $R(+)$-nicotine were reported to be equally toxic after i.p. administration in rats and guinea pigs. However, when guinea pigs were injected s.c., $S(−)$-nicotine was twice as lethal as $R(+)$-nicotine [7, 57]. In comparison, $S(−)$-nicotine was shown to be 7 times more toxic than $R(+)$-nicotine in rats injected intravenously [58]. In other studies, $S(−)$-nicotine was reported to be as toxic as, or slightly more toxic than, (±)-nicotine when administered intraperitoneally in rats, intravenously in rabbits, and intraperitoneally or intravenously in cats [59, 60]. In addition, the former study claimed that synergism resulted when the levorotatory and racemic forms of nicotine were mixed in certain proportions; however, few animals/group were used and no statistics were calculated to support the claim. Also worthy of note is that the lethality of $S(−)$-nicotine is highly dependent upon species [61]. For example, the oral (p.o.) lethal dose 50% (LD_{50}) of (−)-nicotine in mouse ($LD_{50} = 3.3$ mg/kg) is approximately 3 times more potent than that in dog ($LD_{50} = 9.2$ mg/kg) and over 15 times more potent than that in rat ($LD_{50} = 50$ mg/kg). Thus, mice seem to be particularly sensitive to the toxic effects of (−)-nicotine.

4.2 Pharmacology

The first study to examine the behavioral pharmacology of the optical isomers of nicotine compared their effects in a conditioned avoidance task in rats, a preclinical test that is sometimes used to assess antipsychotic-like effects. Both enantiomers blocked the rats' conditioned avoidance response and S(−)-nicotine was 7 times more potent than R(+)-nicotine [62]. Studies of drug discrimination with S(−)-nicotine as training stimulus have consistently reported that S(−)-nicotine is more potent than R(+)-nicotine. For example, the first enantiomeric potency comparison based on ED_{50} doses was determined from rats trained to discriminate either 0.2 or 0.4 mg/kg of S(−)-nicotine (s.c.) from saline. In both groups of rats, S(−)-nicotine was about 10 times more potent than R(+)-nicotine [63]. Other studies also have reported S(−)-nicotine to be more potent than R(+)-nicotine, although thorough dose-response tests have not always been conducted nor ED_{50} calculations been performed (Table 2). Moreover, only animals trained to discriminate S(−)-nicotine from saline have evaluated the effects of S(−)- and R(+)-nicotine. In those studies, stimulus generalization tests indicated that S(−)-nicotine was, at least, ~10 times more potent than R(+)-nicotine as measured by S(−)-nicotine-like responding; to date, racemic nicotine has not been evaluated. As such, two areas for future drug discrimination studies can be proposed. First, even though R(+)-nicotine is about 10 times less potent than S(−)-nicotine, it is still a very potent drug that has not been employed as a training stimulus. Stereoselective drug effects are not solely a property of a drug, but are related both to the drug and the specific pharmacological (biological and/or behavioral) activity being examined; i.e. different methods/assays can afford dissimilar results. Therefore, R(+)-nicotine should be studied as a discriminative stimulus in animals and the results of stimulus generalization and antagonism tests compared to known results already obtained with S(−)-nicotine as training stimulus. Second, racemic nicotine also should be targeted for study as a training drug. (±)-Nicotine is a mixture of equal amounts of its two

Table 2 S(−)-Nicotine as a discriminative stimulus and comparative effects of nicotine stereoisomers

Species and S(−)-nicotine		Potency ratio	
Training dose	ED_{50} dose	S > R	Reference
Rat [0.2 mg/kg; (s.c.)]	(−)-Nicotine (0.083 mg/kg) (+)-Nicotine (0.764 mg/kg)	9.2	Meltzer et al. [63]
Rat [0.4 mg/kg; (s.c.)]	(−)-Nicotine (0.129 mg/kg) (+)-Nicotine (1.318 mg/kg)	10.2	Meltzer et al. [63]
Rat [0.4 mg/kg; (s.c.)]	(−)-Nicotine (0.11 mg/kg) (+)-Nicotine (ED_{50} not stated)	ND[a]	Romano et al. [64]
Squirrel monkey [0.032 or 0.065 mg/kg; (i.v.)]	(−)-Nicotine (0.015 mg/kg) (+)-Nicotine (0.44 mg/kg)	29	Takada et al. [44]
Rat [0.1 mg/kg; (s.c.)][b]	(−)-Nicotine (0.036 mg/kg) (−)-Nicotine (0.054 mg/kg) (+)-Nicotine (ED_{50} not stated)	10–20[c]	Goldberg et al. [65]
Rat [0.3 mg/kg; (i.p.)]	(−)-Nicotine (ED_{50} not stated) (+)-Nicotine (ED_{50} not stated)	3–10[c]	Brioni et al. [66]

[a]Isomer potency ratio not determined
[b]Two groups of rats were trained to discriminate 0.1 mg/kg of (−)-nicotine from saline
[c]Estimated potency ratio

enantiomers (Fig. 1) and comparative drug discrimination results of (±)-, $S(-)$- and R (+)-nicotine, when taken together, could (a) indicate a role for complex stereochemical effects of nicotine and (b) provide a unique perspective on mechanisms of action of nicotine.

5 Tobacco Alkaloids and Nicotine Metabolism

Chemical constituents in tobacco leaf exceed 4,000 and smoke from a burning cigarette contains over 7,000 substances from many chemical classes (e.g., [67–70]). The tobacco leaf contains many alkaloids and in fresh *Nicotina tabacum* (the leaf species most commonly used for the production of cigarette tobacco) the average alkaloid mixture typically consists of 93% $S(-)$-nicotine, 3.9% $S(-)$-anatabine, 2.4% $S(-)$-nornicotine, and 0.5% $S(-)$-anabasine (e.g., [4, 5, 71]; Fig. 2; but see [72]). In comparison, a typical *tobacco cigarette* contains approximately 1.5% of $S(-)$-nicotine, which constitutes ≥95% of alkaloid content [73]. In addition, some of the alkaloid content of tobacco leaf is decomposed during drying and fermentation, leading to substances such as myosmine, $S(-)$-cotinine and others (e.g., [74–77]). There is little doubt, however, that $S(-)$-nicotine is the major component responsible for the appeal of tobacco-based products.

In the body, nicotine is extensively metabolized and is susceptible to a significant first-pass effect during which 80–90% of it is metabolized by the liver. Also, the lung is able to metabolize nicotine, but to a much lesser degree [78, 79]. In humans, about 70–80% of nicotine is converted to the primary metabolite (−)-cotinine, a lactam derivative (Fig. 2). As mentioned earlier, (−)-cotinine also is a minor alkaloid found in tobacco leaf and is often used as a biomarker to detect tobacco use because of its relatively long half-life compared to that of (−)-nicotine. Another primary metabolite of nicotine is nicotine N'-oxide, although only about 4–7% of (−)-nicotine absorbed by smokers is metabolized to this product [80, 81]. Lastly, $S(-)$-nornicotine is a minor metabolite of nicotine and, as

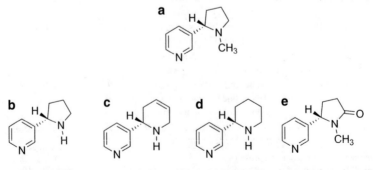

Fig. 2 Structural comparison of $S(-)$-nicotine (**a**) and related tobacco alkaloids $S(-)$-nornicotine (**b**), $S(-)$-anatabine (**c**), $S(-)$-anabasine (**d**), and $S(-)$-cotinine (**e**)

mentioned previously, is considered a minor alkaloid of tobacco (Fig. 2). Interestingly, however, in some varieties of tobacco, $S(-)$-nornicotine concentration exceeds that of $S(-)$-nicotine [82].

In drug discrimination studies, the activity and potency of metabolites have been shown to be important considerations in evaluations of the stimulus properties of drugs (for review, see [34]). Table 3 reviews the results of tobacco alkaloids and/or nicotine metabolites after their administration to animals trained to discriminate $(-)$-nicotine from saline. As can be seen, only one study has convincingly demonstrated $(-)$-

Table 3 $S(-)$-Nicotine as a discriminative stimulus: results of stimulus generalization and stimulus antagonism tests with racemic mixtures or stereoisomers of tobacco alkaloids and/or $(-)$-nicotine metabolites

Training dose of $(-)$-nicotine (route)	Species	Result[a]	Reference
\multicolumn{4}{c}{$S(-)$-Cotinine}			
0.2 mg/kg (s.c.)	Rat	PG (36%; s.c.)	Rosecrans et al. [83]
0.2 mg/kg (s.c.)	Rat	PG (47%; i.v.t.)[b]	Rosecrans et al. [83]
0.1 mg/kg (s.c.)	Rat	G[c]	Goldberg et al. [65]
0.032 or 0.065 mg/kg (i.v.)	Squirrel monkey	G	Takada et al. [44]
0.4 mg/kg (s.c.)	Rat	PG (74%; i.v.t.)	Rosecrans and Chance [48]
\multicolumn{4}{c}{$(\pm)$-Nornicotine}			
0.1 mg/kg (i.p.)	Rat	PG (76%)	Desai et al. [84]
0.32 mg/kg (i.p.)	Mouse	G	Caine et al. [85]
\multicolumn{4}{c}{$S(-)$-Nornicotine}			
0.1 mg/kg (s.c.)	Rat	G	Goldberg et al. [65]
0.032 or 0.065 mg/kg (i.v.)	Squirrel monkey	G	Takada et al. [44]
\multicolumn{4}{c}{$R(+)$-Nornicotine}			
0.1 mg/kg (s.c.)	Rat	G	Goldberg et al. [65]
\multicolumn{4}{c}{$(\pm)$-Anabasine}			
0.4 mg/kg; (s.c.)	Rat	G	Romano et al. [64]
0.032 or 0.065 mg/kg (i.v.)	Squirrel monkey	G	Takada et al. [44]
0.3 mg/kg (i.p.)	Rat	PG (~75%)	Brioni et al. [66]
0.32 mg/kg (i.p.)	Mouse	PG (~75%)	Caine et al. [85]
\multicolumn{4}{c}{$S(-)$-Anabasine}			
0.1 mg/kg (s.c.)	Rat	G	Stolerman et al. [55]
0.2 mg/kg (s.c.)	Rat	PG (~60%)	Stolerman et al. [55]
0.4 mg/kg (s.c.)	Rat	PG (~57%)	Stolerman et al. [55]
0.4 mg/kg (s.c.)	Rat	NA	Stolerman et al. [55]
0.4 mg/kg (s.c.)	Rat	PG (~60%)	Pratt et al. [50]
\multicolumn{4}{c}{$(\pm)$Anatabine}			
0.32 mg/kg (i.p.)	Mouse	G	Caine et al. [85]

[a]*PG* partial generalization, *G* stimulus generalization *NA* no stimulus antagonism
[b]*i.v.t.* intraventricular route of administration
[c]Cotinine sample was reported to be significantly contaminated with $(-)$-nicotine

nicotine stimulus generalization to (−)-cotinine and that occurred only at relatively high doses of drug [44]. Thus, S(−)-cotinine, the major metabolite of nicotine, does not appear to exert a significant role in the stimulus properties of (−)-nicotine. However, this conclusion does not rule out the possibility that (−)-cotinine may exert stimulus properties of its own that differ from those of (−)-nicotine and/or that occur at much lower doses of (−)-cotinine than doses that produced (−)-nicotine-like responding.

In comparison, S(−)-nicotine stimulus generalization occurred not only to "natural" S(−)-nornicotine, a minor metabolite of nicotine, but also to R(+)- and (±)-nornicotine (Table 3); however, these drugs were not as potent as (−)-nicotine [44, 65, 84, 85]. In the Goldberg et al. [65] study, S(−)- and R(+)-nornicotine produced dose response functions and ED_{50} values that were nearly identical, but unfortunately racemic nornicotine was not tested. Nevertheless, these data are of added interest. That is, nornicotine is a chiral substance and it would not be unusual to expect that one of its stereoisomers would (predominately) exhibit the targeted pharmacologic activity [i.e., (−)-nicotine-like responding] and that its antipode would be less potent, inactive, or exhibit a different type of biological/behavioral activity. In this study, however, the optical isomers of nornicotine did not exhibit any of the latter outcomes. Specifically, both isomers produced equally potent percent (−)-nicotine-like responding. This suggests that each isomer of nornicotine would contribute equally to (an expected) (−)-nicotine-like response that would be produced by (±)-nornicotine. Therefore, S(−)-, R(+)- and (±)-nornicotine should be evaluated in future DD studies of (−)-nicotine to explore what might be complex steric interactions in regard to their production of S(−)-nicotine-like responding.

Also of interest are the effects of anabasine and anatabine. (−)-Nicotine-trained animals exhibited very high partial or complete generalization to racemic anabasine but mostly partial generalization to "natural" S(−)-anabasine, except in animals trained to discriminate 0.1 mg/kg of (−)-nicotine from saline (Table 3). These results suggest that the untested R(+)-isomer of anabasine could produce, to some degree, marked (−)-nicotine-like effects; this possibility should be evaluated in future studies. Lastly, (−)-nicotine-trained mice generalized completely to (±)-anatabine but, unfortunately, "natural" S(−)-anatabine and its enantiomer were not tested (Table 3).

An important advisory from studies of the previously mentioned drugs is that results from racemic mixtures and optical isomers are not interchangeable. For example, the "natural" alkaloid substances in tobacco are reported to be stereoisomers that exhibit the S-configuration and levorotatory rotation. The three forms of drug [(±)-, S(−)- and R(+)-forms)] should be viewed as separate chemical entities and results obtained with one substance should not be used as substitute data for the (untested) other two drugs. For example, definitive conclusions about the activity/potency of an untested S(−)-enantiomer should not be drawn from results obtained from its racemic mixture or R(+)-isomer.

In summary, (−)-nicotine stimulus generalization tests of tobacco alkaloids and nicotine metabolites indicated that nicotine-like stimulus effects were produced by S(−)-cotinine, S(−)-nornicotine, and S(−)-anabasine, but these drugs were clearly less potent than $(-)$-nicotine. The results strongly support the idea that $(-)$-nicotine is the most pharmacologically potent alkaloid in tobacco and that its stimulus effects are not due to the effects of a (more) potent metabolite. This does not, however, rule out the possibility that these other tobacco alkaloids and the metabolites of $(-)$-nicotine could exert some other kind of activity or that they may produce interactive effects in combination with $(-)$-nicotine.

6 Mechanism of Action

Historically, $(-)$-nicotine has facilitated our knowledge of the cholinergic nervous system (e.g., [13, 86–88]). It is now well established that nicotine binds to nicotinic acetylcholine receptors (nAChRs) at the cellular level and is the prototype drug used to classify nAChRs. These receptors belong to the super-family of ligand-gated ion channels that also includes $GABA_A$, $GABA_C$, glycine, and $5\text{-}HT_3$ receptors [89–91]. In the mammalian brain, nAChRs are composed of $\alpha2$–$\alpha7$ and $\beta2$–$\beta4$ subunits with distribution patterns that appear to be distinct or to overlap (e.g., [92–94]). These subunits surround an ion channel and receptor binding by an agonist [e.g., $(-)$-nicotine] causes a closed (i.e., rest) conformation of the subunits to change to an open conformation, which allows inflow of sodium ions, and, consequently, produces cell depolarization (e.g., [95, 96]). $(-)$-Nicotine activates all brain nAChR subtypes, but binds preferentially and with high affinity to $\alpha4\beta2$ nAChRs (e.g., [12]). Moreover, these subunits are thought to modulate the release of other neurotransmitters and this has led to the idea that nAChRs, at least in part, are located presynaptically (e.g., [97–100]). For example, $(-)$-nicotine may increase dopamine activity at some brain sites such as the nucleus accumbens, an area thought to be important to drugs of abuse (e.g., [14, 101, 102]; but see [16, 103]).

In drug discrimination studies, Schechter and Rosecrans [40] provided very strong, if not the strongest, evidence for the conclusion that the stimulus effects of $(-)$-nicotine were mediated centrally. In this study, rats trained to discriminate 0.4 mg/kg of $(-)$-nicotine from saline did not generalize (recognize) the administration of nicotine isomethonium iodide hydroiodide, a quaternary amine analog of nicotine that produces the peripheral, but not the central effects of nicotine because of poor penetration into the CNS. In addition, Schechter and Rosecrans [39] and Rosecrans et al. [104] reported on a series of studies in which cannulae were implanted into the dorsal hippocampus of control, dopamine (DA)-depleted or norepinephrine (NE)-depleted rats. These three groups of rats had previously been trained to discriminate 0.4 mg/kg of $(-)$-nicotine (s.c.) from saline and this discrimination was maintained fully in the control group but was somewhat lessened (but still maintained) in both the DA- and NE-depleted animals after peripheral administration of the training treatments (following surgery). However, when rats were injected with $(-)$-nicotine (1 µg) bilaterally into the hippocampus, the discrimination was markedly weakened in NE-depleted rats and was not observed in DA-depleted rats. Taken together, these results suggested that the reduced "strength" of the stimuli in catecholamine-depleted animals provided support, at least in part, for the involvement of DA and NE in the stimulus properties of (−)-nicotine. Lastly, (−)-nicotine also may alter other neuronal systems that are related to substance use and abuse, such as opioid, glutamate, serotonin, and glucocorticoid (e.g., [97–99]).

Other drug discrimination studies of (−)-nicotine have been performed to determine its mechanisms of action. For example, stimulus antagonism tests of nicotine have been studied by three general approaches and the results of such studies are summarized in Tables 4 and 5. In one approach, doses of a receptor antagonist are combined with the training dose of nicotine to determine whether the stimulus effect can be blocked. If a drug is an effective antagonist of (−)-nicotine, then a dose related antagonism will occur in the animals' percentage of (−)-nicotine-appropriate responding (i.e., lever pressing does not stop, but occurs on the saline-designated lever). In a second technique, the dose response of (−)-nicotine is determined in both the presence and absence of a constant dose of the receptor antagonist. If the antagonism is competitive, then the dose response of nicotine will shift in a rightward and parallel manner. In a third tactic, various doses of (−)-nicotine are combined with various doses of a receptor antagonist. This method generates a series of nicotine/antagonist dose response curves and provides the most comprehensive picture of the interactions between the drugs. All three of these approaches have been employed to evaluate putative receptor antagonists of (−)-nicotine stimuli.

The results of antagonism tests typically fall into one of three categories: (a) complete antagonism (i.e., saline- or vehicle-like responding); (b) partial antagonism (i.e., 40 to ~70% drug-appropriate responding), and (c) no antagonism (i.e., ≥80% drug-appropriate responding). In tests that result in no stimulus antagonism, subjects respond ≥80% on the nicotine-designated lever after administration of doses of a receptor antagonist in combination with the training dose of nicotine. Such results indicate that percent (−)-nicotine-appropriate responding is still like that of the dose of training drug and that the receptor antagonist does not interfere with the neurochemical mechanisms that are important for the discrimination. In cases of partial stimulus antagonism, subjects respond "moderately" on the drug-designated lever after administration of doses of a receptor antagonist in combination with doses of the training drug. Such results indicate that drug-appropriate responding is still somewhat like the stimulus effect(s) of the dose of training drug but also somewhat "saline-like." However, the saline-designated lever is also the default lever and subjects will press it under the saline (i.e., inert) condition or if the combination of drugs produces a stimulus effect(s) that is sufficiently dissimilar from that of the dose of training drug. Consequently, this type of data can be most difficult to interpret. Lastly, in cases of complete stimulus antagonism, subjects respond in a manner that is appropriate for the vehicle condition after administration of an appropriate dose of receptor antagonist in combination with the training dose of nicotine – but see discussion below of third state hypothesis.

Over the past 45 years, (−)-nicotine has been the subject of numerous attempts to block its stimulus effects and such tests have indicated quite clearly that nicotine

Table 4 $S(-)$-Nicotine as a discriminative stimulus: antagonism by mecamylamine

Training dose of (−)-nicotine (route)	Species	Result[a]	Reference
0.2 mg/kg (s.c.)	Rat	A	Morrison and Stephenson [35]
0.4 mg/kg (s.c.)	Rat	A	Schechter and Rosecrans [36, 37]
0.4 mg/kg (s.c.)	Rat	A	Schechter and Rosecrans [38]
0.2 mg/kg (s.c.)	Rat	A	Hirschhorn and Rosecrans [53]
0.4 mg/kg (s.c)	Rat	A	Hirschhorn and Rosecrans [53]
0.2 mg/kg (s.c.)	Rat	A	Chance et al. [105]
0.2 mg/kg (s.c.)	Rat	A	Meltzer et al. [63]
0.4 mg/kg (s.c.)	Rat	A	Meltzer et al. [63]
0.4 mg/kg (s.c.)	Rat	A	Romano et al. [64]
0.4 mg/kg (s.c.)	Rat	A	Stolerman et al. [106]
0.1 mg/kg (s.c.)	Rat	A	Stolerman et al. [55]
0.5 mg/kg (p.o.)	Rat	A[b]	Craft and Howard [107]
0.4 mg/kg (s.c.)	Rat	A	Stolerman et al. [108]
0.4 mg/kg (s.c.)[c]	Rat	A	Stolerman and Garcha [54]
0.5 mg/kg (s.c.)	Rat	A[d]	Miyata et al. [109]
0.5 mg/kg (s.c.)	Rat	A[d]	Ando et al. [110]
0.5 mg/kg (s.c.)	Rat	NA[e]	Ando et al. [110]
0.4 mg/kg (s.c.)	Rat	A	James et al. [28]
0.3 mg/kg (s.c.)	Rat	A	Brioni et al. [66]
0.2 mg/kg (s.c.)	Rat	A	Chandler and Stolerman [111]
1.6 mg/kg (s.c.)	Mouse	A	Stolerman et al. [43]
0.1 mg/kg (s.c.)	Rat	A	Gasior et al. [112]
0.4 mg/kg (s.c.)	Rat	A	Gasior et al. [112]
0.02 mg/kg (nasal spray)	Human	A	Perkins et al. [42]
0.4 mg/kg (s.c)	Rat	A	Mansbach et al. [113]
0.4 mg/kg (s.c.)	Rat	A	Wiley et al. [114]
0.6 mg/kg (s.c.)	Rat	A	Young and Glennon [115]
0.4 mg/kg (s.c.)	Rat	A	Zaniewska et al. [116]
0.4 mg/kg (i.p.)	Rat	A	Paterson et al. [117, 118]
0.32 mg/kg (s.c.)	Rat	A	Jutkiewicz et al. [119]
1.78 mg/kg (s.c.)	Rat	A	Jutkiewicz et al. [119]
1.78 mg/kg (s.c.)	Rhesus monkey	A	Cunningham et al. [120]
0.56 mg/kg (s.c.)	Mouse	A	Cunningham and McMahon [121]
1.0 mg/kg (s.c.)	Mouse	A	Cunningham and McMahon [121]
1.78 mg/kg (s.c.)	Mouse	A	Cunningham and McMahon [121]

[a]A stimulus antagonism, NA no stimulus antagonism
[b]Mecamylamine administered orally

(continued)

Table 4 (continued)

[c]Three groups of rats trained at 0.4 mg/kg s.c. of (−)-nicotine with different pre-session injection intervals

[d]Mecamylamine injected into nucleus accumbens blocked the systemic administration of (−)-nicotine

[e]Mecamylamine injected into ventral tegmental area or dorsal hippocampus did not block the systemic administration of (−)-nicotine

Table 5 *S*(−)-Nicotine as a discriminative stimulus: results of antagonism tests with dihydro-β-erythrodine (DHβE) and methyllycaconitine (MLA)

Training dose of (−)-nicotine (route)	Species	Result[a]	Reference
DHβE			
0.1 mg/kg (s.c.)	Rat	A	Stolerman et al. [122]
0.4 mg/kg (s.c.)	Rat	A	Stolerman et al. [122]
0.4 mg/kg (s.c.)	Mouse	A	Gommans et al. [123]
0.2 mg/kg (s.c.)	Rat	A	Shoaib et al. [124]
0.4 mg/kg (s.c.)	Rat	A	Zaniewska et al. [116]
0.4 mg/kg (i.p.)	Rat	A	Paterson et al. [117, 118]
0.32 mg/kg (s.c.)	Rat	A	Jutkiewicz et al. [119]
1.78 mg/kg (s.c.)	Rat	A	Jutkiewicz et al. [119]
1.78 mg/kg (s.c.)	Rhesus monkey	NA	Cunningham et al. [120]
0.56 mg/kg (s.c.)	Mouse	A	Cunningham and McMahon [121]
1.0 mg/kg (s.c.)	Mouse	NA	Cunningham and McMahon [121]
1.78 mg/kg (s.c.)	Mouse	A	Cunningham and McMahon [121]
0.4 mg/kg (s.c)	Rat	A	Lee et al. [125]
MLA			
0.3 mg/kg (i.p.)	Rat	NA	Brioni et al. [66]
0.4 mg/kg (s.c.)	Mouse	NA	Gommans et al. [123]
0.4 mg/kg (s.c.)	Rat	NA	Zaniewska et al. [116]
0.8 mg/kg (s.c.)	Mouse	PA (~50%)	Quarta et al. [126]
0.4 mg/kg (i.p.)	Rat	NA	Paterson et al. [117, 118]
0.4 mg/kg (s.c)	Rat	NA	Lee et al. [125]

[a]*A* stimulus antagonism, *NA* no stimulus antagonism, *PA* partial antagonism

exerts its stimulus effect, at least in part, through an interaction at nicotinic receptors in brain and, in particular, at a subtype of nicotinic receptor termed $\alpha 4\beta 2$ receptors. This conclusion is based on the fact that the stimulus effects of nicotine are convincingly blocked by (a) mecamylamine, a voltage dependent noncompetitive channel blocker at nicotinic receptors (Fig. 3; Table 4) and (b) dihydro-β-erythrodine (DHβE), a nicotinic receptor antagonist that shows high affinity for the nAChR $\alpha 4\beta 2$ subunit (Fig. 3; Table 5) but not by methyllycaconitine (MLA), a $\alpha 7$ nicotinic receptor antagonist (Table 5).

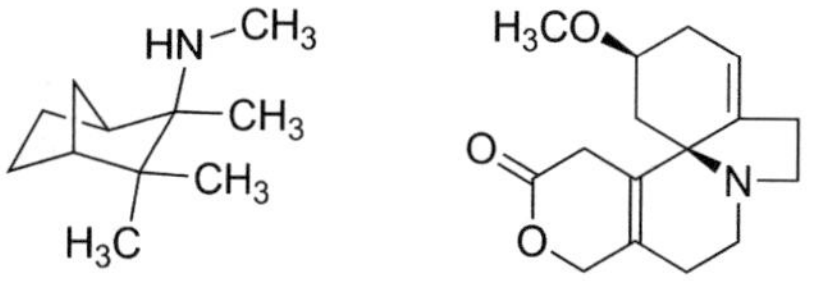

Fig. 3 Structures of nicotinic receptor antagonists mecamylamine (*left*) and dihydro-β-erythroidine (DHβE; *right*)

6.1 Mecamylamine

Mecamylamine (Inversine®, Vecamyl®; Fig. 3) was developed over 60 years ago and marketed as a ganglionic blocker for the treatment of hypertension (e.g., [127]). However, it is rarely used today for this purpose because of parasympathetic and sympathetic ganglia-related adverse effects (e.g., blurred vision, dry mouth, and dizziness). In addition, mecamylamine can produce CNS effects that include tremor, mental confusion, seizures, mania, and depression but the mechanisms by which these effects are produced are unclear. Also, mecamylamine is sometimes used as an anti-addictive drug to help people stop smoking tobacco products (e.g., [128, 129]). However, an early study of mecamylamine in human smokers reported an increased rate (30%) of smoking, which was regarded as evidence for self-titration of (−)-nicotine [130].

Biochemical and pharmacological studies have characterized mecamylamine as a nonselective, voltage dependent and noncompetitive receptor antagonist of neuronal nAChRs and it is often referred to as a "nicotine receptor antagonist." As such, mecamylamine probably exerts its effects via interaction with sites distinct from nAChR agonist binding sites and, therefore, does not compete with (−)-nicotine for binding. For example, some biochemical studies suggest that mecamylamine is a channel blocker that inhibits most neuronal nAChRs (e.g., [131–133]). Table 4 shows that mecamylamine antagonism of (−)-nicotine discriminative stimuli has been consistently demonstrated in many studies. In general, these studies employed mecamylamine at 1–3 mg/kg to block the stimulus effects of nicotine (0.1–1.78 mg/kg). In addition, some of these studies have confirmed the non-competitive antagonism character of mecamylamine because the antagonism effects were not always reversed or surmounted by higher doses of (−)-nicotine (e.g., [106]).

6.2 DHβE (Dihydro-β-Erythroidine)

DHβE (Fig. 3) is an alkaloid found in plant seeds of *Erythrina* and is a competitive nAChR receptor antagonist with a preference for neuronal β2 subtypes. For example, DHβE (at nM concentrations) blocks α4β2 and α3β2 nAChRs but is much less potent at α3β4 and α7 nAChRs expressed in Xenopus oocytes (e.g., [134–137]). DHβE interacts reversibly with nAChRs at, or close to, the agonist binding site(s), stabilizes the receptor in a conformation with the channel closed and prevents access for receptor agonists; however, this blockade is surmountable with increased agonist [e.g., (−)-nicotine] concentrations (e.g., [123]). DHβE has been employed in stimulus antagonism studies of (−)-nicotine to assess involvement of the α4β2 nAChR subunit. Research results summarized in Table 5 indicate that DHβE effectively blocked the stimulus effects of (−)-nicotine in rats or mice (but see exceptions reported by [120, 121]). In addition, and in contrast to mecamylamine, DHβE antagonism of the stimulus effect of (−)-nicotine was reversed (competitively) by increased doses of (−)-nicotine (e.g., [122]).

6.3   Methyllycaconitine (MLA)

Methyllycaconitine (MLA) is an alkaloid found in many plant species of *Delphinium* (larkspurs) and is generally toxic to animals (e.g., [138]). Its biochemical pharmacology indicates that it is a relatively potent competitive receptor antagonist that is selective for α7 nAChRs (e.g., [139–141]). MLA has been used in stimulus antagonism studies of (−)-nicotine to assess potential involvement of α7 nAChRs. Table 5 presents results of MLA/(−)-nicotine combination studies and shows that MLA failed to alter the stimulus effects of (−)-nicotine in rats or mice (but see partial antagonism reported by Quarta et al. [126]).

6.3.1   Summary and Analysis of Antagonism Results

Mecamylamine and DHβE have repeatedly been shown to produce complete antagonism of the stimulus effects of (−)-nicotine. However, the resultant responding on the saline-designated lever may, but does not necessarily, indicate that the combination of substances produced an inert effect. The possibility exists that the effects of either (or both) receptor antagonist in combination with the training dose(s) of nicotine are not like those of the training dose of nicotine nor like the vehicle (i.e., inert) condition. In such cases, the saline-designated lever would have served as the "default" response [a.k.a. "transfer test over-inclusiveness" (see [142]) or 'third-state hypothesis" (see [143])]. Thus, results of antagonism tests only indicated that the stimulus effects produced by the combination of antagonists and training doses of (−)-nicotine are dissimilar from those produced by the training dose of (−)-nicotine (alone). In this regard, there is some drug discrimination data that suggest the need for tests to determine if a "third state hypothesis" explanation could account for results when mecamylamine (or DHβE) is combined with (−)-nicotine. For example, Garcha and Stolerman [144] trained rats to discriminate mecamylamine (3.5 mg/kg s.c.) from saline. The mecamylamine stimulus generalized completely to the ganglion blockers pentolinium and pempidine but not to hexamethonium, trimetaphan, or chlorisondamine. Mecamylamine stimulus generalization also did not occur to (−)-nicotine, atropine, or scopolamine. In antagonism tests, (−)-nicotine failed to block the stimulus effects of mecamylamine. In another study, Cunningham et al. [145] trained rhesus monkeys to discriminate 5.6 mg/kg of mecamylamine from saline and reported that the mecamylamine stimulus was not blocked by (−)-nicotine nor did (−)-nicotine produce mecamylamine-like responding. Mecamylamine stimulus generalization did occur, however, to the peripherally mediated nicotinic receptor antagonist hexamethonium (at a relatively high dose), a quaternary drug that does not readily penetrate into the CNS. It should be noted that hexamethonium, at relatively low doses, does not block the stimulus effects of (−)-nicotine but when administered at high doses has occasionally been reported to attenuate nicotine-like responding; probably the result of penetration into the CNS of a small proportion of the administered dose of drug (e.g., [35, 38, 64, 106, 146]).

Taken together with previous studies (Table 4), results indicate that (a) mecamylamine, by itself, can exert discriminative stimulus effects, (b) cross stimulus generalization does not occur between mecamylamine and (−)-nicotine regardless of which drug was used as training stimulus and (c) asymmetric antagonism occurred between (−)-nicotine and mecamylamine; i.e. mecamylamine blocked the discriminative stimulus effects of nicotine but not vice versa. The latter results could be accounted for by the possibility of dose-dependent effects of mecamylamine such that different doses produced qualitatively or mechanistically different pharmacological effects. That is, relatively low doses (i.e., 1–3 mg/kg) of mecamylamine are used routinely in studies to block the stimulus effects of (−)-nicotine. However, both Garcha and Stolerman [144] and Cunningham et al. [145] reported that low training doses (1–3 mg/kg) of mecamylamine were not sufficient to produce stable discrimination learning and, thus, higher doses (3.5 mg/kg and 5.6 mg/kg, respectively) of drug had to be employed. This suggests that different doses of mecamylamine might produce, to some degree, dissimilar pharmacological (i.e., stimulus) effects. Follow-up studies should evaluate the argument that the dose of mecamylamine can determine its "distinctiveness" in drug discrimination studies. Other studies should be designed to test the third state hypothesis to ascertain if subjects can be trained to discriminate a *drug mixture* of a (reported) dose of mecamylamine that blocked a (reported) dose of (−)-nicotine from saline vehicle. If animals cannot learn to discriminate the drug mixture from saline, then this would be evidence that the drug combination likely exerts an inert stimulus effect in the animals; i.e. negative data that would argue against a third state effect to explain mecamylamine antagonism of (−)-nicotine. On the other hand, if animals can discriminate the drug mixture from saline, then this suggests that the combination of drugs produces a stimulus effect that is unlike (−)-nicotine and, consequently, results in responding on the saline-designated (default) lever in (−)-nicotine-trained animals. Alternatively, or additionally, a three-choice operant conditioning procedure could be employed to examine the stimulus effects of (−)-nicotine versus mecamylamine versus saline vehicle to assess qualitative and/or quantitative similarities/differences in actions. Thus, if a group of subjects can be trained to discriminate a dose of mecamylamine versus a dose of (−)-nicotine versus saline, then this could lead to a more precise characterization of the stimulus properties of the treatment conditions. The interested reader is referred to Frey and Winter [143] and Glennon and Young [34] for further discussion of third state hypothesis.

7 Pharmacotherapies for Smoking Cessation

> Giving up smoking is the easiest thing in the world. I know because I've done it thousands of times.
> – Mark Twain

When a nicotine-dependent user tries to quit nicotine consumption, cessation of use is typically followed by a withdrawal period that may last for months and includes symptoms that can lead an individual to relapse. (−)-Nicotine withdrawal symptoms might begin within a few hours after the last nicotine product, and include irritability/anger/stress/anxiety, sleep disturbances, depressed mood, craving, cognitive and attention deficits, and increased appetite. Symptoms may last a few days or persist for months or longer. Unfortunately, most smokers relapse within just a few days, and less than 10% of those who try to quit on their own achieve more than a year of abstinence; thus, quitting the nicotine product often requires multiple attempts (e.g., [147]).

One of the most common smoking-cessation treatments is nicotine-replacement therapy (NRT), when smokers simply substitute (−)-nicotine inhaled via cigarettes with, ironically, "safer formulations" of (−)-nicotine. In fact, nicotine itself was the first pharmacological agent approved by the U.S. Food and Drug Administration (FDA) for use in smoking cessation therapy. NRT formulations include gum, patch, inhaler, spray, lozenge, and e-cigarette. These products can maintain reinforcement effects of (−)-nicotine at (gradually) reduced doses and concomitantly reduce withdrawal symptoms associated with cessation. The goal is to gradually diminish the body's ingestion of nicotine, establish remission, and sustain it long enough for the ex-smoker to develop "coping" strategies to avoid relapse (for review, see [148]). These nicotine delivery systems are thought to be equally effective, with about 20% of those that received therapy not smoking at 1 year and up to 10% remaining non-smokers if treatment is continued [148, 149]. Other products that have been (are) promoted for cessation of (−)-nicotine consumption include (−)-lobeline (unapproved by the FDA) and the FDA approved products bupropion (Zyban®) and varenicline (Chantix®).

7.1 (−)Lobeline

(−)Lobeline (Fig. 4) is a natural substance found in, for example, "Indian tobacco" (*Lobelia inflata*) and "Devil's tobacco" (*Lobelia tupa*). Lobeline binds with high affinity to α4β2 nAChRs and displays mixed receptor agonist/antagonist actions

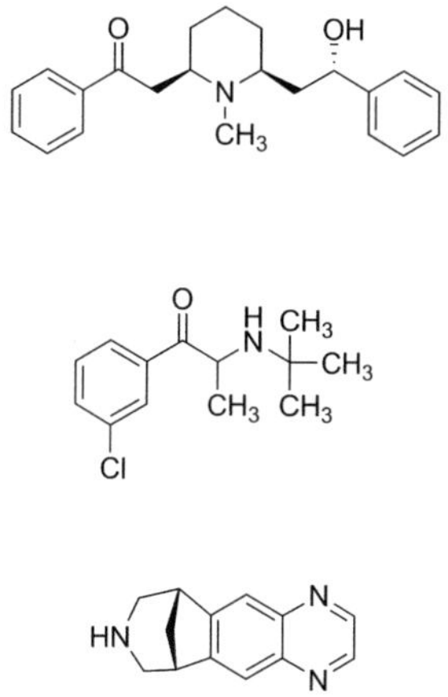

Fig. 4 Structures of pharmacotherapies that have been [(−)-lobeline (*top*)] or are promoted [bupropion (*middle*) and varenicline (*bottom*)] for cessation of (−)-nicotine consumption

(e.g., [150–154]). At one time, lobeline was promoted (but not FDA approved) for smoking cessation in several products: CigArrest®, Bantron®, and Nicoban®. However, the FDA removed these products from the US marketplace in 1993 until such time that they could be shown to be efficacious in scientific studies of smoking cessation (see [155–157]). In drug discrimination studies, (−)-lobeline has been utilized only as a test drug. For example, in rats and squirrel monkeys trained to discriminate cocaine, S(+)-methamphetamine or (−)-nicotine from vehicle, stimulus generalization to lobeline occurred only in relatively "low dose" cocaine-trained rats at a relatively short pre-session injection interval [158]. Importantly, however, (−)-nicotine-trained rats have consistently shown that (−)-lobeline does not produce marked (−)-nicotine-like responding, which supports the argument of unproven efficacy of lobeline as a substitute for nicotine-like effects in smoking cessation treatment (Table 6). Lobeline has, however, been shown to antagonize partially the stimulus effects of (−)-nicotine and S(+)-methamphetamine ([64, 160]; but see [66]). Surprisingly, and unfortunately, there does not appear to be any reports in the literature of (−)-lobeline as a training stimulus in drug discrimination studies. Such studies could prove informative and might reveal important similarities and differences in stimulus effects between (−)-lobeline, (−)-nicotine, cocaine and S(+)-methamphetamine.

7.2 Bupropion

Bupropion [a.k.a. amfebutamone, (RS)-2-(tert-Butylamino)-1-(3-chlorophenyl)propan-1-one, 3-Chloro tert-butylcathinone, 3-Chloro-N-tert-butyl-β-ketoamphetamine; Fig. 4] is a phenylaminoketone or cathinone derivative that is a weak central nervous system (CNS) stimulant. It is prescribed as medication for the treatment of depression (Wellbutrin®) and/or as an adjunct in smoking cessation therapy (Zyban®). In fact, the application of bupropion for smoking cessation was first noted serendipitously by

Table 6 Effects of (−)-lobeline as a test drug in drug discrimination studies

Training drug	Species	Result[a]	Reference
Cocaine[b]	Rat	G	Cunningham et al. [158]
Cocaine[c]	Rat	NG	Desai et al. [159]
S(+)Methamphetamine[d]	Rat	NG, PA	Miller et al. [160]
S(+)Methamphetamine	Squirrel monkey	NG (~30%)	Desai and Bergman [161]
(−)-Nicotine	Rat	NG	Schechter and Rosecrans [40]
(−)-Nicotine	Rat	NG (34%; s.c.)	Rosecrans et al. [83]
(−)-Nicotine	Rat	PG (48%; i.v.t.)	Rosecrans et al. [83]
(−)-Nicotine[e]	Rat	NG, PA	Romano et al. [64]
(−)-Nicotine	Rat	NG (~36%)	Reavill et al. [162]
(−)-Nicotine	Rat	NG, NA	Brioni et al. [66]

[a]G stimulus generalization, NG no stimulus generalization, PA partial stimulus antagonism, NA no stimulus antagonism
[b]Separate groups of rats trained to discriminate either 1.6 or 5 mg/kg of cocaine from saline. Both cocaine training stimuli generalized to lobeline but only with a relatively short (10 min) pre-session injection interval
[c]No generalization in rats trained to discriminate 8.9 mg/kg of cocaine from saline
[d](−)-Lobeline reduced %S(+)-Methamphetamine-appropriate responding from 100% to approximately 60%
[e]T-maze study. (−)-Nicotine stimulus did not generalize to lobeline, but lobeline did produce partial antagonism of the (−)-nicotine stimulus

clinical observations that depressed patients who received the drug decreased their smoking of tobacco (e.g., [163]). Follow-up studies showed that the administration of bupropion, in combination with counseling, produced comparable efficacy to NRTs at the 1-year benchmark for smoking cessation (e.g., [149, 164–166]). Its mechanisms of action (for both indications), however, are not known with certainty but bupropion may produce indirect agonist effects, at least in part, via relatively weak norepinephrine/dopamine reuptake inhibition (NDRI) (e.g., [167–170]). Other studies have reported that bupropion blocked the acute effects of (−)-nicotine in a number of behavioral assays in mice (e.g., [171, 172]). In drug discrimination studies, bupropion has received limited attention as a training agent, but much attention as a test drug (Tables 7, 8, and 9).

To date, only three drug discrimination studies have employed bupropion as training drug and each of these studies used rats as subjects. Jones et al. [173] reported the first drug discrimination study of bupropion and demonstrated that 20 mg/kg, but not 5 or 10 mg/kg, of bupropion was effective as training dose. Two other studies used 17 or 40 mg/kg of bupropion as dose of training drug [174, 175]. Table 7 summarizes the results of these studies and indicates quite clearly that bupropion stimulus generalization occurred to other CNS stimulants and several catecholamine reuptake inhibitors. These drugs included (±)- and S(+)-amphetamine, caffeine, cocaine, methylphenidate, mazindol, SKF 82958, nomifensine, WIN 35428 and GBR 12909 (vanoxerine). In addition, bupropion stimuli produced (high) partial generalization to a number of direct or indirect

Table 7 Bupropion as a discriminative stimulus: drugs that produced complete bupropion-like stimulus effects (i.e., $\geq 80\%$ bupropion-appropriate responding)[a]

Mechanism of action	Drug class or test drug
$(\pm)$-Amphetamine (Benzadrine®)	Stimulant
$S(+)$-Amphetamine (Adderall®)	Stimulant
Benocyclidine (BTCP)	Stimulant
	Dopamine Reuptake Inhibitor
Benzylpiperazine	Designer Drug (Stimulant)
Caffeine	Stimulant
Cocaine	Stimulant
EXP-561[b]	Norepinephrine/Dopamine/Serotonin
	Reuptake Inhibitor
GBR 12909 (Vanoxerine)	Dopamine Reuptake Inhibitor
GBR 12935[c]	Dopamine Reuptake Inhibitor
LU 17-133[d]	Norepinephrine/Dopamine Uptake Inhibitor
Mazindol (Mazanor®, Sanorex®)	Stimulant
	Norepinephrine/Dopamine/Serotonin
	Reuptake Inhibitor
Methylphenidate (Ritalin®)	Stimulant
Nomifensine (Merital®, Alival®)	Antidepressant
	Norepinephrine/Dopamine Reuptake Inhibitor
RU 24213[e]	Dopamine D_2 Receptor Agonist
SKF 82958[f]	Stimulant
	Dopamine D_1/D_5 Receptor Agonist
Viloxazine (Vivalan®, Vivarint®, Vicilan®)[g]	Antidepressant/Stimulant
	Norepinephrine Reuptake Inhibitor
WIN 35428 (β-CFT)[h]	Stimulant
	Dopamine Reuptake Inhibitor

[a]Data from rats trained to discriminate 20 mg/kg of bupropion [173], 40 mg/kg of bupropion [174] or 17 mg/kg of bupropion [175] from saline vehicle
[b]4-Phenylbicyclo[2.2.2]octan-1-amine
[c]1-(2-(Diphenylmethoxy)ethyl)-4-(3-phenylpropyl)piperazine
[d]$(\pm)$-Trans-4-[3-(3,4-dichlorophenyl)-indan-1-yl]-1-piperazineethanol
[e]N-n-propyl-N-phenylethyl-4(3-hydroxyphenyl)ethylamine
[f]3-Allyl-6-chloro-1-phenyl-1,2,4,5-tetrahydro-3-benzazepine-7,8-diol
[g]Jones et al. [173] reported complete bupropion generalization to viloxazine but Blitzer and Becker [174] reported bupropion (high) partial generalization to viloxazine
[h]$(-)$-2-β-Carbomethoxy-3-β-(4-fluorophenyl)tropane

receptor agonists of catecholamine systems such as indatraline, pergolide, SKF 38393, and viloxazine (Table 8). Table 8 also indicates that bupropion stimulus generalization did not occur to (a) antidepressants from other chemical classes such as amitriptyline, mianserin, desipramine or zimelidine, or (b) substances from other drug classes such as the anxiolytic diazepam (but see partial generalization to chlordiazepoxide in [173]), the hallucinogen LSD, the analgesic morphine, or the sedative pentobarbital [173–175]. Curiously, however, the stimulus effects of

Table 8 Bupropion as a discriminative stimulus: drugs that produced partial or no bupropion-like responding[a]

Test drug	Highest % of bupropion-appropriate responding	Drug class or mechanism of action
Amitriptyline (Elavil®)	~30	Antidepressant
Benztropine (Cogentin®)	41	Anti-Parkinson
Bromocriptine (Parlodel®, Cycloset®)	~20	Monoamine receptor agent
Chlordiazepoxide (Librium®)	~60	Antianxiety
Clonidine (Catapres®)	~20	α_2-Adrenergic/imidazoline Receptor agonist
Desipramine (Norpramin®, Pertofrane®)	~35	Antidepressant
Diazepam (Valium®)	~20	Antianxiety
Imipramine (Tofranil®)	~50	Antidepressant
Indatraline (LU 19-005)	71	Non-selective monoamine Transporter inhibitor
Isoproterenol (Isuprel®)	~15	Non-selective β-adrenergic Receptor agonist
LSD	~20	Hallucinogen
Mianserin (Tolvon®, Norval®)	~35	Antidepressant/anxianxiety
Morphine (MScontin®, Oramorph®)	~15	Analgesic
Nisoxetine	~20	Norepinephrine reuptake inhibitor
Nortriptyline (Aventyl®, Sensova®)	~60	Antidepressant
(−)-NPA[b]	~63	Dopamine D_2 agonist
Pentobarbital (Nembutal®)	0	Sedative/hypnotic
Pergolide (Permax®, Prascend®)	69	Non-selective dopamine/serotonin agonist
Phenethylamine	~50	Monoamine neuromodulator
(−)Quinpirole	57	Dopamine D_2/D_3 receptor agonist
Quipazine	~35	Non-selective serotonin receptor agent
Scopolamine (Transdermscop®)	~40	Anti-motion sickness Muscarinic receptor antagonist
SCH 23390[c]	<5	Dopamine D_1 receptor antagonist
SKF 38393[d]	67	Dopamine D_1/D_5 receptor partial agonist
SKF 75670[e]	52	Dopamine D_1 agent
SKF 77434[f]	65	Dopamine D_1 receptor partial agonist
Spiperone (Spiropitan®)	~40	Antipsychotic
Thyrotropin-Releasing Hormone (TRH)	~20	Hypothalamic hormone
Viloxazine (Vivalan®, Vivarint®)[g]	~70	Antidepressant/stimulant Norepinephrine reuptake inhibitor

(continued)

Table 8 (continued)

Test drug	Highest % of bupropion-appropriate responding	Drug class or mechanism of action
Zimelidine	<5	Antidepressant
		Selective serotonin reuptake inhibitor

[a]Results from Jones et al. [173], Blitzer and Becker [174] and Terry and Katz [175]
[b]R(−)-10,11-dihydroxy-N-n-propylnorapomorphine
[c]7-Chloro-3-methyl-1-phenyl-1,2,4,5-tetrahydro-3-benzazepin-8-ol
[d]1-Phenyl-2,3,4,5-tetrahydro-1H-3-benzazepine-7,8-diol
[e]7,8-Dihydroxy-3-methyl-1-phenyl-2,3,4,5-tetrahydro-1H-3-benzazepine
[f]3-Allyl-1-phenyl-1,2,4,5-tetrahydro-3-benzazepine-7,8-diol
[g]Jones et al. [173] reported complete bupropion generalization to viloxazine but Blitzer and Becker [174] reported bupropion (high) partial generalization to viloxazine

bupropion were not blocked by monoamine receptor antagonists such as haloperidol, thioridazine, thiothixene, propranolol, phenoxybenzamine, and cyproheptadine but were blocked partially by spiperone (which also generalized partially) and completely by SCH 23390 [174, 175]. The general lack of antagonism of the stimulus effects of bupropion, especially by catecholamine receptor antagonists, was rather unexpected and suggests that additional (stimulus antagonism and/or generalization) studies be undertaken to more fully elucidate its mechanism of action.

In comparison, bupropion has received extensive evaluation as a test drug in studies that used animals trained to discriminate S(+)-amphetamine, clenbuterol, cocaine, ethanol, GBR 12909, imipramine, isoproterenol, MDMA ("Ecstasy"), S(+)-methamphetamine, methylphenidate, mirtazapine, (−)-nicotine, oxazepam, rimonabant or Δ^9-THC from vehicle (Table 9). The results of these studies showed clearly that S(+)-amphetamine-, cocaine-, or GBR 12909-trained animals generalized to bupropion, which indicates cross-generalization between bupropion and these drugs regardless of which drug was used as training stimulus (e.g., [173–176, 188]). However, in one study, Rush et al. [179] trained humans to discriminate S(+)-amphetamine from vehicle and reported only partial generalization to bupropion, but the lack of complete stimulus generalization may be related to administered doses of bupropion. That is, the dose range for the antidepressant effect of bupropion in humans is considered to be 300–750 mg and it is within this dose range where concerns have been raised about the occurrence of seizures and the possibility of drug-induced psychotic symptoms, the latter a noted risk when prescribing CNS stimulants (e.g., [168, 169, 203]). On the other hand, the anti-smoking doses of bupropion are stated to be 150 and 300 mg; doses above 300 mg are not recommended and, in fact, are discouraged [204]. However, doses up to 300–400 mg of bupropion usually do not exert marked CNS stimulant effects in humans. Miller and Griffith [205], for example, reported that the effects of bupropion up to 400 mg produced little resemblance to S(+)-amphetamine. They speculated that because bupropion is such a

Table 9 Results of bupropion as a test drug in animals trained to discriminate various training drugs from saline vehicle

Training drug	Species	Result[a]	Reference
S(+)-Amphetamine	Pigeon	G	Evans and Johanson [176]
S(+)-Amphetamine	Rhesus monkey	G	de la Garza and Johanson [177]
S(+)-Amphetamine	Rhesus monkey	G	Kamien and Woolverton [178]
S(+)-Amphetamine	Human	PG (40%)	Rush et al. [179]
S(+)-Amphetamine	Rat	G	Bondarev et al. [180]
S(+)-Amphetamine	Rat	G	Heal et al. [181]
S(+)-Methamphetamine	Rhesus monkey	G	Banks et al. [182]
Clenbuterol	Rat	NG	Makhay and O'Donnell [183]
Cocaine	Rat	G	Lamb and Griffiths [184]
Cocaine	Rhesus monkey	G	Kleven et al. [185]
Cocaine	Pigeon	G	Johanson and Barrett [186]
Cocaine	Rat	G	Broadbent et al. [187]
Cocaine	Rat	G	Baker et al. [188]
Cocaine	Rat	G	Quinton et al. [189]
Cocaine	Rat	G	Paterson et al. [117, 118]
Cocaine	Rat	G	Awasaki et al. [190]
Ethanol	Rat (P)[b]	PG (50%)	McMillan et al. [191]
Ethanol	Rat (NP)[b]	G	McMillan et al. [191]
GBR 12909 (Vanoxerine)	Squirrel monkey	G	Melia and Spealman [192]
Imipramine	Pigeon	G	Zhang and Barrett [193]
Isoproterenol	Rat	G	Crissman and O'Donnell [194]
MDMA ("Ecstasy")	Rat	NG	Mori et al. [195]
S(+)-Methamphetamine	Pigeon	G	Sasaki et al. [196]
S(+)-Methamphetamine	Rat	G	Munzar and Goldberg [197]
Methylphenidate	Rat	G	Mori et al. [195]
Mirtazapine	Rat	PG (40%)	Dekeyne and Millan [198]
(−)-Nicotine	Rat	G	Young and Glennon [115]
(−)-Nicotine	Rat	G	Wiley et al. [114]
(−)-Nicotine	Rat	PG (70%)	Desai et al. [159]
(−)-Nicotine	Rat	NG	Shoaib et al. [199]
(−)-Nicotine	Mouse	PG (70%)	Damaj et al. [200]
(−)-Nicotine	Rhesus monkey	NG (23%)	Cunningham et al. [120]
Oxazepam	Pigeon	NG	de la Garza et al. [201]
Rimonabant	Rhesus monkey	PG (45%)	Schulze et al. [202]
Δ^9-THC	Rhesus monkey	NG	Schulze et al. [202]

[a]G stimulus generalization, PG partial generalization, NG no stimulus generalization
[b]P ethanol preferring rats, NP non-ethanol preferring rats

weakly potent CNS stimulant they were possibly examining only the "lower portion" of the dose response curve for bupropion; however, if the dose(s) was elevated, then S(+)-amphetamine-like effects would have occurred. In the Rush et al. [179], only "lower" doses of 50–400 mg of bupropion were examined in human subjects trained to discriminate the effect of 20 mg of S(+)-amphetamine from placebo. The administration of these doses of bupropion in substitution tests produced low to moderate levels of (+)-amphetamine-like responding, accompanied by subject-rated drug effects such as "alert/energetic," "elated," and "vigorous" that somewhat overlapped with those of S(+)-amphetamine. As such, complete S(+)-amphetamine stimulus generalization to bupropion may have occurred if relatively higher doses of bupropion had been administered (for further discussion, see [115]).

Table 9 also reveals that animals trained to discriminate (−)-nicotine from saline produced (high) partial generalization or complete generalization to bupropion; however, two studies reported no (−)-nicotine stimulus generalization to bupropion. Although future studies will be needed to explain the apparent discrepancy in results, evidence does suggest that (−)-nicotine and bupropion can share, to some degree, a similar stimulus effect. Consequently, bupropion may help some people refrain from smoking because it produces effects that serve as a suitable substitute for (−)-nicotine in the individual who is motivated to quit smoking. Lastly, the issue of bupropion as pharmacotherapy for cessation of nicotine consumption may be complicated by the actions of bupropion metabolites (e.g., [180, 200]). For example, chemical pathways of bupropion metabolism include hydroxylation of its *tertiary*-butyl group (with or without subsequent cyclization) and/or reduction of its carbonyl group to an alcohol (e.g., [206]). Moreover, species differences are known in the metabolism of bupropion (e.g., [207]). In humans, two major metabolites are a phenylmorpholinol, hydroxy-bupropion (BW 306U), and an aminoalcohol, *threo*hydrobupropion [sometimes referred to as *threo*dihydrobupropion, also known as *R,R*-2-(*tert*-butylamino)-1-(3-chlorophenyl)propanol or BW A494U] [206, 208–210]. Another human metabolite, although formed in lesser amounts than the others, is *erythro*hydrobupropion. The stimulus effects of bupropion metabolites (isomers) were examined in (−)-nicotine-trained rats and (+)- and (−)-threohydrobupropion substituted partially [180]. In contrast, *R,R*-hydroxybupropion produced vehicle-like responding in (−)-nicotine animals but, when given in combination with the training dose of (−)-nicotine, resulted in an attenuated nicotine-like effect. On the other hand, *S,S*-hydroxy-bupropion partially (66%) substituted for (−)-nicotine [180]. In another study, Damaj et al. [200] reported similar results. That is, a (−)-nicotine stimulus in mice partially generalized to *S,S*-hydroxybupropion but did not generalize to *R,R*-hydroxybupropion. Taken together, these results appear to indicate that some bupropion metabolites probably play a role(s) in the complex actions of this drug. Moreover, data also suggested that it is unlikely that any one metabolite (or isomer) is chiefly responsible for the stimulus actions of bupropion. In this regard, there does not appear to be any reports in the literature of bupropion metabolites (or isomers) being tested in bupropion-trained animals. Such data might give the best indication of the role each metabolite exerts in the stimulus properties of bupropion. In addition, bupropion stimulus generalization tests should be conducted with other pharmacotherapies used for smoking cessation, such as (−)-lobeline or varenicline, to assess potential similarities and/or differences in effects between these drugs.

7.3 Varenicline

Varenicline (Chantix®; Fig. 4) is prescribed as an adjunct medication in smoking cessation therapy and is thought to exert its effects as a partial agonist at α4β2 nAChRs and as a full agonist at α7 nAChRs [211, 212]. In drug discrimination studies, (−)-nicotine-trained rodents (rat or mouse) or non-human primates (rhesus monkeys) displayed a high degree of partial generalization or complete substitution to varenicline (Table 10). These data indicate that (−)-nicotine and varenicline share a similar stimulus effect. Thus, varenicline may assist people to refrain from smoking because it produces effects that serve as a suitable substitute for nicotine in the individual who wants to quit smoking. In comparison, S(+)-methamphetamine-trained animals generalized partially to varenicline, whereas cocaine-trained monkeys did not. Lastly, a search of the literature did not reveal any reports of varenicline as a training stimulus in drug discrimination studies. Such studies could provide important insights into the characterization of stimulus effects of varenicline in comparison to the degree of varenicline-like responses that might be produced by (−)-nicotine, cocaine, and/or S(+)-methamphetamine.

8 Summary and Conclusions

S(−)-nicotine is the pivotal reason that individuals persist in their consumption of smoke and smokeless nicotine-based products. The psychoactive effects of (−)-nicotine have been characterized as both "stimulant" and "calming." These effects are probably influenced by the mental status and expectations of the user. Thus,

Table 10 Varenicline as a test drug in animals trained to discriminate cocaine, S(+)-amphetamine or (−)-nicotine from saline

Training drug	Species	Result[a]	Reference
Cocaine	Rhesus monkey	NG	Gould et al. [213]
S(+)Methamphetamine	Rat	PG (~60%)	Desai and Bergman [161]
S(+)Methamphetamine	Squirrel monkey	PG (~50–65%)	Desai and Bergman [161]
S(+)Methamphetamine	Rhesus monkey	PG (~35–40%)	Banks et al. [182]
(−)Nicotine	Rat	PG (60%)	Smith et al. [214]
(−)Nicotine	Rat	G	Rollema et al. [215]
(−)Nicotine	Rat	PG (63%)	LeSage et al. [216]
(−)Nicotine	Rat	G	Paterson et al. [117, 118]
(−)Nicotine	Rat	G	Jutkiewicz et al. [119]
(−)Nicotine	Rhesus monkey	G	Cunningham et al. [120]
(−)Nicotine	Rat	G	Le Foll et al. [217]
(−)Nicotine	Mouse	PG (~50–70%)	Cunningham and McMahon [121]
(−)Nicotine	Mouse	PG (71%)	Rodriguez et al. [218]

[a]G stimulus generalization, PG partial generalization, NG no stimulus generalization

consumers of nicotine may experience alertness or relaxation and these effects could form the basis of the claim that "(-)-nicotine is a drug for all seasons." (−)-Nicotine produces effects that appear to be mediated primarily through $\alpha4\beta2$ nicotinic acetylcholine receptors (nAChRs) with subsequent influence on other neurotransmitters systems (e.g., dopamine). This chapter has described the basic methodology and usefulness of drug discrimination procedures to study the effects of (−)-nicotine. In this assay, subjects are tasked to identify the effects of (−)-nicotine versus saline vehicle. (−)-Nicotine can serve as a discriminative stimulus in non-human animal and human subjects. The model exhibits stability, sensitivity and displays several advantages over acute behavioral techniques to study in vivo pharmacological effects of this drug. Once established, the (−)-nicotine stimulus can be demonstrated to be dose related, time dependent and stereoselective: $S(−)$-nicotine is more potent than $R(+)$-nicotine in the production of (−)-nicotine-appro-priate responding. Tests of stimulus generalization (substitution) have been conducted to determine the similarity of effects produced by a test drug to those produced by the training dose of nicotine. Such tests have shown that other "natural" tobacco alkaloids and metabolites of (−)-nicotine can produce nicotine-like effects, but these drugs are less potent than (−)-nicotine. Stimulus antagonism (blockade) tests confirm that the (−)-nicotine stimulus is mediated via brain $\alpha4\beta2$ nAChR subtype receptors. This conclusion is based on reports that nicotine stimuli are blocked by (a) mecamylamine, a noncompetitive channel blocker at nAChRs and (b) dihydro-β-erythrodine, a nicotinic $\alpha4\beta2$ subunit receptor antagonist, but not by (c) methyllycaconitine, an $\alpha7$ nAChR receptor antagonist. In other studies, (−)-nicotine stimuli appear to share a marked degree of effects with bupropion (Zyban®) and varenicline (Chantix®), pharmacotherapies prescribed for smoking cessation. However, studies have not been conducted with the latter two drugs as training drugs to assess the effects of (−)-nicotine in stimulus generalization tests. Results from such studies would determine if cross generalization occurs between the drugs and could elucidate more clearly the relationships between the stimulus effects of (−)-nicotine versus those of the prescribed treatment medications. Overall, the application of drug discrimination procedures to study the effects of (−)-nicotine has achieved much success and progress. At this point in time, however, the model should be directed and applied to the issue of maintenance of abstinence from nicotine-based products. Thus, when an individual tries to quit nicotine consumption, cessation of use is typically followed by a withdrawal period that, unfortunately, usually leads to relapse; withdrawal symptoms include anxiety, depression, craving, cognitive and attention deficits. Moreover, just about every form of nicotine cessation therapy that has been employed typically demonstrates high immediate success rates, but high relapse rates almost certainly follow. The discriminative stimulus model demonstrates specificity and strength. These attributes could prove useful in the invention of new pharmacotherapies to assist the individual who desires to end their use of nicotine. For example, Harris et al. [219] trained rats to discriminate pentylenetetrazol [a.k.a. metrazol (PTZ)] from saline and suggested that the basis for the discrimination was PTZ-induced anxiety. In support of this argument, they reported that their animals showed a PTZ-like response when they were in nicotine "withdrawal." That is, their

PTZ-trained rats were administered high doses of nicotine over a 3-week period and subsequently responded on the PTZ-appropriate lever 24 h after the cessation of dosing. These investigators suggested that rats in nicotine withdrawal may be experiencing "anxiety" as measured by their PTZ partial generalization response. This application of the model could be an important finding: the potential to measure subjective effects during withdrawal from (−)-nicotine. Follow-up studies could exploit the finding and evaluate targeted drugs as potential antagonists of the PTZ-like withdrawal response. This approach may be able to identify candidate drugs to assist people who want to cease their consumption of (−)-nicotine-based products.

Acknowledgements The authors gratefully acknowledge the helpful comments and suggestions provided by Dr. John R. James.

References

1. Efron DH (1967) Ethnopharmacologic search for psychoactive drugs. In: Proceedings of a symposium held in San Francisco, CA, 28–30 Jan 1967. Public Health Service Publication Number 1645. U.S. Department of Health, Education and Welfare, Washington, DC
2. Larson PS, Haagard HB, Silvette H (1961) Tobacco. Experimental and clinical studies: a comprehensive account of the world literature. Williams & Wilkins, Baltimore
3. Brecher EM (1972) Licit and illicit drugs. Little, Brown, Boston
4. Leete E, Mueller ME (1982) Biomimetic synthesis of anatabine from 2,5-dihydropyridine produced by the oxidative decarboxylation of baikiain. J Am Chem Soc 104:6440–6444
5. Leete E, Slattery SA (1976) Incorporation of [2-^{14}C]-and [6-^{14}C]nicotinic acid into the tobacco alkaloids. Biosynthesis of anatabine and α,β-dipyridyl. J Am Chem Soc 98:6326–6330
6. Pinner A (1895) Über Nicotine. Des Konstitutions des Alkaloids. Ber Dtsch Chem Ges 28:456
7. Pictet A, Crepieux P (1895) Über Phenyl- and Pyrrylpyrroles and die constitution des nicotines. Ber Dtsch Chem Ges 28:1904
8. Spath E, Bretschneider H (1928) Eine neue synthese des Nicotines. Ber Dtsch Chem Ges 61:327
9. Hudson CS, Neuberger A (1950) The stereochemical formulas of hydroxypyrolines and some related substances. J Org Chem 15:24
10. Griffiths RR, Henningfield JE, Bigelow GE (1982) Human cigarette smoking: manipulation of number of puffs per bout, interbout interval and nicotine dose. J Pharmacol Exp Ther 220:256–265
11. Matta SG, Balfour DJ, Benowitz NL, Boyd RT, Buccafusco JJ, Caggiula AR, Craig CR, Collins AC, Damaj MI, Donny EC, Gardiner PS, Grady SR, Heberlein U, Leonard SS, Levin ED, Lukas RJ, Markou A, Marks MJ, McCallum SE, Parameswaran N, Perkins KA, Picciotto MR, Quik M, Rose JE, Rothenfluh A, Schafer WR, Stolerman IP, Tyndale RF, Wehner JM, Zirger JM (2007) Guidelines on nicotine dose selection for in vivo research. Psychopharmacology 190:269–319
12. Marubio LM, del Mar Arroyo-Jimenez M, Cordero-Erausquin M, Léna C, Le Novère N, de Kerchove, d'Exaerde A, Huchet M, Damaj MI, Changeux JP (1999) Reduced antinociception in mice lacking neuronal nicotinic receptor subunits. Nature 398:805–810

13. Lukas RJ, Changeux JP, Le Novère N, Albuquerque EX, Bslfour DJ, Berg DK, Bertrand D, Chiappinelli VA, Clarke PB, Collins AC, Dani JA, Grady SR, Kellar KJ, Lindstrom JM, Marks MJ, Quik M, Taylor PW, Wonnacott S (1999) International Union of Pharmacology. XX. Current status of the nomenclature for nicotinic acetylcholine receptors and their subunits. Pharmacol Rev 51:397–401
14. Corrigall WA, Coen KM, Adamson KL (1994) Self-administered nicotine activates the mesolimbic dopamine system through the ventral tegmental area. Brain Res 653:278–284
15. Reavill C, Stolerman IP (1987) Interaction of nicotine with dopaminergic mechanisms assessed through drug discrimination and rotational behaviour in rats. J Psychopharmacol 1:264–273
16. Domino EF (2002) Conflicting evidence for the dopamine release theory of nicotine/tobacco dependence. Nihon Shinkei Seishin Yakurigaku Zasshi 22:181–184
17. Morrison CF, Lee PN (1968) A comparison of the effects of nicotine and physostigmine on a measure of activity in the rat. Psychopharmacologia 13:210–221
18. Pradhan SN (1970) Effects of nicotine on several schedules of behavior in rats. Arch Int Pharmacodyn Ther 183:127–138
19. Philibin SD, Vann RE, Varvel SA, Covington 3rd HE, Rosecrans JA, James JR, Robinson SE (2005) Differential behavioral responses to nicotine in Lewis and Fischer-344 rats. Pharmacol Biochem Behav 80:87–92
20. Prus AJ, Vann RE, Rosecrans JA, James JR, Pehrson AL, O'Connell MM, Philibin SD, Robinson SE (2008) Acute nicotine reduces and repeated nicotine increases spontaneous activity in male and female Lewis rats. Pharmacol Biochem Behav 91:150–154
21. Rosecrans JA, Schechter MD (1972) Brain area nicotine levels in male and female rats of two strains. Arch Int Pharmacodyn Ther 196:46–54
22. Rosecrans JA (1971) Effects of nicotine on behavioral arousal and brain 5-hydroxytryptamine function in female rats selected for differences in activity. Eur J Pharmacol 14:29–37
23. Hendry JS, Rosecrans JA (1982) Effects of nicotine on conditioned and unconditioned behaviors in experimental animals. Pharmacol Ther 17:431–454
24. Johnston LM (1942) Tobacco smoking and nicotine. Lancet 2:742
25. Volle RL, Koelle RB (1975) Ganglionic stimulating and blocking agents. In: Goodman LS, Gilman A (eds) The pharmacological basis of therapeutics, 5th edn. Macmillan, New York, pp. 565–574
26. Jarvik ME (1979) Tolerance to the effects of tobacco. NIDA Res Monogr 23:150–157
27. Danielson K, Truman P, Kivell BM (2011) The effects of nicotine and cigarette smoke on the monoamine transporters. Synapse 65:866–879
28. James JR, Villanueva HF, Johnson JH, Arezo S, Rosecrans JA (1994) Evidence that nicotine can acutely desensitize central nicotinic acetylcholinergic receptors. Psychopharmacology 114:456–462
29. Sershen H, Toth E, Lajha A, Vizi ES (1995) Nicotine effects on presynaptic receptor interactions. Ann N Y Acad Sci 757:238–244
30. Rosecrans JA (1989) Nicotine as a discriminative stimulus: a neurobiobehavioral approach to studying central cholinergic mechanisms. J Subst Abus 1:287–300
31. Altman JL, Albert J, Milstein SL, Greenberg I (1976) Drugs as discriminative events in humans. Psychopharmacol Commun 2:327–330
32. Kallman WM, Kallman MJ, Harry GJ, Woodson PP, Rosecrans JA (1978) Nicotine as a discriminative stimulus in human subjects. In: Colpaert FC, Slangen JL (eds) Drug discrimination: applications in CNS pharmacology. Elsevier Biomedical Press, Amsterdam, pp. 211–218
33. Perkins KA (2011) Nicotine discrimination in humans. In: Glennon RA, Young R (eds) Drug discrimination: application to medicinal chemistry and drug studies. Wiley, New York, pp. 463–481. Chapter 15

34. Glennon RA, Young R (2011) Drug discrimination: practical considerations. In: Glennon RA, Young R (eds) Drug Discrimination: Application to medicinal chemistry and drug studies. Wiley, New York, pp. 41–128

35. Morrison CF, Stephenson JA (1969) Nicotine injections as the conditioned stimulus in discrimination learning. Psychopharmacologia 15:351–360

36. Schechter MD, Rosecrans JA (1971) C.N.S. effect of nicotine as the discriminative stimulus for the rat in a T-maze. Life Sci 10:821–832

37. Schechter MD, Rosecrans JA (1971) Behavioral evidence for two types of cholinergic receptors in the C.N.S. Eur J Pharmacol 15:375–378

38. Schechter MD, Rosecrans JA (1972) Effect of mecamylamine on discrimination between nicotine- and arecoline-produced cues. Eur J Pharmacol 17:179–182

39. Schechter MD, Rosecrans JA (1972) Nicotine as a discriminative stimulus in rats depleted of norepinephrine or 5-hydroxytryptamine. Psychopharmacologia 24:417–429

40. Schechter MD, Rosecrans JA (1972) Nicotine as a discriminative cue in rats: inability of related drugs to produce a nicotine-like cueing effect. Psychopharmacologia 27:379–387

41. Schechter MD, Rosecrans JA (1973) D-amphetamine as a discriminative cue: drugs with similar stimulus properties. Eur J Pharmacol 21:212–216

42. Perkins KA, Sanders M, Fonte C, Wilson AS, White W, Stiller R, McNamara D (1999) Effects of central and peripheral nicotinic blockade on human nicotine discrimination. Psychopharmacology 142:158–164

43. Stolerman IP, Naylor C, Elmer GI, Goldberg SR (1999) Discrimination and self-administration of nicotine by inbred strains of mice. Psychopharmacology 141:297–306

44. Takada K, Swedberg MD, Goldberg SR, Katz JL (1989) Discriminative stimulus effects of intravenous l-nicotine and nicotine analogs or metabolites in squirrel monkeys. Psychopharmacology 99:208–212

45. Varvel SA, James JR, Bowen S, Rosecrans JA, Karan LD (1999) Discriminative stimulus (DS) properties of nicotine in the C57BL/6 mouse. Pharmacol Biochem Behav 63:27–32

46. Kamien JB, Bickel WK, Hughes JR, Higgins ST, Smith BJ (1993) Drug discrimination by humans compared to nonhumans: current status and future directions. Psychopharmacology 111:259–270

47. Rosecrans JA (1987) Noncholinergic mechanisms involved in the behavioral and stimulus effects of nicotine, and relationships to the process of nicotine dependence. In: Martin WR, Van Loon GR, Iwamoto ET, Davis L (eds) Tobacco smoking and nicotine: a neurobiological approach. Plenum, New York, pp. 125–139

48. Rosecrans JA, Chance WT (1977) Cholinergic and non-cholinergic aspects of the discriminative stimulus properties of nicotine. In: Lal H (ed) Discriminative stimulus properties of drugs. Plenum Press, New York, pp. 155–185

49. Extance K, Goudie AJ (1981) Inter-animal olfactory cues in operant drug discrimination procedures in rats. Psychopharmacology 73:363–371

50. Pratt JA, Stolerman IP, Garcha HS, Giardini V, Feyerabend C (1983) Discriminative stimulus properties of nicotine: further evidence for mediation at a cholinergic receptor. Psychopharmacology 81:54–60

51. White FJ, Appel JB (1981) A neuropharmacological analysis of the discriminative stimulus properties of fenfluramine. Psychopharmacology 73:110–115

52. Chance WT, Murfin D, Krynock GM, Rosecrans JA (1977) A description of the nicotine stimulus and tests of its generalization to amphetamine. Psychopharmacology 55:19–26

53. Hirschhorn ID, Rosecrans JA (1974) Studies on the time course and the effect of cholinergic and adrenergic receptor blockers on the stimulus effect of nicotine. Psychopharmacologia 40:109–120

54. Stolerman IP, Garcha HS (1989) Temporal factors in drug discrimination: experiments with nicotine. J Psychopharmacol 3:88–97

55. Stolerman IP, Garcha HS, Pratt JA, Kumar R (1984) Role of training dose in discrimination of nicotine and related compounds by rats. Psychopharmacology 84:413–419

56. Aceto MD, Tucker SM, Ferguson GS, Hinson JR (1986) Rapid and brief tolerance to (+)- and (−)-nicotine in unanesthetized rats. Eur J Pharmacol 132:213–218
57. Hicks CS, Sinclair DA (1947) Toxicities of the optical isomers of nicotine and nornicotine. Aust J Exp Biol Med Sci 25:83–86
58. Aceto MD, Martin BR, Uwaydah IM, May EL, Harris LS, Izazola-Conde C, Dewey WL, Bradshaw TJ, Vinck WC (1979) Optically pure (+)-nicotine from (+/−)-nicotine and biological comparisons with (−)-nicotine. J Med Chem 22:174–177
59. Macht DI (1929) Pharmacological synergism of stereoisomers. Proc Natl Acad Sci 15:63–70
60. Macht DI, Davis ME (1934) Toxicity of alpha- and beta-nicotines and nornicotines, an inquiry into chemopharmacodynamic relationships. J Pharmacol Exp Ther 50:93–99
61. RTECS (1986) Registry of toxic effects of chemical substances. NIOSH 3A:3060–3424
62. Domino EF (1965) Some comparative pharmacological actions of (−)-nicotine, its optical isomer, and related compounds. In: von Euler US (ed) Tobacco alkaloids and related compounds, Wenner-Gren center international symposium series, vol 4. Pergamon, New York, London, pp. 303–314
63. Meltzer LT, Rosecrans JA, Aceto MD, Harris LS (1980) Discriminative stimulus properties of the optical isomers of nicotine. Psychopharmacology 68:283–286
64. Romano C, Goldstein A, Jewell NP (1981) Characterization of the receptor mediating the nicotine discriminative stimulus. Psychopharmacology 74:310–315
65. Goldberg SR, Risner ME, Stolerman IP, Reavill C, Garcha HS (1989) Nicotine and some related compounds: effects on schedule-controlled behaviour and discriminative properties in rats. Psychopharmacology 97:295–302
66. Brioni JD, Kim DJ, O'Neill AB (1996) Nicotine cue: lack of effect of the alpha 7 nicotinic receptor antagonist methyllycaconitine. Eur J Pharmacol 301:1–5
67. Hoffmann D, Hoffmann I (1998) Tobacco smoke components. Beiträge zur Tabakforsch ung International 18:49–52
68. Rodgman A, Perfetti TA (2009) The chemical components of tobacco and tobacco smoke. CRC Press, Taylor & Francis, Boca Raton
69. Dwoskin LP, Teng L, Buxton ST, Ravard A, Deo N, Crooks PA (1995) Minor alkaloids of tobacco release [3H]dopamine from superfused rat striatal slices. Eur J Pharmacol 276:195–199
70. Löfroth G (1989) Environmental tobacco smoke: overview of chemical composition and genotoxic components. Mutat Res 45:133–144
71. Felpin F-X, Girard S, Vo-Thanh G, Robins RJ, Villiéras J, Lebreton J (2001) Efficient enantiomeric synthesis of pyrrolidine and piperidine alkaloids from tobacco. J Org Chem 66:6305–6312
72. Armstrong DW, Wang X, Lee J-T, Liu Y-S (1999) Enantiomeric composition of nornicotine, anatabine, and anabasine in tobacco. Chirality 11:82–84
73. Benowitz NL, Hukkanen J, Jacob III P (2009) Nicotine chemistry, metabolism, kinetics and biomarkers. Handb Exp Pharmacol 192:29–60
74. Armstrong DW, Wang X, Ercal N (1998) Enantiomeric composition of nicotine in smokeless tobacco, medicinal products, and commercial reagents. Chirality 10:587–591
75. Jacob III P, Yu L, Shulgin AT, Benowitz NL (1999) Minor tobacco alkaloids as biomarkers for tobacco use: comparison of users of cigarettes, smokeless tobacco, cigars, and pipes. Am J Public Health 89:731–736
76. Leete E (1965) Biosynthesis of alkaloids. Science 147:1000–1006
77. Pool WF, Godin CS, Crooks PA (1985) Nicotine racemization during nicotine smoking. Toxicologist 5:232
78. Armitage AK, Dollery CT, George CF, Houseman TH, Lewis PJ, Turner DM (1975) Absorption and metabolism of nicotine from cigarettes. Br Med J 4:313–316
79. Turner DM, Armitage AK, Briant RH, Dollery CT (1975) Metabolism of nicotine by the isolated perfused dog lung. Xenobiotica 5:539–551

80. Benowitz NL, Jacob 3rd P, Fong I, Gupta S (1994) Nicotine metabolic profile in man: comparison of cigarette smoking and transdermal nicotine. J Pharmacol Exp Ther 268:296–303
81. Byrd GD, Cnang KM, Greene JM, de Bethizy JD (1992) Evidence for urinary excretion of glucuronide conjugates of nicotine, cotinine, and trans-3′-hydroxycotinine in smokers. Drug Metab Dispos 20:192–197
82. Schmeltz I, Hoffmann D (1977) Nitrogen containing compounds in tobacco and tobacco smoke. Chem Rev 77:295–311
83. Rosecrans JA, Spencer RM, Krynock GM, Chance WT (1978) Discriminative stimulus properties of nicotine and nicotine-related compounds. In: Bättig K (ed) Behavioral effects of nicotine. S. Karger, Basel, pp. 70–82
84. Desai RI, Barber DJ, Terry P (1999) Asymmetric generalization between the discriminative stimulus effects of nicotine and cocaine. Behav Pharmacol 10:647–656
85. Caine SB, Collins GT, Thomsen M, Wright C, Lanier RK, Mello NK (2014) Nicotine-like behavioral effects of the minor tobacco alkaloids nornicotine, anabasine, and anatabine in male rodents. Exp Clin Psychopharmacol 22:9–22
86. Dale HH (1914) The action of certain esters and ethers of choline, and their relation to muscarine. J Pharmacol Exp Ther 6:147
87. Langley JN (1905) On the reaction of cells and of nerve-endings to certain poisons, chiefly as regards the reaction of striated muscle to nicotine and to curari. J Physiol 33:374–413
88. Langley JN (1914) The antagonism of curari and nicotine in skeletal muscle. J Physiol 48:73–108
89. Changeux JP (2012) The nicotinic acetylcholine receptor: the founding father of the pentameric ligand-gated ion channel superfamily. J Biol Chem 287:40207–40215
90. Lindstrom J, Anand R, Gerzanich V, Peng X, Wang F, Wells G (1996) Structure and function of neuronal nicotinic acetylcholine receptors. Prog Brain Res 109:125–137
91. Wolstenholme AJ (2012) Glutamate-gated chloride channels. J Biol Chem 287:40232–40238
92. Gotti C, Zoli M, Clementi F (2006) Brain nicotinic acetylcholine receptors: native subtypes and their relevance. Trends Pharmacol Sci 27:482–491
93. Hurst R, Rollema H, Bertrand D (2013) Nicotinic acetylcholine receptors: from basic science to therapeutics. Pharmacol Ther 137:22–54
94. McGehee DS, Role LW (1995) Physiological diversity of nicotinic acetylcholine receptors expressed by vertebrate neurons. Annu Rev Physiol 57:521–546
95. Corringer PJ, Le Novère N, Changeux JP (2000) Nicotinic receptors at the amino acid level. Annu Rev Pharmacol Toxicol 40:431–458
96. Miyazawa A, Fujiyoshi Y, Stowell M, Unwin N (1999) Nicotinic acetylcholine receptor at 4.6 A resolution: transverse tunnels in the channel wall. J Mol Biol 288:765–786
97. Dani JA, De Biasi M (2001) Cellular mechanisms of nicotine addiction. Pharmacol Biochem Behav 70:439–446
98. Kenny PJ, Markou A (2001) Neurobiology of the nicotine withdrawal syndrome. Pharmacol Biochem Behav 70:531–549
99. Malin DH (2001) Nicotine dependence: studies with a laboratory model. Pharmacol Biochem Behav 70:551–559
100. Wonnacott S (1997) Presynaptic nicotinic ACh receptors. Trends Neurosci 20:92–98
101. Di Chiara G, Imperato A (1988) Drugs abused by humans preferentially increase synaptic dopamine concentrations in the mesolimbic system of freely moving rats. Proc Natl Acad Sci 1988(85):5274–5278
102. Schechter MD, Meehan SM (1992) Further evidence for the mechanisms that may mediate nicotine discrimination. Pharmacol Biochem Behav 41:807–812
103. Corrigall WA, Coen KM (1994) Dopamine mechanisms play at best a small role in the nicotine discriminative stimulus. Pharmacol Biochem Behav 48:817–820

104. Rosecrans JA, Chance WT, Schechter MD (1976) The discriminative stimulus properties of nicotine, d-amphetamine and morphine in dopamine depleted rats. Psychpharmacol Commun 2:349–356
105. Chance WT, Kallman MD, Rosecrans JA, Spencer RM (1978) A comparison of nicotine and structurally related compounds as discriminative stimuli. Br J Pharmacol 63:609–616
106. Stolerman IP, Pratt JA, Garcha HS, Giardini V, Kumar R (1983) Nicotine cue in rats analysed with drugs acting on cholinergic and 5-hydroxytryptamine mechanisms. Neuropharmacology 22:1029–1037
107. Craft RM, Howard JL (1988) Cue properties of oral and transdermal nicotine in the rat. Psychopharmacology 96:281–284
108. Stolerman IP, Kumar R, Reavill C (1988) Discriminative stimulus effects of cholinergic agonists and the actions of their antagonists. Psychopharmacol Ser 1988(4):32–43
109. Miyata H, Ando K, Yanagita T (1991) Studies on the involvement of the nucleus accumbens in the discriminative effects of nicotine in rats. Nihon Yakurigaku Zasshi 98:389–397
110. Ando K, Miyata H, Hironaka N, Tsuda T, Yanagita T (1993) The discriminative effects of nicotine and their central sites in rats. Yakubutsu Seishin Kodo 13:129–136
111. Chandler CJ, Stolerman IP (1997) Discriminative stimulus properties of the nicotinic agonist cytisine. Psychopharmacology 129:257–264
112. Gasior M, Shoaib M, Yasar S, Jaszyna M, Goldberg SR (1999) Acquisition of nicotine discrimination and discriminative stimulus effects of nicotine in rats chronically exposed to caffeine. J Pharmacol Exp Ther 288:1053–1073
113. Mansbach RS, Chambers LK, Rovetti CC (2000) Effects of the competitive nicotinic antagonist erysodine on behavior occasioned or maintained by nicotine: comparison with mecamylamine. Psychopharmacology 148:234–242
114. Wiley JL, Lavecchia KL, Martin BR, Damaj MI (2002) Nicotine-like discriminative stimulus effects of bupropion in rats. Exp Clin Psychopharmacol 10:129–135
115. Young R, Glennon RA (2002) Nicotine and bupropion share a similar discriminative stimulus effect. Eur J Pharmacol 443:113–118
116. Zaniewska M, McCreary AC, Przegaliński E, Filip M (2006) Evaluation of the role of nicotinic acetylcholine receptor subtypes and cannabinoid system in the discriminative stimulus effects of nicotine in rats. Eur J Pharmacol 540:96–106
117. Paterson NE, Fedolak A, Oliver B, Hanania T, Ghavami A, Caldarone B (2010) Psychostimulant-like discriminative stimulus and locomotor sensitization properties of the wake-promoting agent modafinil in rodents. Pharmacol Biochem Behav 95:449–456
118. Paterson NE, Min W, Hackett A, Lowe D, Hanania T, Caldarone B, Ghavami A (2010) The high-affinity nAChR partial agonists varenicline and sazetidine-A exhibit reinforcing properties in rats. Prog Neuro-Psychopharmacol Biol Psychiatry 34:1455–1464
119. Jutkiewicz EM, Brooks EA, Kynaston AD, Rice KC, Woods JH (2011) Patterns of nicotinic receptor antagonism: nicotine discrimination studies. J Pharmacol Exp Ther 339:194–202
120. Cunningham CS, Javors MA, McMahon LR (2012) Pharmacologic characterization of a nicotine-discriminative stimulus in rhesus monkeys. J Pharmacol Exp Ther 341:840–849
121. Cunningham CS, McMahon LR (2013) Multiple nicotine training doses in mice as a basis for differentiating the effects of smoking cessation aids. Psychopharmacology 228:321–333
122. Stolerman IP, Chandler CJ, Garcha HS, Newton JM (1997) Selective antagonism of behavioural effects of nicotine by dihydro-beta-erythroidine in rats. Psychopharmacology 129:390–397
123. Gommans J, Stolerman IP, Shoaib M (2000) Antagonism of the discriminative and aversive stimulus properties of nicotine in C57BL/6J mice. Neuropharmacology 39:2840–2847
124. Shoaib M, Zubaran C, Stolerman IP (2000) Antagonism of stimulus properties of nicotine by dihydro-beta-erythroidine (DHβE) in rats. Psychopharmacology 149:140–146
125. Lee JY, Choi MJ, Choe ES, Lee YJ, Seo JW, Yoon SS (2016) Differential discriminative-stimulus effects of cigarette smoke condensate and nicotine in nicotine-discriminating rats. Behav Brain Res 306:197–201

126. Quarta D, Naylor CG, Barik J, Fernandes C, Wonnacott S, Stolerman IP (2009) Drug discrimination and neurochemical studies in alpha7 null mutant mice: tests for the role of nicotinic alpha7 receptors in dopamine release. Psychopharmacology 203:399–410

127. Ford RV, Madison JC, Moyer JH (1956) Pharmacology of mecamylamine. Am J Med Sci 232:129–143

128. Rose JE, Behm FM, Westman EC (1998) Nicotine-mecamylamine treatment for smoking cessation: the role of pre-cessation therapy. Exp Clin Psychopharmacol 6:331–343

129. Shytle RD, Penny E, Silver AA, Goldman J, Sanberg PR (2002) Mecamylamine (Inversine): an old antihypertensive with new research directions. J Hum Hypertens 16:453–457

130. Stolerman IP, Goldfarb T, Fink R, Jarvik ME (1973) Influencing cigarette smoking with nicotine antagonists. Psychopharmacologia 28:2472–2459

131. Arias HR, Targowska KM, Feuerbach D, Sullivan CJ, Maciejewski R, Jozwiak K (2010) Different interaction between tricyclic antidepressants and mecamylamine with the human alpha3beta4 nicotinic acetylcholine receptor ion channel. Neurochem Int 56:642–649

132. Kaiser SA, Soliakov L, Harvey SC, Luetje CW, Wonnacott S (1998) Differential inhibition by alpha-conotoxin-MII of the nicotinic stimulation of [3H]dopamine release from rat striatal synaptosomes and slices. J Neurochem 70:1069–1076

133. Papke RL, Wecker L, Stitzel JA (2010) Activation and inhibition of mouse muscle and neuronal nicotinic acetylcholine receptors expressed in Xenopus oocytes. J Pharmacol Exp Ther 333:501–518

134. Chavez-Noriega LE, Crona JH, Washburn MS, Urrutia A, Elliott KJ, Johnson EC (1997) Pharmacological characterization of recombinant human neuronal nicotinic acetylcholine receptors h alpha 2 beta 2, h alpha 2 beta 4, h alpha 3 beta 2, h alpha 3 beta 4, h alpha 4 beta 2, h alpha 4 beta 4 and h alpha 7 expressed in Xenopus oocytes. J Pharmacol Exp Ther 280:346–356

135. Harvey SC, Maddox FN, Luetje CW (1996) Multiple determinants of dihydro-beta-erythroidine sensitivity on rat neuronal nicotinic receptor alpha subunits. J Neurochem 67:1953–1959

136. Papke RL, Dwoskin LP, Crooks PA, Zheng G, Zhang Z, McIntosh JM, Stokes C (2008) Extending the analysis of nicotinic receptor antagonists with the study of alpha6 nicotinic receptor subunit chimeras. Neuropharmacology 54:1189–1200

137. Verbitsky M, Rothlin CV, Katz E, Elgoyhen AB (2000) Mixed nicotinic-muscarinic properties of the alpha9 nicotinic cholinergic receptor. Neuropharmacology 39:2515–2524

138. Harbourne JB, Baxter H (1993) Phytochemical dictionary. Taylor & Francis, London, p. 153

139. Absalom NL, Quek G, Lewis TM, Qudah T, von Arenstorff I, Ambrus JI, Harpsøe K, Karim N, Balle T, McLeod MD, Chebib M (2013) Covalent trapping of methyllycaconitine at the α4-α4 interface of the α4β2 nicotinic acetylcholine receptor: antagonist binding site and mode of receptor inhibition revealed. J Biol Chem 288:26521–26532

140. Capelli AM, Castelletti L, Chen YH, Van der Keyl H, Pucci L, Oliosi B, Salvagno C, Bertani B, Gotti C, Powell A, Mugnaini M (2011) Stable expression and functional characterization of a human nicotinic acetylcholine receptor with α6β2 properties: discovery of selective antagonists. Br J Pharmacol 163:313–329

141. Mogg AJ, Whiteaker P, McIntosh JM, Marks M, Collins AC, Wonnacott S (2002) Methyllycaconitine is a potent antagonist of alpha-conotoxin-MII-sensitive presynaptic nicotinic acetylcholine receptors in rat striatum. J Pharmacol Exp Ther 302:197–204

142. Overton DA (1974) Experimental methods for the study of state-dependent learning. Fed Proc 33:1800–1813

143. Frey LG, Winter JC (1978) Current trends in the study of drugs as discriminative stimuli. In: Ho BT, Richards III DW, Chute DL (eds) Drug discrimination and state dependent learning. Academic Press, New York, pp. 35–45

144. Garcha HS, Stolerman IP (1993) Discriminative stimulus effects of the nicotine antagonist mecamylamine in rats. J Psychopharmacol 7:43–51

145. Cunningham CS, Moerke MJ, McMahon LR (2014) The discriminative stimulus effects of mecamylamine in nicotine-treated and untreated rhesus monkeys. Behav Pharmacol 25:296–305

146. Kumar R, Reavill C, Stolerman IP (1987) Nicotine cue in rats: effects of central administration of ganglion-blocking drugs. Br J Pharmacol 90:239–246

147. Hughes JR, Gulliver SB, Fenwick JW, Valliere WA, Cruser K, Pepper S, Shea P, Solomon LJ, Flynn BS (1992) Smoking cessation among self-quitters. Health Psychol 11:331–334

148. Raw M, McNeill A, West R (1998) Smoking cessation guidelines for health professionals. A guide to effective smoking cessation interventions for the health care system. Health education authority. Thorax 53(Suppl 5 Pt 1):S1–19

149. Britton J, Jarvis MJ (2000) Bupropion: a new treatment for smokers. Nicotine replacement treatment should also be available on the NHS. BMJ 321:65–66

150. Damaj MI, Patrick GS, Creasy KR, Martin BR (1997) Pharmacology of lobeline, a nicotinic receptor ligand. J Pharmacol Exp Ther 282:410–419

151. Dwoskin LP, Crooks PA (2001) Competitive neuronal nicotinic receptor antagonists: a new direction for drug discovery. J Pharmacol Exp Ther 298:395–402

152. Kaniaková M, Lindovský J, Krůšek J, Adámek S, Vyskočil F (2011) Dual effect of lobeline on α4β2 rat neuronal nicotinic receptors. Eur J Pharmacol 658:108–113

153. Miller DK, Crooks PA, Dwoskin LP (2000) Lobeline inhibits nicotine-evoked [(3)H]dopamine overflow from rat striatal slices and nicotine-evoked (86)Rb(+) efflux from thalamic synaptosomes. Neuropharmacology 39:2654–2662

154. Miller DK, Harrod SB, Green TA, Wong MY, Bardo MT, Dwoskin LP (2003) Lobeline attenuates locomotor stimulation induced by repeated nicotine administration in rats. Pharmacol Biochem Behav 74:279–286

155. Department of Health and Human Services, Food and drug Administration (1993) Smoking deterrent drug products for over-the-counter human use. Federal Register 58(103):31236–31241

156. Sachs DP (1986) Cigarette smoking. Health effects and cessation strategies. Clin Geriatr Med 2:337–362

157. U.S Department of Health and Human Services (1988) The health consequences of smoking: nicotine addiction. A report of the Surgeon General. Office on Smoking and Health, Maryland

158. Cunningham CS, Polston JE, Jany JR, Segert IL, Miller DK (2006) Interaction of lobeline and nicotinic receptor ligands with the discriminative stimulus properties of cocaine and amphetamine. Drug Alcohol Depend 84:211–222

159. Desai RI, Barber DJ, Terry P (2003) Dopaminergic and cholinergic involvement in the discriminative stimulus effects of nicotine and cocaine in rats. Psychopharmacology 167:335–343

160. Miller DK, Crooks PA, Teng L, Witkin JM, Munzar P, Goldberg SR, Acri JB, Dwoskin LP (2001) Lobeline inhibits the neurochemical and behavioral effects of amphetamine. J Pharmacol Exp Ther 296:1023–1034

161. Desai RI, Bergman J (2014) Drug discrimination in methamphetamine-trained rats: effects of cholinergic nicotinic compounds. J Pharmacol Exp Ther 335:807–816

162. Reavill C, Walther B, Stolerman IP, Testa B (1990) Behavioural and pharmacokinetic studies on nicotine, cytosine and lobeline. Neuropharmacology 29:619–624

163. Balfour DJ (2001) The pharmacology underlying pharmacotherapy for tobacco dependence: a focus on bupropion. Int J Clin Pract 55:53–57

164. Harrison C (2001) Bupropion may not be as good as editorial implies. Br Med J 322:431

165. Hurt RD, Sachs DPL, Glover ED, Offord KP, Johnston JA, Dale LC, Khayrallah MA, Schroeder DR, Glover PN, Sullivan CR, Croghan IT, Sullivan PM (1997) A comparison of sustained-release bupropion and placebo for smoking cessation. N Engl J Med 337:1195–1202

166. Jorenby DE, Leischow SJ, Nides MA, Rennard SI, Johnston JA (1999) A controlled trial of sustained-release bupropion, a nicotine patch, or both for smoking cessation. N Engl J Med 340:685–691
167. Butz RF, Welch RM, Findlay JWA (1982) Relationship between bupropion disposition and dopamine uptake inhibition in rats and mice. J Pharmacol Exp Ther 221:676–685
168. Dufrensne RL, Weber SS, Becker RE (1984) Bupropion hydrochloride. Drug Intell Clin Pharm 18:957–964
169. Dufrensne RL, Becker RE, Blitzer R, Wagner RL, Lal H (1985) Safety and efficacy of bupropion, a novel antidepressant. Drug Dev Res 6:39–45
170. Ferris RM, Maxwell RA, Cooper BR, Soroko FE (1982) Neurochemical and neuropharmacological investigations into the mechanisms of action of bupropion. HCI—a new atypical antidepressant agent. Adv Biochem Psycholpharmacol 31:277–286
171. Fryer JD, Lukas RJ (1999) Noncompetitive functional inhibition at diverse, human nicotinic acetylcholine receptor subtypes by bupropion, phencylidine and ibogaine. J Pharmacol Exp Ther 288:88–92
172. Slemmer JE, Martin BR, Damaj MI (2000) Bupropion is a nicotinic antagonist. J Pharmacol Exp Ther 295:321–327
173. Jones CN, Howard JL, McBennett ST (1980) Stimulus properties of antidepressants in the rat. Psychopharmacology 67:111–118
174. Blitzer RD, Becker RE (1985) Characterization of the bupropion cue in the rat: lack of evidence for a dopaminergic mechanism. Psychopharmacology 85:173–177
175. Terry P, Katz JL (1997) Dopaminergic mediation of the discriminative stimulus effects of bupropion in rats. Psychopharmacology 134:201–212
176. Evans SM, Johanson CE (1987) Amphetamine-like effects of anorectics and related compounds in pigeons. J Pharmacol Exp Ther 241:817–825
177. de la Garza R, Johanson CE (1987) Discriminative stimulus properties of intragastrically administered d-amphetamine and pentobarbital in rhesus monkeys. J Pharmacol Exp Ther 243:955–962
178. Kamien JB, Woolverton WL (1989) A pharmacological analysis of the discriminative stimulus properties of d-amphetamine in rhesus monkeys. J Pharmacol Exp Ther 248:938–946
179. Rush CR, Kollins SH, Pazzaglia PJ (1998) Discriminative-stimulus and participant-rated effects of methylphenidate, bupropion, and triazolam in d-amphetamine-trained humans. Exp Clin Psychopharmacol 6:32–44
180. Bondarev ML, Bondareva TS, Young R, Glennon RA (2003) Behavioral and biochemical investigations of bupropion metabolites. Eur J Pharmacol 474:85–93
181. Heal DJ, Frankland AT, Gosden J, Hutchins LJ, Prow MR, Luscombe GP, Buckett WR (1992) A comparison of the effects of sibutramine hydrochloride, bupropion and methamphetamine on dopaminergic function: evidence that dopamine is not a pharmacological target for sibutramine. Psychopharmacology 107:303–309
182. Banks ML, Smith DA, Blough BE (2016) Methamphetamine-like discriminative stimulus effects of bupropion and its two hydroxyl metabolites in male rhesus monkeys. Behav Pharmacol 27:196–203
183. Makhay MM, O'Donnell JM (1999) Effects of antidepressants in rats trained to discriminate the beta-2 adrenergic agonist clenbuterol. Pharmacol Biochem Behav 63:319–324
184. Lamb RJ, Griffiths RR (1990) Self-administration in baboons and the discriminative stimulus effects in rats of bupropion, nomifensine, diclofensine and imipramine. Psychopharmacology 102:183–190
185. Kleven MS, Anthony EW, Woolverton WL (1990) Pharmacological characterization of the discriminative stimulus effects of cocaine in rhesus monkeys. J Pharmacol Exp Ther 254:312–317
186. Johanson CE, Barrett JE (1993) The discriminative stimulus effects of cocaine in pigeons. J Pharmacol Exp Ther 267:1–8

187. Broadbent J, Gaspard TM, Dworkin SI (1995) Assessment of the discriminative stimulus effects of cocaine in the rat: lack of interaction with opioids. Pharmacol Biochem Behav 51:379–385
188. Baker LE, Riddle EE, Saunders RB, Appel JB (1993) The role of monoamine uptake in the discriminative stimulus effects of cocaine and related compounds. Behav Pharmacol 4:69–79
189. Quinton MS, Gerak LR, Moerschbaecher JM, Winsauer PJ (2006) Effects of pregnanolone in rats discriminating cocaine. Pharmacol Biochem Behav 85:385–392
190. Awasaki Y, Nojima H, Nishida N (2011) Application of the conditioned taste aversion paradigm to assess discriminative stimulus properties of psychostimulants in rats. Drug Alcohol Depend 118:288–294
191. McMillan DE, Li M, Shide DJ (1999) Differences between alcohol-preferring and alcohol-nonpreferring rats in ethanol generalization. Pharmacol Biochem Behav 64:415–419
192. Melia KF, Spealman RD (1991) Pharmacological characterization of the discriminative-stimulus effects of GBR 12909. J Pharmacol Exp Ther 258:626–632
193. Zhang L, Barrett JE (1991) Imipramine as a discriminative stimulus. J Pharmacol Exp Ther 259:1088–1093
194. Crissman AM, O'Donnell JM (2002) Effects of antidepressants in rats trained to discriminate centrally administered isoproterenol. J Pharmacol Exp Ther 302:606–611
195. Mori T, Uzawa N, Kazawa H, Watanabe H, Mochizuki A, Shibasaki M, Yoshizawa K, Higashiyama K, Suzuki T (2014) Differential substitution for the discriminative stimulus effects of 3,4-methylenedioxymethamphetamine and methylphenidate in rats. J Pharmacol Exp Ther 350:403–411
196. Sasaki JE, Tatham TA, Barrett JE (1995) The discriminative stimulus effects of methamphetamine in pigeons. Psychopharmacology 120:303–310
197. Munzar P, Goldberg SR (2000) Dopaminergic involvement in the discriminative-stimulus effects of methamphetamine in rats. Psychopharmacology 148:209–216
198. Dekeyne A, Millan MJ (2009) Discriminative stimulus properties of the atypical antidepressant, mirtazapine, in rats: a pharmacological characterization. Psychopharmacology 203:329–341
199. Shoaib M, Sidhpura N, Shafait S (2003) Investigating the actions of bupropion on dependence-related effects of nicotine in rats. Psychopharmacology 165:405–412
200. Damaj MI, Grabus SD, Navarro HA, Vann RE, Warner JA, King LS, Wiley JL, Blough BE, Lukas RJ, Carroll FI (2010) Effects of hydroxymetabolites of bupropion on nicotine dependence behavior in mice. J Pharmacol Exp Ther 334:1087–1095
201. de la Garza R, Evans S, Johanson CE (1987) Discriminative stimulus properties of oxazepam in the pigeon. Life Sci 40:71–79
202. Schulze DR, Carroll FI, McMahon LR (2012) Interactions between dopamine transporter and cannabinoid receptor ligands in rhesus monkeys. Psychopharmacology 222:425–438
203. Golden RN, James SP, Sherer MA, Rudorfer MV, Sack DA, Potter WZ (1985) Psychoses associated with bupropion treatment. Am J Psychiatry 142:1459–1462
204. Glaxo Wellcome (2001) Zyban® (bupropion hydrochloride) sustained-release tablets. Product information
205. Miller L, Griffith J (1983) A comparison of bupropion, dextramphetamine, and placebo in mixed-substance abusers. Psychopharmacology 80:199–205
206. Schroeder DH (1983) Metabolism and kinetics of bupropion. J Clin Psychiatry 44:79–81
207. Horst WD, Preskorn SH (1998) Mechanisms of action and clinical characteristics of three atypical antidepressants: venlafaxine, nefazodone, bupropion. J Affect Disord 51:237–254
208. Rotzinger S, Bourin M, Akimoto Y, Coutts RT, Baker GB (1999) Metabolism of some "second"- and "fourth"-generation antidepressants: iprindole, viloxazine, bupropion, mianserin, maprotiline, trazodone, nefazodone, and venlafaxine. Cell Mol Neurobiol 19:427–442

209. Suckow RF, Smith TM, Perumal AS, Cooper TB (1986) Pharmacokinetics of bupropion and metabolites in plasma and brain of rats, mice, and guinea pigs. Drug Metab Dispos 14:692–697

210. Welch RM, Lai AA, Schroeder DH (1987) Pharmacological significance of the species differences in bupropion metabolism. Xenobiotica 17:287–298

211. Coe JW, Brooks PR, Vetelino MG, Wirtz MC, Arnold EP, Huang J, Sands SB, Davis TI, Lebel LA, Fox CB, Shrikhande A, Heym JH, Schaeffer E, Rollema H, Lu Y, Mansbach RS, Chambers LK, Rovetti CC, Schulz DW, Tingley 3rd FD, O'Neill BT (2005) Varenicline: an alpha4beta2 nicotinic receptor partial agonist for smoking cessation. J Med Chem 48:3474–3477

212. Mihalak KB, Carroll FI, Lueje CW (2006) Varenicline is a partial agonist at alpha4beta2 and a full agonist at alpha7 neuronal nicotinic receptors. Mol Pharmacol 70:801–805

213. Gould RW, Czoty PW, Nader SH, Nader MA (2011) Effects of varenicline on the reinforcing and discriminative stimulus effects of cocaine in rhesus monkeys. J Pharmacol Exp Ther 339:678–686

214. Smith JW, Mogg A, Tafi E, Peacey E, Pullar IA, Szekeres P, Tricklebank M (2007) Ligands selective for α4β2 but not α3β4 or α7 nicotinic receptors generalise to the nicotine discriminative stimulus in the rat. Psychopharmacology 190:157–170

215. Rollema H, Chambers LK, Coe JW, Glowa J, Hurst RS, Lebel LA, Lu Y, Mansbach RS, Mather RJ, Rovetti CC, Sands SB, Schaeffer E, Schulz DW, Tingley 3rd FD, Williams KE (2007) Pharmacological profile of the alpha4beta2 nicotinic acetylcholine receptor partial agonist varenicline, an effective smoking cessation aid. Neuropharmacology 52:985–994

216. LeSage MG, Shelley D, Ross JT, Carroll FI, Corrigall WA (2009) Effects of the nicotinic receptor partial agonists varenicline and cytisine on the discriminative stimulus effects of nicotine in rats. Pharmacol Biochem Behav 91:461–467

217. Le Foll B, Chakraborty-Chtterjee M, Lev-Ran S, Barnes C, Pushparaj A, Gamaleddin I, Yan Y, Khaled M, Goldberg SR (2012) Varenicline decreases nicotine self-administration and cue-induced reinstatement of nicotine-seeking behaviour in rats when a long pretreatment time is used. Int J Neuropsychopharmacol 15:1265–1274

218. Rodriguez JS, Cunningham CS, Moura FB, Ondachi P, Carroll FI, McMahon LR (2014) Discriminative stimulus and hypothermic effects of some derivatives of the nAChR agonist epibatidine in mice. Psychopharmacology 231:4455–4466

219. Harris CM, Emmett-Oglesby MW, Robinson NG, Lal H (1986) Withdrawal from chronic nicotine substitutes partially for the interoceptive stimulus produced by pentylenetetrazol (PTZ). Psychopharmacology 90:85–89

Cross-Species Translational Findings in the Discriminative Stimulus Effects of Ethanol

Daicia C. Allen, Matthew M. Ford, and Kathleen A. Grant

Abstract The progress on understanding the pharmacological basis of ethanol's discriminative stimulus effects has been substantial, but appears to have plateaued in the past decade. Further, the cross-species translational efforts are clear in laboratory animals, but have been minimal in human subject studies. Research findings clearly demonstrate that ethanol produces a compound stimulus with primary activity through GABA and glutamate receptor systems, particularly ionotropic receptors, with additional contribution from serotonergic mechanisms. Further progress should capitalize on chemogenetic and optogenetic techniques in laboratory animals to identify the neural circuitry involved in mediating the discriminative stimulus effects of ethanol. These infrahuman studies can be guided by in vivo imaging of human brain circuitry mediating ethanol's subjective effects. Ultimately, identifying receptors systems, as well as where they are located within brain circuitry, will transform the use of drug discrimination procedures to help identify possible treatment or prevention strategies for alcohol use disorder.

Keywords Alcohol • Drug discrimination • Ethanol • Interspecies • Translational

D.C. Allen
Department of Behavioral Neurosciences, Oregon Health & Science University, Portland, OR 97239, USA

M.M. Ford and K.A. Grant (✉)
Department of Behavioral Neurosciences, Oregon Health & Science University, Portland, OR 97239, USA

Division of Neuroscience, Oregon National Primate Research Center, Beaverton, OR 97006, USA
e-mail: grantka@ohsu.edu

© Springer International Publishing AG 2017
Curr Topics Behav Neurosci (2018) 39: 95–112
DOI 10.1007/7854_2017_2
Published Online: 24 March 2017

Contents

1   Introduction

Although reports of state dependent learning using alcohol go back to the 1950s [1], the study of the pharmacological basis of the ethanol discriminative stimulus began in earnest in the early 1970s. The literature reviewed here comprises reports in which ethanol was used as a training stimulus as well as manuscripts that used ethanol in substitution tests for other drugs used as training stimuli. Studies that report only subjective effects or reinstatement procedures are not included in this review. In general, given that the cumulative dataset encompasses over four decades of research, the volume of studies addressing the discriminative stimulus effects of ethanol is not large. Figure 1 depicts both the cumulative publications (Fig. 1a) and the yearly publication rate (Fig. 1b) reporting ethanol trained as a discriminative stimulus as well as ethanol substitution tests in other drug discriminations. Most of the substitution tests using ethanol were either to control for nonspecific drug effects (i.e., as a negative control for discriminative stimuli other than ethanol) or to test for cross-generalization. These contributions have been at a low, but fairly consistent rate over time (Fig. 1b). Overall, for studies that trained an ethanol discrimination there has been about 4–5 publications/year, but the timeframe of 1990–2005 was clearly the most productive (Fig. 1a). Publications have fallen off considerably since 2005. This trend is remarkably similar to the trend encompassing the entire drug discrimination literature as recently reviewed [2]. The reason for this decline in the use of ethanol discrimination in understanding the behavioral pharmacology of ethanol is not readily obvious. Given the utility of drug discrimination as an in vivo pharmacological assay, there clearly remain many important questions that can be addressed with an ethanol discrimination preparation, particularly in the context of recent advancements in brain-region specific targeting and genetic manipulations. Neurobiological approaches, including chemogenetic and optogenetic manipulations, have the potential to greatly expand our knowledge of dose-dependent mechanisms of ethanol in the brain and improve cross-species translational cohesion of the interoceptive effects of ethanol. In general, animal models of human behavior ultimately strive to provide data that inform the human condition. In alcohol discrimination procedures, this emphasis is on pharmacological variables and receptor mechanisms that mediate the stimulus effects and a rational approach to pharmacotherapeutic development for alcohol use disorders. However, specificity in terms of

a

Yearly publications with ethanol discrimination

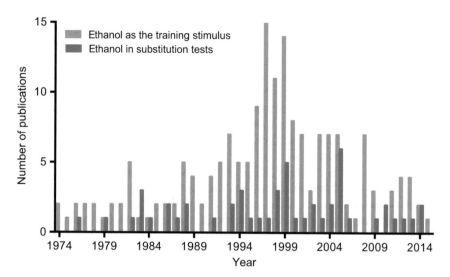

b

Cumulative publications with ethanol discrimination

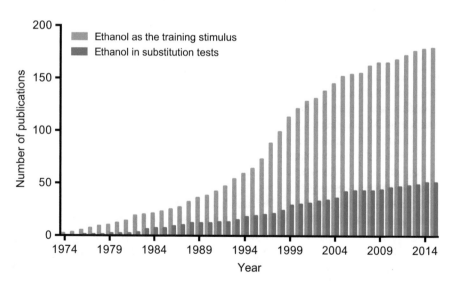

Fig. 1 Cumulative (**a**) and yearly (**b**) publication rate for ethanol as a training stimulus and in substitution tests

discrete neural circuitry to target and lower adverse off target effects has not been achieved, but represents an important avenue forward.

Although circuitry and genomic approaches are only just beginning to be applied to ethanol discriminations, ethanol is one substance that has a relatively strong record in cross-species translational studies. Ethanol has been trained as a discriminative stimulus in pigeons, mice, rats, gerbils, monkeys, and humans, although not in a proportional fashion. Indeed, the contribution to the literature can be rank ordered by rodents (89%), monkeys (7%), pigeons (2%), and humans (2%). Somewhat perplexing, given the legal and historical status of alcohol, there are far fewer human subject studies of ethanol discrimination than there are of stimulant, opiate, or cannabinoid discriminations (reviewed in Bolin et al. [2] and in the section below).

Given the low rate of new data on ethanol discrimination in recent years, an in-depth review of the basic pharmacology underlying ethanol's discriminative stimulus effects would add little to available reviews on the subject [3–6]. Instead, this review focuses on more recent developments in refining the specific action of ethanol at each receptor target. Specifically, the receptor systems that are primarily identified in rodents will be compared and contrasted with findings from monkeys and humans in order to highlight whether this information has been translated across species. A major conclusion is that in order for ethanol discrimination studies to advance pharmacotherapies for alcohol use disorders, a new approach is needed. Specifically, the future of translational ethanol discrimination studies must focus on region specific receptor mechanisms and how this fits into a cohesive understanding of brain circuitry. Over the last four decades, ethanol drug discrimination studies have established three primary receptor targets involved in ethanol's discriminative stimulus effects: $GABA_A$, NMDA, and $5\text{-}HT_{1B/2C}$ systems. There has also been some evidence for a secondary, modulatory role of both the opioid [7–11] and acetylcholine [12–16] receptor systems, but there is no evidence of direct mediation of ethanol's discriminative stimulus effects at these receptor sites. Ethanol is known to act as a positive modulator at the $GABA_A$ receptor to increase chloride conductance through the channel and decrease cellular excitability [17]. Additionally, ethanol has antagonist activity at the NMDA glutamate receptor, which appears selective for noncompetitive antagonism. Lastly, ethanol has activity at several 5-HT receptor systems, but agonism at the $5\text{-}HT_{1B/2C}$ receptor subtypes is most prominent [4, 6, 18].

Somewhat unique to ethanol, the relative contribution of these stimulus components varies based on training dose magnitude, with $GABA_A$ receptors exerting greatest influence at low to moderate training doses (≤ 1.5 g/kg) and NMDA receptors playing a larger role at higher doses (≥ 1.5 g/kg) in rodents [6, 19, 20]. Similarly, the 5-HT component of the ethanol stimulus complex is most prominent at low to moderate training doses [21]. More recent work expands upon this foundation and emphasizes the selectivity of ethanol at different receptor subtypes and subunits by incorporating novel ligands. To compare data across species, findings from systemic administration are surveyed in the following sections, with an additional section emphasizing recent work with targeted brain-region approaches. At the conclusion, suggestions for future approaches are presented to maximize the utility of ethanol discrimination procedures for pharmacotherapy development.

1.1   Rodents

1.1.1   GABA

The $GABA_A$ receptor complex is integral to many of ethanol's behavioral and physiological effects (e.g., [17, 22]). Consistent with ethanol's action as a positive modulator at the $GABA_A$ receptor, drugs in the benzodiazepine and barbiturate classes, with a similar mechanism to modulate chloride flow through the $GABA_A$ receptor, consistently produce ethanol-like discriminative stimulus effects (reviewed in Grant [4]). More recent work has expanded upon these findings in two primary ways. First, the specific action of ethanol at $GABA_A$ receptors with distinct subunit compositions has been investigated using a combination of genetic knockout and selective ligand approaches. Second, the selective role of neurosteroid activity at the $GABA_A$ receptor has been confirmed, and consistent with the action of neurosteroids as positive allosteric modulators at $GABA_A$, they exhibit ethanol-like discriminative stimulus effects similar to those generated by the benzodiazepine and barbiturate drug classes.

The $GABA_A$ receptor is a pentameric transmembrane receptor, classically made up of two alpha (α) subunits, two beta (β) subunits, and one gamma (γ) subunit. A delta (δ) subunit may substitute for a γ subunit in some receptor isoforms. Ethanol discrimination studies have primarily focused on isolating the role of $\alpha 1$-, $\alpha 4/6$-, and δ-subunit-containing receptors. Specifically, zolpidem, an $\alpha 1$ subunit-preferring benzodiazepine agonist, partially substitutes for ethanol in rats [23], but does not produce ethanol-like stimulus effects in mice [24], suggesting that activity at the $\alpha 1$ subunit is not sufficient to produce ethanol discriminative stimulus effects in rodents. Additionally, ethanol's action at $\alpha 4/6$-subunits has been investigated using Ro 15-4513, an inverse agonist at the benzodiazepine binding site, with some selectivity for the $\alpha 4/6$-subunits. While Ro 15-4513 successfully antagonizes the discriminative stimulus effects of benzodiazepine, the results are mixed for ethanol-trained rodents, with some studies showing antagonism of ethanol's discriminative effects [25, 26], and others showing no antagonism [27, 28]. The mixed effects of Ro 15-4513 as an ethanol antagonist are likely due to the differences in training doses and routes, suggesting that the prominence of the $\alpha 4/6$-subunits in ethanol discrimination is dependent on experimental parameters that might influence BEC. The δ-subunit of the $GABA_A$ receptor complex has also been isolated in ethanol discrimination using a constitutive δ-subunit knockout line of mice, and the results indicated that there were no differences in either the acquisition of ethanol discrimination or the substitution patterns of the $GABA_A$ receptor positive modulators compared to wild-type mice [24]. Therefore the δ-subunit of $GABA_A$ receptors is not necessary for mediating ethanol-like discriminative stimulus effects or for the substitution of benzodiazepines, barbiturates, or neurosteroids. The δ-subunit is thought to be an identifying feature of extrasynaptic $GABA_A$ receptors that mediate tonic inhibitory currents and confer sensitivity to low doses of ethanol [29, 30], and thus, these findings suggest that either non-δ extrasynaptic or synaptic receptors associated with phasic inhibitory currents may be more prominent in producing the discriminative stimulus effects of ethanol.

The steroid binding site on $GABA_A$ receptors and its modulation by neuroactive steroids has received considerable attention because these endogenous compounds respond to stress and are implicated in a number of behavioral disorders [31]. Neuroactive steroids that act at $GABA_A$ receptors do so through binding sites that are distinct from the benzodiazepine and barbiturate sites, and the conformation of the steroid A-ring 3′ and 5′ carbon hydroxyl groups is the key to receptor activation (see Chen et al. [32]). Select neuroactive steroids generalize from an ethanol training stimulus in rodents, including the reduced metabolites of progesterone (allopregnanolone or 3α,5α-P; pregnanolone or 3α,5β-P; and epipregnanolone or 3β,5β-P) and deoxycorticosterone (allotetrahydro-deoxycorticosterone or 3α,5α-THDOC) [33, 34]. Substitution was more prominent at a lower training dose (1 g/kg, i.g.) versus a higher one (2 g/kg, i.g.) [34]. The ethanol route of administration may also play a role in substitution patterns as 3β,5β-P has mixed effects in ethanol discriminations. 3β,5β-P produced no generalization with ethanol trained via an intraperitoneal route [34] but produced complete substitution, as well as potentiation of the ethanol cue, when trained with an intragastric route [33, 35]. Finally, the neurosteroid substitution patterns for ethanol suggest sex differences in sensitivity. For example, in contrast to earlier studies in male rats [33, 34], female rats showed only partial substitution of allopregnanolone and pregnanolone for a 1 g/kg ethanol training dose [36]. This latter finding is consistent with earlier work demonstrating that females were less sensitive to the modulatory effects of allopregnanolone on ethanol drinking behavior when compared to males [37]. Collectively, these and other studies (e.g., [38]) suggest that $GABA_A$ receptors that contain a neurosteroid binding site contribute to the discriminative stimulus effects of ethanol. Similar to barbiturates and benzodiazepines, neuroactive steroids asymmetrically cross-generalize with ethanol, with only partial substitution when ethanol is substituted in pregnanolone-trained rats [39–41] and mice [42]. This asymmetrical cross-generalization likely reflects the inability of pregnanolone and related neuroactive steroids to encompass other aspects of the compound ethanol cue.

1.1.2 Glutamate

The NMDA glutamatergic receptor is also well established in contributing to the discriminative stimulus effects of ethanol, particularly at higher doses in rodents [8]. Consistent with ethanol's known action as an NMDA antagonist at the synapse [17], drug discrimination studies have established that antagonism of the NMDA receptor produces ethanol-like discriminative effects. One of the earliest studies determined that the noncompetitive channel blocker dizocilpine (i.e., MK-801) fully substituted for ethanol in pigeons [43], and this finding has been replicated in rodents, including multiple strains of rats [19, 44–48] and mice [24, 49]. Other NMDA channel blockers such as memantine, phencyclidine (PCP), and ketamine have yielded similar degrees of substitution for ethanol in rats [19, 45, 47]. Often, however, substitution requires doses of the NMDA antagonists that also attenuate response rates [44, 50] to the extent that full substitution by these compounds is precluded [51].

In addition to the channel blocker site, multiple binding sites on the NMDA receptor have been examined, including the glutamate, glycine, and polyamine sites. Overall, ligands for each of these other binding sites have been far less effective in producing ethanol-like stimulus effects, indicating that ethanol's action is most similar to the non-competitive activity at the channel pore. Competitive antagonists at the glutamate site have generalized from ethanol in some cases (CGS 19755) [47], but have only partially substituted in other cases (CPPene, NPC-17742) [44, 51]. Similar results have been found with glycine site antagonists, with some ligands producing full substitution (L701,324) [50, 52], and others not substituting at all (MRZ2-502 and MRZ2-576) [45, 50]. Lastly, polyamine binding site antagonists (eliprodil and arcaine) produce stimulus effects that do not generalize from ethanol [45, 47]. In conclusion, the contribution of the glutamate, glycine, and polyamine binding sites of the NMDA receptor appears minimal in ethanol discrimination, particularly when compared to the channel pore site. However, it is noteworthy that aforementioned studies were all conducted in rats trained to discriminate a low to moderate dose of ethanol (i.e., 1 g/kg), and it is possible that inconsistent findings between studies may be partially attributable to the training dose studied, as previous work indicates that NMDA receptors contribute more predominantly to the ethanol stimulus at higher doses (>1.5 g/kg) in rodents [6, 19, 52].

In addition to the NMDA receptor, recent studies have begun to examine the metabotropic glutamate system (mGluR1, mGluR2/3, and mGluR5) based on findings that the mGluR5 receptor might modulate activity at the $GABA_A$ receptor [53]. Selective mGluR5 antagonist MPEP antagonized the ethanol dose–response function by decreasing the potency for ethanol to substitute for itself [53–55]. An mGluR2/3 agonist also decreased the potency of ethanol discrimination [56], but no effect was observed with any of the mGluR1 antagonists tested [54]. These studies have provided a novel pharmacological target for ethanol's discriminative stimulus effects, although it should be noted that these effects are modulatory in nature, and they are not sufficient to produce ethanol-like effects on their own. Thus, the direct glutamatergic activity of ethanol remains primarily at the NMDA receptor.

1.1.3   Serotonin

The importance of serotonergic neurotransmission in ethanol discriminative stimulus effects was first reported with the observation that pretreatment with a tryptophan hydroxylase inhibitor (p-chlorophenylalanine; which depletes brain 5-HT) reduces compartment choice between ethanol and water to chance levels in rats studied within a shock avoidance-based discrimination paradigm [57]. Since then, there have been several studies to manipulate levels of synaptic 5-HT, through enhancing 5-HT release (fenfluramine), a nonselective 5-HT receptor agonist (5-MeODMT), and selective serotonin uptake inhibitors (SSRIs; fluoxetine and paroxetine). In general, only SSRIs have produced ethanol-like discriminative stimulus effects [58], but this may be mediated through a non-serotonergic mechanism via their augmentation of brain allopregnanolone levels [59], which would be expected to exert positive modulation of $GABA_A$ receptors.

The first 5-HT receptor to be examined in an ethanol discrimination preparation was the 5-HT$_3$ receptor [60], which is an ionotropic receptor, and therefore from the same superfamily of receptors as the GABA$_A$ and NMDA receptors. Although studies in rats have found that 5-HT$_3$ receptor agonists (mCPBG) and antagonists (ICS 205-930) do not generalize from ethanol [61, 62], there is some limited evidence in pigeons that 5-HT$_3$ receptor antagonists (ICS 205-930 and MDL 72222) block the discriminative stimulus effects of low to moderate ethanol doses [63]. These data suggest that contribution of 5-HT$_3$ receptors in producing discriminative stimulus effects of ethanol is likely minimal. This conclusion is also supported by data from transgenic mice that overexpress 5-HT$_3$ receptors and show no differences in their ability to acquire an ethanol discrimination or in the substitution profiles with GABA$_A$ receptor positive modulators and an NMDA receptor antagonist when compared to wild-type mice [64].

In contrast to nonselective or selective 5-HT$_3$ receptor agonists, there is sufficient evidence to indicate a role for agonism at metabotropic 5-HT receptor subtypes in ethanol discrimination. From an initial characterization of several 5-HT receptor agonists in rats, the only compound to yield full substitution for ethanol in rats was TFMPP, a relatively nonselective 5-HT$_1$ agonist with slightly greater affinity for the 1A isoform [65]. This finding with TFMPP was replicated in both male [21] and female [36] rats. Subsequent evaluations of multiple compounds with various 5HT receptor agonist profiles in male rats revealed that CGS 12066B and CP 94,253 (both selective for 5-HT$_{1B}$) or mCPP and RU 24969 (both selective for 5-HT$_{1B/2C}$) fully generalized from ethanol (1 g/kg), whereas 8-OH DPAT (5-HT$_{1A}$) and DOI (5-HT$_{2A}$) did not [66–68]. A parallel set of antagonism studies used subtype selective antagonists to completely block the ethanol-like effects of CP 94,253 and mCPP [67], leading to an overall conclusion that 5-HT$_{1B}$ and 5-HT$_{2C}$ receptors contribute to the ethanol cue. However, there are inconsistencies in the generalizability of 5-HT$_{1B/2C}$ agonists to substitute for ethanol across sex and species, as RU 24969 only partially substituted for ethanol in female rats [36] and mCPP did not generalize from ethanol in mice [64]. Refinement of receptor ligands with increased selectivity for 5-HT$_1$ and 5-HT$_2$ receptor isoforms (e.g., [69, 70]) coupled with a rapid expansion of novel ligand development for 5-HT$_4$ receptors, which also function to regulate neurotransmission in conjunction with 5-HT$_1$ and 5-HT$_2$ receptors [71, 72], should prompt a fresh look at the involvement of metabotropic 5-HT receptors in modulating the discriminative stimulus effects of ethanol.

1.2   Nonhuman Primates

Ethanol discrimination in monkeys has built upon findings from rodents in several key ways. In general, nearly all of the receptor targets of ethanol in monkeys have been taken from the rodent literature and are largely consistent across species. However, there are several important differences between the rodent and the monkey that may inform future clinical work and shed light on potential limitations of smaller laboratory animals in ethanol discrimination. Nonhuman primate studies have primarily focused on ethanol's action at the GABA$_A$ and NMDA receptors, with some work on the opioid system. Additionally, nonhuman primate work has examined other biological variables that may contribute to ethanol's discriminative stimulus effects, such as sex [73–76], age [77], and menstrual cycle [78].

Ethanol's action at the GABA$_A$ receptor is highly selective in nonhuman primates. Specifically, studies in monkeys have examined subunit-selective ligands and antagonists at the GABA$_A$ receptor [75, 79–81], as well as neuroactive steroid activity [74, 78, 82, 83]. Additionally, cross-generalization analysis was possible by studies that trained ethanol-like GABA$_A$ ligands and examined ethanol in substitution tests [79, 84–86]. Similar to rodents, direct agonists at the GABA$_A$ receptor fail to produce ethanol discriminative stimulus effects, but positive allosteric modulators reliably substitute for ethanol [73]. Specifically, positive modulators at the benzodiazepine and barbiturate binding sites produce the most robust ethanol-like effects [73]. In contrast to rodents, however, GABA$_A$ modulators produce full substitution at low and high training doses (1.0–2.0 g/kg), rather than just predominantly at lower doses. Converging evidence from multiple studies suggests that α5 subunit-containing receptors are particularly important in ethanol's discriminative stimulus effects [75, 80, 81], as well as some contribution of the α1 and α2/3 subunits. Alpha-5 and alpha-1 selective agonists substitute for ethanol, but only inverse agonists selective for α5 (L-655,708) and α5 + α4/6 (Ro-154513) are able to antagonize ethanol's discriminative stimulus effects [75, 87]. Ro-154513 is also able to antagonize the substitution of benzodiazepines and barbiturates for ethanol, suggesting a shared action at the GABA$_A$ subunit level [76]. Neuroactive steroids also selectively produce ethanol-like discriminative effects based on their pharmacological effect at the GABA$_A$ receptor. Specifically, 3-alpha-hydroxy metabolites of progesterone such as allopregnanolone and pregnanolone are positive modulators at the GABA$_A$ receptor and produce ethanol-like stimulus effects in male and female monkeys [74, 82, 83]. However, 3-beta-hydroxy metabolites do not reliably substitute for ethanol at any training dose [80]. Several studies in monkeys have trained GABA$_A$ ligands and tested ethanol for substitution. To summarize this work, ethanol only cross-substituted with pentobarbital [85], but did not substitute for midazolam [86] or lorazepam [84]. These data suggest that ethanol's discriminative stimulus effects in the monkey are more similar to barbiturates, as compared to benzodiazepines.

Ethanol's discriminative stimulus effects are also mediated by antagonist activity at the NMDA receptor, and may be modulated by the opioid system. Noncompetitive antagonists at the channel pore MK-801 (or dizocilpine) and PCP produce full substitution for ethanol in male and female monkeys, but (unlike rodents) ketamine has not produced full substitution [76]. NMDA antagonist substitution was most potent and efficacious at a lower training dose, which is also in contrast to studies in rodents suggesting that a higher ethanol training dose conferred greater NMDA antagonism substitution [6] (see Sect. 1.1 above). These data are consistent with rodent data in characterizing ethanol as a compound stimulus in the monkey, with activity at both GABA$_A$ and NMDA receptors. Further, there has been a limited attempt to characterize the role of mu and delta opioid receptors in mediating the ethanol cue in monkeys. This examination found that selective agonists at both the mu (i.e., morphine and fentanyl) and delta (i.e., SNC 80 and SNC 162) receptors did not produce ethanol-like stimulus effects [73, 87], indicating that the opioid system is likely not a primary target in ethanol's discriminative stimulus effects. However, nonselective antagonist naltrexone antagonized the ethanol dose–response relationship [87], suggesting that the opioid system may function as a modulator of the ethanol stimulus, adding to the complex basis of the ethanol cue.

Lastly, nonhuman primate studies have taken advantage of the overlapping physiology between humans and monkeys to examine biological variables that may contribute to ethanol's discriminative stimulus effects. Most notably, a few of the nonhuman primate studies have directly compared male and female subjects in the analysis of GABA$_A$ and NMDA receptor involvement in ethanol's discriminative stimulus effects [73, 76]. Though there are small differences between male and female monkeys, in general the pharmacological basis of the ethanol cue is shared across the sexes. One exception relates to neurosteroid substitution for ethanol, which appears dependent on the phase of the menstrual cycle in female monkeys [78, 83]. In the luteal phase, when progesterone levels are high, allopregnanolone is more potent in its substitution for ethanol, consistent with greater levels of allopregnanolone in the plasma. Lastly, one study examined the effect of age on ethanol discriminative stimulus effects and determined that ethanol served as a relatively weaker stimulus in middle-aged monkeys, despite elevated blood ethanol concentrations relative to when the same monkeys were young adults [77]. Additionally, this study demonstrated that ethanol discrimination was persistent and demonstrated up to 3 years without any intermediate training [77].

1.3   Humans

To our knowledge, there are only five reports of training ethanol as a discriminative stimulus in human subjects [88–92] and one report of ethanol substitution in a nicotine-trained discrimination, in which it did not substitute [93]. These studies primarily demonstrated that ethanol can be trained with equal sensitivity in male and female subjects [88, 91], but the acquisition is sensitive to baseline weekly alcohol intake [89, 90] and ethanol generalization occurs in a dose-dependent manner [88, 89, 92]. The only study to test a compound other than ethanol examined the benzodiazepine lorazepam and found complete substitution [91]. Thus, the only receptor system directly implicated in the basis of an ethanol discrimination in humans is the GABA$_A$ receptor system.

1.4   Neuroanatomical Targets

In the last 20 years, there have been a handful of laboratories that have investigated the neuroanatomical basis of ethanol's discriminative stimulus effects. These studies have been conducted exclusively in rodents and have focused on the GABA and glutamate components of the ethanol cue using intracranial site-specific microinjections. Additionally, some work has been done measuring c-Fos activation after performance of an ethanol discrimination to identify the primary brain regions involved in ethanol's discriminative stimulus effects and the direction (activation or inactivation) of their involvement. A majority of these studies are based on an initial finding that agonism of the GABA$_A$ receptor in the nucleus accumbens (NAc) produced full substitution for ethanol [94]. Since then, GABA$_A$ positive modulators such as pentobarbital and allopregnanolone administered into the NAc core have also produced full ethanol substitution [94–96]. However, ethanol substitution is not blocked by the GABA$_A$ antagonist bicuculline in the NAc indicating that GABA$_A$ receptors within the NAc are sufficient, but not necessary to produce ethanol-like discriminative stimulus effects [94]. This is supported by work demonstrating that NMDA antagonist MK-801 in the NAc also produces full substitution for ethanol [96], and there appears to be some secondary contribution of mGlu5 receptors in the NAc, consistent with systemic administration of these compounds [54]. Thus, it appears that within the NAc, ethanol is acting as a compound cue on GABA and glutamate systems. It is important to note that these findings are highly consistent with ethanol's known action to activate GABA$_A$ and inhibit NMDA activity within the NAc in slice electrophysiology studies [17, 97, 98], resulting in an overall suppression of neuronal firing. This is further supported by c-Fos studies in discrimination-trained rats demonstrating decreased c-Fos activity within the NAc after ethanol [99, 100].

In addition to the NAc, there have also been select studies examining the role of the amygdala, several cortical areas (mPFC, prelimbic, and insula), hippocampus, and thalamus (rhomboid nucleus). In general, these primarily limbic brain regions have been demonstrated to contribute to some extent to ethanol's discriminative stimulus effects. Interestingly, these brain areas appear to have some selectivity for whether they are involved primarily in ethanol's GABAergic or glutamatergic component. Specifically, GABA$_A$ modulation in the amygdala produces ethanol-like effects, but there is no evidence for this brain region in the NMDA component [96, 101]. Conversely, NMDA the antagonist MK-801 in the prelimbic cortex and hippocampus produced full ethanol substitution, but GABA$_A$ agonists did not substitute [96]. The mPFC, insula, and rhomboid thalamus have also been shown to contribute to the GABA component through pharmacological inactivation using a GABA$_A$ + GABA$_B$ cocktail [100]. This fairly limited body of literature raises some important questions that can be addressed with future research. A differential contribution of different brain structures to the compound ethanol cue strongly suggests that our focus should be redirected to understanding sensitive circuitry mediating the discriminative stimulus effects of ethanol. Because the preliminary data on sensitive brain areas (not circuitry per se) is exclusively derived in rodent subjects, replicating and extending these results to the primate brain is needed.

1.5 New Paradigm for Advancing Knowledge and Pharmacotherapeutic Development with Ethanol Discriminations

From a translational perspective, these brain circuitry studies in rodents provide a strong foundation for potential target sites for future work in monkeys and humans. The recent development of chemogenetic or optogenetic approaches, using viral-based molecular targeting strategies, will allow for repeated manipulation of specific brain nuclei to understand their role in mediating the ethanol cue. Additionally, application of fMRI techniques in humans can examine the connectivity patterns of brain activation in mediating the discriminative stimulus effects of ethanol. The combination of human brain mapping and functional testing of identified areas in animal models with molecular targeting approaches will open up a new understanding of how the subjective effects of ethanol are mediated. Overall, although the number of laboratories involved in ethanol discrimination studies appears to be declining, these new technologies are likely to revive interest in knowing how the ethanol cue is mediated and its role in the subjective effects that maintain human alcohol consumption.

References

1. Conger JJ (1956) Alcoholism: theory, problem and challenge. II. Reinforcement theory and the dynamics of alcoholism. Q J Stud Alcohol 17:296–305
2. Bolin BL, Alcorn JL, Reynolds AR, Lile JA, Rush CR (2016) Human drug discrimination: a primer and methodological review. Exp Clin Psychopharmacol 24:214–228
3. Barry H (1991) Distinctive discriminative effects of ethanol. NIDA Res Monogr 116:131–144
4. Grant KA (1994) Emerging neurochemical concepts in the actions of ethanol at ligand-gated ion channels. Behav Pharmacol 5:383–404
5. Hodge CW, Grant KA, Becker HC, Besheer J, Crissman AM, Platt DM, Shannon EE, Shelton KL (2006) Understanding how the brain perceives alcohol: neurobiological basis of ethanol discrimination. Alcohol Clin Exp Res 30:203–213
6. Stolerman IP, Childs E, Ford MM, Grant KA (2011) Role of training dose in drug discrimination: a review. Behav Pharmacol 22:415–429
7. Mhatre M, Holloway F (2003) Micro1-opioid antagonist naloxonazine alters ethanol discrimination and consumption. Alcohol 29:109–116
8. Middaugh LD, Kelley BM, Cuison ER Jr, Groseclose CH (1999) Naltrexone effects on ethanol reward and discrimination in C57BL/6 mice. Alcohol Clin Exp Res 23:456–464
9. Middaugh LD, Kelley BM, Groseclose CH, Cuison ER Jr (2000) Delta-opioid and 5-HT3 receptor antagonist effects on ethanol reward and discrimination in C57BL/6 mice. Pharmacol Biochem Behav 65:145–154
10. Shippenberg TS, Altshuler HL (1985) A drug discrimination analysis of ethanol-induced behavioral excitation and sedation: the role of endogenous opiate pathways. Alcohol 2:197–201
11. Winter JC (1975) The stimulus properties of morphine and ethanol. Psychopharmacologia 44:209–214
12. Bienkowski P, Kostowski W (1998) Discrimination of ethanol in rats: effects of nicotine, diazepam, CGP 40116, and 1-(m-chlorophenyl)-biguanide. Pharmacol Biochem Behav 60:61–69

13. Ford MM, McCracken AD, Davis NL, Ryabinin AE, Grant KA (2012) Discrimination of ethanol-nicotine drug mixtures in mice: dual interactive mechanisms of overshadowing and potentiation. Psychopharmacology (Berl) 224:537–548

14. Ford MM, Davis NL, McCracken AD, Grant KA (2013) Contribution of NMDA glutamate and nicotinic acetylcholine receptor mechanisms in the discrimination of ethanol-nicotine mixtures. Behav Pharmacol 24:617–622

15. Korkosz A, Taracha E, Plaznik A, Wrobel E, Kostowski W, Bienkowski P (2005) Extended blockade of the discriminative stimulus effects of nicotine with low doses of ethanol. Eur J Pharmacol 512:165–172

16. Le Foll B, Goldberg SR (2005) Ethanol does not affect discriminative-stimulus effects of nicotine in rats. Eur J Pharmacol 519:96–102

17. Lovinger DM, Roberto M (2013) Synaptic effects induced by alcohol. Curr Top Behav Neurosci 13:31–86

18. Stolerman IP, Mariathasan EA, White JA, Olufsen KS (1999) Drug mixtures and ethanol as compound internal stimuli. Pharmacol Biochem Behav 64:221–228

19. Grant KA, Colombo G (1993) Discriminative stimulus effects of ethanol: effect of training dose on the substitution of N-methyl-D-aspartate antagonists. J Pharmacol Exp Ther 264:1241–1247

20. Colombo G, Grant KA (1992) NMDA receptor complex antagonists have ethanol-like discriminative stimulus effects. Ann N Y Acad Sci 654:421–423

21. Grant KA, Colombo G (1993) Substitution of the 5-HT1 agonist trifluoromethylphenylpiperazine (TFMPP) for the discriminative stimulus effects of ethanol: effect of training dose. Psychopharmacology (Berl) 113:26–30

22. Breese GR, Criswell HE, Carta M, Dodson PD, Hanchar HJ, Khisti RT, Mameli M, Ming Z, Morrow AL, Olsen RW, Otis TS, Parsons LH, Penland SN, Roberto M, Siggins GR, Valenzuela CF, Wallner M (2006) Basis of the gabamimetic profile of ethanol. Alcohol Clin Exp Res 30:731–744

23. Bienkowski P, Iwinska K, Stefanski R, Kostowski W (1997) Discriminative stimulus properties of ethanol in the rat: differential effects of selective and nonselective benzodiazepine receptor agonists. Pharmacol Biochem Behav 58:969–973

24. Shannon EE, Shelton KL, Vivian JA, Yount I, Morgan AR, Homanics GE, Grant KA (2004) Discriminative stimulus effects of ethanol in mice lacking the gamma-aminobutyric acid type A receptor delta subunit. Alcohol Clin Exp Res 28:906–913

25. Rees DC, Balster RL (1988) Attenuation of the discriminative stimulus properties of ethanol and oxazepam, but not of pentobarbital, by Ro 15-4513 in mice. J Pharmacol Exp Ther 244:592–598

26. Gatto GJ, Grant KA (1997) Attenuation of the discriminative stimulus effects of ethanol by the benzodiazepine partial inverse agonist Ro 15-4513. Behav Pharmacol 8:139–146

27. Hiltunen AJ, Järbe TUC (1988) Effects of Ro 15-4513, alone or in combination with ethanol, Ro 15-1788, diazepam, and pentobarbital on instrumental behaviors of rats. Pharmacol Biochem Behav 31:597–603

28. Middaugh LD, Bao K, Becker HC, Daniel SS (1991) Effects of Ro 15-4513 on ethanol discrimination in C57BL/6 mice. Pharmacol Biochem Behav 38:763–767

29. Carver CM, Reddy DS (2016) Neurosteroid structure-activity relationships for functional activation of extrasynaptic δGABA(A) receptors. J Pharmacol Exp Ther 357:188–204

30. Farrant M, Nusser Z (2005) Variations on an inhibitory theme: phasic and tonic activation of GABA(A) receptors. Nat Rev Neurosci 6:215–229

31. Paul SM, Purdy RH (1992) Neuroactive steroids. FASEB J 6:2311–2322

32. Chen ZW, Manion B, Townsend RR, Reichert DE, Covey DF, Steinbach JH, Sieghart W, Fuchs K, Evers AS (2012) Neurosteroid analog photolabeling of a site in the third transmembrane domain of the β3 subunit of the GABAA receptor. Mol Pharmacol 82:408–419

33. Ator NA, Grant KA, Purdy RH, Paul SM, Griffiths RR (1993) Drug discrimination analysis of endogenous neuroactive steroids in rats. Eur J Pharmacol 241:237–243

34. Bowen CA, Purdy RH, Grant KA (1999) Ethanol-like discriminative stimulus effects of endogenous neuroactive steroids: effect of ethanol training dose and dosing procedure. J Pharmacol Exp Ther 289:405–411
35. Ginsburg BC, Lamb RJ (2005) Alphaxalone and epiallopregnanolone in rats trained to discriminate ethanol. Alcohol Clin Exp Res 29:1621–1629
36. Helms CM, McCracken AD, Heichman SL, Moschak TM (2013) Ovarian hormones and the heterogeneous receptor mechanisms mediating the discriminative stimulus effects of ethanol in female rats. Behav Pharmacol 24:95–104
37. Finn DA, Beckley EH, Kaufman KR, Ford MM (2010) Manipulation of GABAergic steroids: sex differences in the effects on alcohol drinking- and withdrawal-related behaviors. Horm Behav 57:12–22
38. Bienkowski P, Kostowski W (1997) Discriminative stimulus properties of ethanol in the rat: effects of neurosteroids and picrotoxin. Brain Res 753:348–352
39. Engel SR, Purdy RH, Grant KA (2001) Characterization of discriminative stimulus effects of the neuroactive steroid pregnanolone. J Pharmacol Exp Ther 297:489–495
40. Gerak LR, Moerschbaecher JM, Winsauer PJ (2008) Overlapping, but not identical, discriminative stimulus effects of the neuroactive steroid pregnanolone and ethanol. Pharmacol Biochem Behav 89:473–479
41. Vanover KE (2000) Effects of benzodiazepine receptor ligands and ethanol in rats trained to discriminate pregnanolone. Pharmacol Biochem Behav 67:483–487
42. Shannon EE, Porcu P, Purdy RH, Grant KA (2005) Characterization of the discriminative stimulus effects of the neuroactive steroid pregnanolone in DBA/2J and C57BL/6J inbred mice. J Pharmacol Exp Ther 314:675–685
43. Grant KA, Knisely JS, Tabakoff B, Barrett JE, Balster RL (1991) Ethanol-like discriminative stimulus effects of non-competitive n-methyl-d-aspartate antagonists. Behav Pharmacol 2:87–95
44. Shelton KL, Balster RL (1994) Ethanol drug discrimination in rats: substitution with GABA agonists and NMDA antagonists. Behav Pharmacol 5:441–451
45. Hundt W, Danysz W, Hölter SM, Spanagel R (1998) Ethanol and N-methyl-D-aspartate receptor complex interactions: a detailed drug discrimination study in the rat. Psychopharmacology (Berl) 135:44–51
46. Kotlinska J, Liljequist S (1997) The NMDA/glycine receptor antagonist, L-701,324, produces discriminative stimuli similar to those of ethanol. Eur J Pharmacol 332:1–8
47. Sanger DJ (1993) Substitution by NMDA antagonists and other drugs in rats trained to discriminate ethanol. Behav Pharmacol 4:523–528
48. Schechter MD, Meehan SM, Gordon TL, McBurney DM (1993) The NMDA receptor antagonist MK-801 produces ethanol-like discrimination in the rat. Alcohol 10:197–201
49. Shelton KL, Grant KA (2002) Discriminative stimulus effects of ethanol in C57BL/6J and DBA/2J inbred mice. Alcohol Clin Exp Res 26:747–757
50. Bienkowski P, Danysz W, Kostowski W (1998) Study on the role of glycine, strychnine-insensitive receptors (glycineB sites) in the discriminative stimulus effects of ethanol in the rat. Alcohol 15:87–91
51. Shelton KL (2004) Substitution profiles of N-methyl-D-aspartate antagonists in ethanol-discriminating inbred mice. Alcohol 34:165–175
52. Grant KA, Colombo G (1993) Pharmacological analysis of the mixed discriminative stimulus effects of ethanol. Alcohol Alcohol Suppl 2:445–449
53. Besheer J, Hodge CW (2005) Pharmacological and anatomical evidence for an interaction between mGluR5- and GABA(A) alpha1-containing receptors in the discriminative stimulus effects of ethanol. Neuropsychopharmacology 30:747–757
54. Besheer J, Grondin JJ, Salling MC, Spanos M, Stevenson RA, Hodge CW (2009) Interoceptive effects of alcohol require mGlu5 receptor activity in the nucleus accumbens. J Neurosci 29:9582–9591

55. Besheer J, Stevenson RA, Hodge CW (2006) mGlu5 receptors are involved in the discriminative stimulus effects of self-administered ethanol in rats. Eur J Pharmacol 551:71–75
56. Cannady R, Grondin JJ, Fisher KR, Hodge CW, Besheer J (2011) Activation of group II metabotropic glutamate receptors inhibits the discriminative stimulus effects of alcohol via selective activity within the amygdala. Neuropsychopharmacology 36:2328–2338
57. Schechter MD (1973) Ethanol as a discriminative cue: reduction following depletion of brain serotonin. Eur J Pharmacol 24:278–281
58. Maurel S, Schreiber R, De Vry J (1997) Substitution of the selective serotonin reuptake inhibitors fluoxetine and paroxetine for the discriminative stimulus effects of ethanol in rats. Psychopharmacology (Berl) 130:404–406
59. Pinna G, Costa E, Guidotti A (2006) Fluoxetine and norfluoxetine stereospecifically and selectively increase brain neurosteroid content at doses that are inactive on 5-HT reuptake. Psychopharmacology (Berl) 186:362–372
60. Lovinger DM (1991) Ethanol potentiation of 5-HT3 receptor-mediated ion current in NCB-20 neuroblastoma cells. Neurosci Lett 122:57–60
61. Mhatre MC, Garrett KM, Holloway FA (2001) 5-HT 3 receptor antagonist ICS 205-930 alters the discriminative effects of ethanol. Pharmacol Biochem Behav 68:163–170
62. Stefanski R, Bienkowski P, Kostowski W (1996) Studies on the role of 5-HT3 receptors in the mediation of the ethanol interoceptive cue. Eur J Pharmacol 309:141–147
63. Grant KA, Barrett JE (1991) Blockade of the discriminative stimulus effects of ethanol with 5-HT3 receptor antagonists. Psychopharmacology (Berl) 104:451–456
64. Shelton KL, Dukat M, Allan AM (2004) Effect of 5-HT3 receptor over-expression on the discriminative stimulus effects of ethanol. Alcohol Clin Exp Res 28:1161–1171
65. Signs SA, Schechter MD (1986) Nicotine-induced potentiation of ethanol discrimination. Pharmacol Biochem Behav 24:769–771
66. Grant KA, Colombo G, Gatto GJ (1997) Characterization of the ethanol-like discriminative stimulus effects of 5-HT receptor agonists as a function of ethanol training dose. Psychopharmacology (Berl) 133:133–141
67. Maurel S, Schreiber R, De Vry J (1998) Role of 5-HT1B, 5-HT2A and 5-HT2C receptors in the generalization of 5-HT receptor agonists to the ethanol cue in the rat. Behav Pharmacol 9:337–343
68. Szeliga KT, Grant KA (1998) Analysis of the 5-HT2 receptor ligands dimethoxy-4-indophenyl-2-aminopropane and ketanserin in ethanol discriminations. Alcohol Clin Exp Res 22:646–651
69. Gupta S, Villalón CM (2010) The relevance of preclinical research models for the development of antimigraine drugs: focus on 5-HT(1B/1D) and CGRP receptors. Pharmacol Ther 128:170–190
70. Jensen NH, Cremers TI, Sotty F (2010) Therapeutic potential of 5-HT$_{2C}$ receptor ligands. ScientificWorldJournal 10:1870–1885
71. Bureau R, Boulouard M, Dauphin F, Lezoualc'h F, Rault S (2010) Review of 5-HT$_4$R ligands: state of art and clinical applications. Curr Top Med Chem 10:527–553
72. Fink KB, Göthert M (2007) 5-HT receptor regulation of neurotransmitter release. Pharmacol Rev 59:360–417
73. Grant KA, Waters CA, Green-Jordan K, Azarov A, Szeliga KT (2000) Characterization of the discriminative stimulus effects of GABA A receptor ligands in *Macaca fascicularis* monkeys under different ethanol training conditions. Psychopharmacology (Berl) 152:181–188
74. Grant KA, Helms CM, Rogers LSM, Purdy RH (2008) Neuroactive steroid stereospecificity of ethanol-like discriminative stimulus effects in monkeys. J Pharmacol Exp Ther 326:354–361
75. Helms CM, Rogers LSM, Grant KA (2009) Antagonism of the ethanol-like discriminative stimulus effects of ethanol, pentobarbital, and midazolam in cynomolgus monkeys reveals involvement of specific GABA(A) receptor subtypes. J Pharmacol Exp Ther 331:142–152
76. Vivian JA, Waters CA, Szeliga KT, Jordan K, Grant KA (2002) Characterization of the discriminative stimulus effects of N-methyl-D-aspartate ligands under different ethanol training conditions in the cynomolgus monkey (*Macaca fascicularis*). Psychopharmacology (Berl) 162:273–281

77. Helms CM, Grant KA (2011) The effect of age on the discriminative stimulus effects of ethanol and its GABA(A) receptor mediation in cynomolgus monkeys. Psychopharmacology (Berl) 216:333–343

78. Green KL, Azarov AV, Szeliga KT, Purdy RH, Grant KA (1999) The influence of menstrual cycle phase on sensitivity to ethanol-like discriminative stimulus effects of GABA(A)-positive modulators. Pharmacol Biochem Behav 64:379–383

79. Licata SC, Platt DM, Rüedi-Bettschen D, Atack JR, Dawson GR, Van Linn ML, Cook JM, Rowlett JK (2010) Discriminative stimulus effects of L-838,417 (7-tert-butyl-3-(2,5-difluoro-phenyl)-6-(2-methyl-2H-[1,2,4]triazol-3-ylmethoxy)-[1,2,4]triazolo[4,3-b]pyridazine): role of GABA(A) receptor subtypes. Neuropharmacology 58:357–364

80. Platt DM, Duggan A, Spealman RD, Cook JM, Li X, Yin W, Rowlett JK (2005) Contribution of alpha 1GABAA and alpha 5GABAA receptor subtypes to the discriminative stimulus effects of ethanol in squirrel monkeys. J Pharmacol Exp Ther 313:658–667

81. Helms CM, Rogers LS, Waters CA, Grant KA (2008) Zolpidem generalization and antagonism in male and female cynomolgus monkeys trained to discriminate 1.0 or 2.0 g/kg ethanol. Alcohol Clin Exp Res 32:1197–1206

82. Grant KA, Azarov A, Bowen CA, Mirkis S, Purdy RH (1996) Ethanol-like discriminative stimulus effects of the neurosteroid 3 alpha-hydroxy-5 alpha-pregnan-20-one in female *Macaca fascicularis* monkeys. Psychopharmacology (Berl) 124:340–346

83. Grant KA, Azarov A, Shively CA, Purdy RH (1997) Discriminative stimulus effects of ethanol and 3 alpha-hydroxy-5 alpha-pregnan-20-one in relation to menstrual cycle phase in cynomolgus monkeys (*Macaca fascicularis*). Psychopharmacology (Berl) 130:59–68

84. Ator NA, Griffiths RR (1997) Selectivity in the generalization profile in baboons trained to discriminate lorazepam: benzodiazepines, barbiturates and other sedative/anxiolytics. J Pharmacol Exp Ther 282:1442–1457

85. Massey BW, Woolverton WL (1994) Discriminative stimulus effects of combinations of pentobarbital and ethanol in rhesus monkeys. Drug Alcohol Depend 35:37–43

86. McMahon LR, France CP (2005) Combined discriminative stimulus effects of midazolam with other positive GABAA modulators and GABAA receptor agonists in rhesus monkeys. Psychopharmacology (Berl) 178:400–409

87. Platt DM, Bano KM (2011) Opioid receptors and the discriminative stimulus effects of ethanol in squirrel monkeys: mu and delta opioid receptor mechanisms. Eur J Pharmacol 650:233–239

88. Duka T, Stephens DN, Russel C, Tasker R (1998) Discriminative stimulus properties of low doses of ethanol in humans. Psychopharmacology (Berl) 136:379–389

89. Duka T, Jackson A, Smith DC, Stephens DN (1999) Relationship of components of an alcohol interoceptive stimulus to induction of desire for alcohol in social drinkers. Pharmacol Biochem Behav 64:301–309

90. Jackson A, Stephens D, Duka T (2001) A low dose alcohol drug discrimination in social drinkers: relationship with subjective effects. Psychopharmacology (Berl) 157:411–420

91. Jackson A, Stephens D, Duka T (2005) Gender differences in response to lorazepam in a human drug discrimination study. J Psychopharmacol 19:614–619

92. Kelly TH, Stoops TH, Perry AS, Prendergast MA, Rush CR (1997) Clinical neuropharmacology of drugs of abuse: a comparison of drug-discrimination and subject-report measures. Behav Cogn Neurosci Rev 2:227–260

93. Perkins K (2009) Discriminative stimulus effects of nicotine in humans. Handb Exp Pharmacol 192:369–400

94. Hodge CW, Aiken AS (1996) Discriminative stimulus function of ethanol: role of GABAA receptors in the nucleus accumbens. Alcohol Clin Exp Res 20:1221–1228

95. Hodge CW, Nannini MA, Olive MF, Kelley SP, Mehmert KK (2001) Allopregnanolone and pentobarbital infused into the nucleus accumbens substitute for the discriminative stimulus effects of ethanol. Alcohol Clin Exp Res 25:1441–1447

96. Hodge CW, Cox AA (1998) The discriminative stimulus effects of ethanol are mediated by NMDA and GABA(A) receptors in specific limbic brain regions. Psychopharmacology (Berl) 139:95–107

97. Nie Z, Madamba SG, Siggins GR (1994) Ethanol inhibits glutamatergic neurotransmission in nucleus accumbens neurons by multiple mechanisms. J Pharmacol Exp Ther 271:1566–1573

98. Nie Z, Madamba SG, Siggins GR (2000) Ethanol enhances gamma-aminobutyric acid responses in a subpopulation of nucleus accumbens neurons: role of metabotropic glutamate receptors. J Pharmacol Exp Ther 293:654–661

99. Besheer J, Schroeder JP, Stevenson RA, Hodge CW (2008) Ethanol-induced alterations of c-Fos immunoreactivity in specific limbic brain regions following ethanol discrimination training. Brain Res 1232:124–131

100. Jaramillo AA, Randall PA, Frisbee S, Besheer J (2016) Modulation of sensitivity to alcohol by cortical and thalamic brain regions. Eur J Neurosci 44:2569–2580

101. Besheer J, Cox AA, Hodge CW (2003) Coregulation of ethanol discrimination by the nucleus accumbens and amygdala. Alcohol Clin Exp Res 27:450–456

Discriminative Stimulus Effects of Abused Inhalants

Keith L. Shelton

Abstract Inhalants are a loosely organized category of abused compounds defined entirely by their common route of administration. Inhalants include volatile solvents, fuels, volatile anesthetics, gasses, and liquefied refrigerants, among others. They are ubiquitous in modern society as ingredients in a wide variety of household, commercial, and medical products. Persons of all ages abuse inhalants but the highest prevalence of abuse is in younger adolescents. Although inhalants have been shown to act upon a host of neurotransmitter receptors, the stimulus effects of the few inhalants which have been trained or tested in drug discrimination procedures suggest that their discriminative stimulus properties are mediated by a few key neurotransmitter receptor systems. Abused volatile solvent inhalants have stimulus effects that are similar to a select group of $GABA_A$ positive modulators comprised of benzodiazepines and barbiturates. In contrast the stimulus effects of nitrous oxide gas appear to be at least partially mediated by uncompetitive antagonism of NMDA receptors. Finally, volatile anesthetic inhalants have stimulus effects in common with both $GABA_A$ positive modulators as well as competitive NMDA antagonists. In addition to a review of the pharmacology underlying the stimulus effects of inhalants, the chapter also discusses the scientific value of utilizing drug discrimination as a means of functionally grouping inhalants according to their abuse-related pharmacological properties.

Keywords 1,1,1-trichloroethane · Abuse · Drug discrimination · Inhalant · Isoflurane · Nitrous oxide · Toluene · Trichloroethylene · Volatile vapor

K.L. Shelton (✉)
Department of Pharmacology and Toxicology, Virginia Commonwealth University, 410 North
12th Street, Room 746, P.O. Box 980613, Richmond, VA 23298-0613, USA
e-mail: klshelto@vcu.edu

© Springer International Publishing Switzerland 2016
Curr Topics Behav Neurosci (2018) 39: 113–140
DOI 10.1007/7854_2016_22
Published Online: 5 October 2016

Contents

1 Introduction

Inhalant abuse is a major worldwide public health problem. In the USA approximately one million adults, 18 years and older, used an inhalant in 2013 (SAMSA). This is roughly equal to the reported number of methamphetamine users and almost twice that reported to have used heroin. A number of other drugs including marijuana, prescription drugs, and cocaine are abused at higher rates than inhalants in adults. However, in younger adolescents, inhalant abuse is far more prevalent. Lifetime use of inhalants in 8th graders in 2014 was estimated at 10.8%, which ranked below only marijuana among illicit drugs [1]. The demographics of inhalant abuse are particularly troubling given the possibility that repeated inhalant exposure during the vulnerable adolescent development period may have long-lasting effects, which may not be immediately apparent. Chronic inhalant abuse can produce profound toxic effects to many organs including the liver, bone marrow, heart, and brain [2–5]. These risks and other factors make understanding the neurochemical effects underlying the abuse-related effects of inhalants an important priority. Unfortunately several challenges complicate this task, some of which may be overcome through the use of drug-discrimination procedures.

Inhalants stand alone as being the only major classification of abused drugs that is based solely on a shared route of abuse rather than established similarities in pharmacological actions. Most inhalants are volatile liquids possessing vapor pressures that permit them to readily form vapors at room temperature. Also included are products such as propane and butane as well as various liquefied refrigerants that are compressed into a liquid form which exist as gasses at atmospheric pressure. Finally, the anesthetic adjunct gas nitrous oxide is also considered an inhalant within the drug abuse research community.

Many inhalants are consumer or industrial products comprised of mixtures of various volatile compounds, which may or may not share common pharmacological properties. For instance, gasoline is a mixture composed of toluene, hexane, xylene, octane, and ethanol along with perhaps a dozen additional minor constituents. This chemical complexity makes studies on the abuse-related effects of consumer products problematic. Therefore, almost all research has focused on a small number of individual chemicals commonly present in abused consumer products [6, 7]. Among these, the aromatic hydrocarbon toluene likely has the highest abuse rate [8]. Toluene is present in gasoline, pain thinners, wood coloring stains, spray paints, and cleaning products. The addictive nature of toluene is strikingly illustrated by the precipitous drop in gasoline abuse in areas of Australia where a toluene-free gasoline was introduced by BP to combat rampant abuse [9]. Unfortunately, other commonly abused volatile and gaseous chemicals have received far less attention in the scientific literature than toluene.

Given limited resources a complete understanding of the abuse-related effects of every common inhalant is unrealistic. Instead a more reasonable goal might be to thoroughly explore the actions of a lesser number of inhalants that are known to have differing pharmacological actions. These inhalants could then serve as reference standards with which to compare others. However, it has yet to be adequately established whether pharmacologically distinct subgroups of inhalants even exist. This is one area in which the use of drug discrimination may be particularly helpful given the ability of drug discrimination to compare and contrast drugs using a relevant, abuse-related endpoint.

In the absence of sufficient data to permit inhalants to be segregated by pharmacological activity, other proposed classification systems of inhalants have been suggested [10]. For instance, many inhalants are volatile hydrocarbon chemicals with common structural characteristics. Examples include aromatic hydrocarbons (toluene, xylene), chlorinated hydrocarbons (1,1,1-trichloroethane, trichloroethylene), and halogenated hydrocarbons (isoflurane, sevoflurane). At this time the classification of inhalants according to chemical structure has limited pharmacological value as the importance structural characteristics which determine pharmacological mechanisms of action of inhalants have not been defined as has been the case with some other classes of drugs [11, 12]. Therefore, the most widely accepted system of inhalant classification at present is to group them according to their intended usage. The scientific consensus has generally chosen to utilize the three categories proposed by Balster et al. [10] and Balster [13]. These categories include (1) volatile solvents, fuels, and anesthetics; (2) volatile alkyl nitrites; and (3) nitrous oxide. However, as with any framework some entities are not easily categorized. For instance, liquefied compressed refrigerants such as chloroethane and 1,1,1,2-tetrafluoroethane (R134a) do not fit into any of these categories. Broadly speaking, inhalants grouped according to the categories proposed by Balster do appear to bear some similarities in observable behavioral effects but the categories are probably too inclusive to serve as a reliable indicator of pharmacological actions. The system has, however, proven helpful as an interim solution to a challenging problem.

2 Unique Methodological Aspects of Inhalant Discrimination Studies

The technical issues related to drug-discrimination studies with inhalants have been discussed in greater detail elsewhere and will only be touched upon here (for review see [14]). Briefly, methods previously developed to produce consistent and reproducible inhalant exposures to examine other endpoints have been adapted for use in drug discrimination, the most common being the static exposure procedure. The exposures are "static" in the sense that a test subject and a measured volume of a volatile liquid inhalant chemical are confined in a sealed chamber of a fixed volume for a specified period of time. Provided the liquid inhalant can be completely volatilized, the ideal gas law is used to calculate inhalant chamber concentration without resorting to complicated quantitative assessment methods.

Static exposure chamber concentration and exposure duration are in most respects analogous to drug dose and pretreatment time in traditional drug discrimination studies. However, unlike injected drugs, after the cessation of exposure inhalant blood levels immediately begin to decline as a result of elimination via exhalation. This results in pharmacological effects that are quite labile compared to drugs administered by other routes. Under common drug discrimination training exposure conditions designed to mimic the short duration of inhalation typical of abuse, the majority of toluene within the bloodstream is eliminated unchanged by exhalation with very little undergoing metabolism. In mice trained to discriminate 10 min exposure to 6,000 ppm toluene vapor from 10 min exposure to air, toluene-lever selection declined to less than 50% within 10 min after the cessation of exposure [15] but it required 60 min before its stimulus effects had completely disappeared. In contrast, nitrous oxide is not metabolized and is rapidly eliminated entirely by exhalation. In mice trained to discriminate 10 min exposure to 60% nitrous oxide, drug-appropriate responding had returned to levels near those produced by the air vehicle within only 5 min following the cessation of exposure [16]. Therefore, when inhalant discrimination training and tests are conducted outside of the inhalant exposure environment, the rapid diminution of stimulus effects necessitates very brief operant sessions, generally 2–5 min in duration. In some cases limiting the duration of the training or test session is insufficient to capture the stimulus effects of an inhalant. For example, in order to train the nitrous oxide discrimination just discussed it was necessary to employ specially designed operant conditioning/dynamic inhalant exposure chambers rather than using static exposure chambers. These dynamic chambers allowed continuous gas exposure for the duration of operant training and test sessions.

3 Inhalants as Drug Discrimination Training Stimuli

It has long been known that inhalants can serve as stimuli to control behavior [17] as well as disrupt behavior controlled by exteroceptive stimuli [18]. However it was not until the late 1980s that an inhalant was first used as the training drug in a traditional drug discrimination experiment [19]. The two initial studies one in mice and a second in rats focused specifically on training the interoceptive stimulus properties of toluene utilized administration by intraperitoneal (i.p.) injection [19, 20] rather than inhalation. Although not consistent with the manner in which it is normally abused, toluene is often administered by injection in toxicology studies [21]. Further, while route may have a major impact in some types of experiments, few studies have noted that administration route alters drug-discrimination results [22]. For instance, ethanol trained by i.p. administration in one laboratory produced comparable cross-substitution results with ethanol trained by oral gavage dosing in other laboratories [23–26]. The relative insensitivity of the drug discrimination procedure to administration route has been shown to apply to other drugs including smoke inhalation and intravenous administration of phency-clidine and cocaine [27–29] and even to self-administered versus experimenter-administered ethanol exposure [30, 31]. Among inhalants, only toluene has been compared across administration route where it has been demonstrated inhaled toluene will substitute following i.p. training in mice [19] and conversely, injected toluene cross generalizes in mice trained to discriminate inhaled toluene [32].

In 2006 our laboratory began to actively explore training inhalants as discriminative stimuli by establishing a discrimination between 10 min of exposure to 6,000 ppm inhaled toluene vapor or air [32]. The odor threshold of toluene in humans is approximately 3 ppm, which is orders of magnitude below what we believed was necessary to produce centrally mediated discriminative stimulus effects. It was a significant concern that the odor of toluene and/or its pronounced effects on the trigeminal system might serve as a preferential discriminative stimulus over its centrally mediated subjective effects [33, 34]. Indeed the exteroceptive stimulus properties of odorants have long been used as cues to control behavioral outcomes [35]. To lessen the possibility of odor cues controlling behavior, the toluene and air exposures were conducted in different experimental chambers than the discrimination training with a short temporal break between exposure and placement in the operant training chamber. Under these conditions it is required a mean of 26 sessions to train the 6,000 ppm toluene versus air discrimination, significantly faster than with an i.p. route of administration.

Several control tests were conducted to assess whether the toluene cue was due to interoceptive CNS effects, exteroceptive odor stimuli, or a compound cue composed of both CNS effects and odor. First, based on the premise that i.p. administered toluene was likely to have lower perceived odor than inhaled toluene, the ability of i.p. injected toluene to substitute for the inhaled toluene training condition was examined. When injected via the i.p. route, toluene dose-dependently and fully substituted for inhaled toluene, a result consistent with the prior mouse study, which had demonstrated that inhaled toluene would substitute for an i.p. injected toluene training stimulus [36]. As a second control for potential olfactory stimulus effects it was demonstrated that a brief 1 min of exposure to 6,000 ppm toluene vapor did not substitute for the longer 10 min exposure training condition. Lastly, ethylbenzene, another aromatic hydrocarbon with a strong but distinctive odor, produced nearly identical levels of partial substitution for toluene regardless of whether it was administered IP or by inhalation; whereas the vapor anesthetic isoflurane produced only vehicle-appropriate responding. Interestingly, this study also suggested a reason why the prior experiments using 100 mg/kg i.p. toluene as a training stimulus required such protracted training [19, 20, 36]. Specifically, a much higher dose of 560 mg/kg i.p. toluene was required to produce full substitution in 6,000 ppm toluene vapor-trained mice. This outcome suggests that the stimulus effects of 100 mg/kg i.p. toluene are quite weak and would therefore be expected to require more extended training than the stronger 6,000 ppm inhaled toluene stimulus [37–39].

Although this experiment supported the contention that the olfactory effects of toluene were not alone sufficient to elicit toluene-appropriate responding, it did not rule out the possibility of a contributory role of odor. A subsequent experiment that again trained mice to discriminate between 10 min of 6,000 ppm toluene and air exploited toluene's inhaled pharmacokinetic properties to more thoroughly explore the role of exteroceptive versus interoceptive cue control over behavior [40]. It has been demonstrated that during extended duration exposure to toluene vapor, blood toluene levels rise rapidly for the first hour but do not entirely plateau until several hours later [41, 42]. Therefore, within the first hour of continuous exposure, toluene blood concentration can be manipulated independently from toluene odor intensity by increasing or decreasing exposure duration. Using this strategy we demonstrated across a range of exposure concentrations that 20 min of toluene exposure produced mean toluene-lever selection much greater than that produced by 10 min of exposure. Further, toluene blood concentration as quantified by gas chromatography almost perfectly predicted toluene-lever selection, a finding that has also been extended to the inhalant 1,1,1-trichloroethane [15]. Finally, toluene administered by i.p. injection had an additive effect on the stimulus effects of inhaled toluene. As a whole, these studies supported the conclusion that the stimulus effects of inhaled toluene were governed by the concentration of the drug in the bloodstream at the time of testing rather than simply by the strength of its odor. This uncoupling of stimulus effects from odor lends additional support to the conclusion that the training stimulus of inhaled toluene is primarily, if not exclusively, CNS mediated.

While toluene has been the inhalant most frequently trained as a discriminative stimulus, other inhalants have also served as training stimuli. Two different chlorinated hydrocarbon vapors 1,1,1-trichloroethane [15, 43, 44] and trichloroethylene [45] have been shown to be effective training stimuli in mice. A discrimination based on a behaviorally active, subanesthetic concentration of 6,000 ppm of the volatile anesthetic isoflurane has also been established in mice [46]. Finally, two studies in mice have reported that the discriminative stimulus effects of 60% nitrous oxide gas can be trained [16, 47]. In every case, the concentration response functions of the training inhalants were indistinguishable from those produced by drugs trained using i.p., subcutaneous or oral gavage routes of administration. The success of these studies strongly supports the conclusion that any inhalant with sufficiently robust CNS effects can serve as an effective discriminative stimulus under the proper training conditions.

4 Pharmacological Characterization of Inhalant Discriminative Stimuli

The neurochemical and abuse-related behavioral effects of only a small number of inhalants have been explored in any detail. As previously mentioned the largest number of published reports has focused on toluene. Less attention has been given to other volatile solvents such as 1,1,1-trichloroethane and trichloroethylene. Due to their clinical importance, a fairly large literature base is available on the anesthesia-related neurochemical and behavioral effects of volatile anesthetics such as halothane, isoflurane, and sevoflurane as well as nitrous oxide gas [48]. Conversely, a number of additional inhalants with documented instances of abuse such as chlorofluorocarbons, haloalkanes, butane, propane, and nitrites have been largely neglected. As a whole the accumulated literature convincingly demonstrates that not all inhalants act upon the same receptor target or targets [for review see [6, 49].

Ligand-gated ion channels including GABA$_A$, NMDA, glycine, nicotinic acetylcholine, and 5-HT$_3$ receptors appear to be particularly sensitive targets of inhalants in both in vitro and in vivo assays. The function of voltage gated ion channels are also altered by inhalants [50]. Finally, evidence exists that inhalants interact with g-protein coupled receptors including dopamine and opioid receptors. The subsequent sections of this chapter will briefly review the literature regarding the effects of inhalants on specific receptors and studies exploring whether these mechanisms are also involved in transducing the discriminative stimulus effects of individual inhalants. A summary of the studies in which probe drugs have been tested in subjects trained to discriminate inhalants is presented in Table 1.

4.1 Stimulus Effects of Inhalants: GABA$_A$ Receptors

GABA$_A$ receptors are the most abundant inhibitory ion channel receptors in the CNS and play a critical role in maintaining inhibitory tone [53]. GABA$_A$ receptors are ligand-gated chloride channel receptors composed of 5 subunits [54]. The majority of GABA$_A$ receptors have a binding sites for GABA itself as well as allosteric modulatory sites selective for benzodiazepines, barbiturates, and GABA positive neurosteroids, among others. Although many inhalants act on GABA$_A$ receptors, most [55, 56] but not all [57] studies suggest that their effects are not the

Table 1 Maximal percentage drug-lever selection of cross-test drugs in subjects trained to discriminate various inhalants from vehicle

Test drugs and mechanisms	Training drugs				
	Toluene	1,1,1-Trichloroethane	Trichloroethylene	Isoflurane	Nitrous oxide
GABA receptors					
Classical benzodiazepines	47[a] 50[b] 66[b] 72[j]	66[c] 62[d] 62[d]	48[e]	71[f]	27[h]
Zaleplon	26[b]	28[d]		74[f]	
Barbiturates	66[a] 43[b] 24[b] 34[j]	68[d]	70[e]	70[f]	10[h]
Gaboxadol		1[d]			4[h]
Muscimol				8[f]	22[h]
Neurosteroids	14[b] 8[j]				
Tiagabine		11[d]			
Valproic acid	58[b]	39[d]		85[f]	33[h]
NMDA receptors					
Uncompetitive antagonists	20[b]	13[c] 14[c] 14[d]	45[e]	44[f]	55[h] 50[h] 40[g]
CGS-19755	21[b]	25[d]	35[e]	98[f]	9[h]
L701,324	18[b]	10[d]	1[e]	24[f]	1[h]
Opioid receptors					
Morphine		4[c]			33[h]
U50,488	20[j]		22[e]	14[f]	11[h]
SNC80					10[h]
Nicotinic receptors					
Nicotine		22[c]			
Mecamylamine		1[c]			1[i]
Serotonin receptors					
8-OH-DPAT					4[h]
mCPP	19[j]		3[e]	10[f]	21[h]
MDL-72222	8[j]		2[j]		
Multiple mechanisms and other					
Ethanol	44[j]	23[d]	67[e]	52[f]	55[h] 52[g]
GHB				30[f]	
Telazol		38[d]			
Chlorpromazine	14[b]				
D-Amphetamine					1[g]
L-NAME					2[g]

[a]Reference [20]
[b]Reference [51]
[c]Reference [43]
[d]Reference [44]
[e]Reference [45]
[f]Reference [46]
[g]Reference [16]
[h]Reference [47]
[i]Reference [52]
[j]Shelton, K.L. unpublished data

results of actions at established drug binding sites. Toluene and to a lesser extent 1,1,1-trichloroethane and trichloroethylene all increase GABA-stimulated currents in GABA$_A$ receptors expressed in oocytes, but do not alter steady state GABA$_A$ receptor mediated currents. This indicates they are positive modulators rather than direct GABA$_A$ receptor agonists [58]. Toluene enhances GABA$_A$-mediated inhibitory postsynaptic currents in rat prefrontal cortex neurons [59]. Toluene, 1,1,1-trichloroethane, and trichloroethylene increase GABA$_A$ receptor mediated inhibition in hippocampal pyramidal neurons [60]. Repeated, acute exposure to toluene alters GABA$_A$ subunit expression profiles in the striatum, ventral tegmental area, and nucleus accumbens [61]. Nitrous oxide [56, 62–65], as well as volatile anesthetics, also positively modulates GABA$_A$-receptor mediated effects [66–69]. At the behavioral level, toluene attenuates convulsions induced by the GABA$_A$ receptor antagonist pentylenetetrazol [70] and demonstrates locomotor cross-sensitization with diazepam [71]. Toluene also has anxiolytic-like effects [72] similar to benzodiazepines as well as increases footshock-suppressed operant responding in mice [70] in a manner comparable to GABA$_A$-positive modulators [73].

The role of GABA$_A$ receptors in the discriminative stimulus effects of several inhalants has been examined using both cross tests of GABA$_A$ receptor ligands in animals trained to discriminate inhalants as well as cross tests of inhalants in animals trained to discriminate drugs with GABAergic mechanisms of action. In toluene-trained animals, classical non-selective benzodiazepines including midazolam, oxazepam, diazepam, and chlordiazepoxide produced partial substitution ranging from a low of 47% toluene-lever selection in rats trained to discriminate 100 mg/kg i.p. toluene from vehicle [20] up to 72% toluene-lever responding in mice trained to discriminate 2,000 ppm inhaled toluene from air [51]. Barbiturates also produce toluene-like stimulus effects under some conditions. The short-acting barbiturate methohexital elicited 66% drug-lever selection in rats trained to discriminate i.p. toluene from vehicle [20] but a much lower 24% drug-lever selection in mice trained to discriminate toluene vapor from air [51]. In mice trained to discriminate pentobarbital, toluene vapor produced greater than 85% drug-lever responding in 8 of 10 subjects [36]. In the converse training situation, pentobarbital produced greater than 80% drug-lever selection in mice trained to discriminate an extremely low dose of 100 mg/kg i.p. toluene from vehicle [19] but only 43% drug-lever selection in mice trained to discriminate a more abuse-relevant concentration of 2,000 ppm toluene vapor from air [51].

Barbiturates and benzodiazepines often cross-substitute for one another, therefore it is not surprising that both will produce some toluene-like discriminative stimulus effects. However, it does not appear that all drugs that positively modulate GABA$_A$ receptors will elicit toluene-like discriminative stimulus effects in that the GABA-positive neurosteroid allopregnanolone failed to substitute for toluene [51]. Likewise, zalaplon which preferentially binds to the benzodiazepine site in alpha 1 subunit containing GABA$_A$ receptors also failed to evoke toluene-lever responding [51]. This latter finding suggests that the benzodiazepine-like stimulus effects of toluene may not involve alpha 1 subunit containing GABA$_A$ receptors.

Classical benzodiazepines also act on $GABA_A$ receptors containing alpha 2, 3, and 5 subunits. However, selective ligands for these additional subunits have not been examined in toluene-trained mice.

The data are suggestive of a GABAergic involvement in the stimulus effects of toluene but as noted in several studies, benzodiazepines and barbiturates produced less than complete generalization making this conclusion somewhat tentative. If the same data sets are reanalyzed to take into account different sensitivities of individual subjects to the toluene-like stimulus effect of GABAergic positive modulators the data appear more convincing. For instance, at least one dose of oxazepam fully substituted for 100 mg/kg i.p. toluene in 4 of 5 rats tested [20]. Similarly, 88% of mice exhibited greater than 75% toluene-lever selection at one or more test dose of midazolam [51]. The results with barbiturates when analyzed in the same manner are less consistent. At least one dose of methohexital produced full substitution in 6 of 8 rats [20] and 4 of 5 mice [19] trained to discriminate 100 mg/kg i.p. toluene from vehicle. However in mice trained to discriminate 2,000 ppm inhaled toluene from air, one or more doses of methohexital produced full substitution in only 28% of mice and no dose of pentobarbital fully substituted in any of the subjects [51]. Unfortunately it is difficult to adequately equate these three studies given the differences in training conditions. Of these differences perhaps the most relevant was the observations that in some subjects it required in excess of 100 training sessions to establish a discrimination between 100 mg/kg i.p. toluene and vehicle whereas a mean of 65 sessions was necessary to train the discrimination between 2,000 ppm toluene vapor and air. It is therefore likely that the 100 mg/kg i.p. toluene dose was a fairly weak stimulus. Lower training doses may in some cases result in less specific discriminative stimuli, which could have been responsible for the greater degree of barbiturate lever selection in the earlier versus latter study.

While benzodiazepines substitute fairly consistently in animals trained to discriminate toluene from vehicle, in mice trained to discriminate diazepam from vehicle, toluene vapor exposure exhibited only a very low level of partial substitution [74]. This pattern of asymmetrical substitution is in many respects similar to that reported with ethanol where $GABA_A$ positive modulators as well as NMDA antagonists substitute in ethanol-trained animals more consistently than does ethanol in subjects trained to discriminate $GABA_A$ positive modulators or NMDA antagonists (for review see [75]). The findings with ethanol have been suggested to be a result of its actions on multiple receptors, which attributes ethanol with drug mixture-like properties in discrimination studies. When a drug mixture is trained as a stimulus, individual components of that mixture are sufficient to elicit mixture-appropriate responding [76–78]. However, when a single component of a drug mixture is trained as a stimulus and the mixture then tested the additional component(s) within the mixture may overshadow the common stimulus component and prevent full substitution [79]. Given that toluene, like ethanol, may interact with multiple receptors, a similar phenomenon may be taking place. However, this interpretation is speculation only as any additional components that might underlie the discriminative stimulus effect of toluene are presently unidentified.

As with toluene, classical benzodiazepines elicit partial substitution (i.e., less than 80% drug-lever appropriate responding) in mice trained to discriminate chlorinated hydrocarbons from air. Midazolam produced 66 and 62% drug-lever selection in two different studies in mice trained to discriminate 1,1,1-trichloroethane from air [43, 44]. Similarly, diazepam produced 62% drug-lever selection in 1,1,1-trichloroethane trained mice [44]. Midazolam produced a less robust 48% drug-lever selection in mice trained to discriminate trichloroethylene from air [45]. While classical benzodiazepines produced meaningful substitution for chlorinated hydrocarbons, the alpha 1 subunit preferring benzodiazepine site ligand zaleplon produced little 1,1,1-trichloroethane lever selection [44]. Interestingly, both 1,1,1-trichloroethane and trichloroethylene appear to have somewhat more robust barbiturate-like stimulus effects than does toluene. Pentobarbital elicited 68% drug-lever selection in mice trained to discriminate 1,1,1-trichloroethane from air [44] and methohexital produced 70% drug-lever responding in mice trained to discriminate trichloroethylene vapor from air [45]. In cross-substitution testing, 1,1,1-trichloroethane produced full substitution (>80%) in mice trained to discriminate 15 mg/kg pentobarbital from vehicle as well as mice trained to discriminate 20 mg/kg pentobarbital from vehicle [80] suggesting symmetry in stimulus effects between barbiturates and chlorinated hydrocarbons. As with toluene, not all positive GABA$_A$ modulators produced substitution in animals trained to discriminate chlorinated hydrocarbons as neither the extrasynaptic GABA$_A$ receptor agonist gaboxadol nor the GABA reuptake inhibitor tiagabine substituted for 1,1,1-trichloroethane [44]. The apparent differences between toluene and chlorinated hydrocarbons in their barbiturate-like stimulus effects may be due to fundamentally different receptor actions but it is equally possible they are the consequence of differential training stimulus intensities which are difficult to compare across studies.

The discriminative stimulus effects of the volatile anesthetics also appear to be partially the result of positive GABA$_A$ receptor modulation. The volatile halogenated anesthetic methoxyflurane produced full substitution in mice trained to discriminate diazepam from vehicle [74] as did halothane in mice trained to discriminate pentobarbital from vehicle [80]. The converse is also true in that both midazolam and pentobarbital produced fairly robust substitution in mice trained to discriminate isoflurane [46]. Interestingly, zaleplon produced the same level of isoflurane-lever selection as did midazolam and pentobarbital suggesting that positive modulation of alpha 1 subunit containing GABA$_A$ receptors alone is sufficient to produce isoflurane-like stimulus effects. Similar to the other inhalants previously discussed, the GABA-positive stimulus effects of isoflurane did not extend to all drugs that facilitate GABA$_A$ neurotransmission, as the direct GABA$_A$ agonist muscimol produced only vehicle-appropriate responding in isoflurane-trained mice [46]. Lastly, of those inhalants which have been examined to date, only the anesthetic gas nitrous oxide appears to be completely devoid of GABA$_A$ positive modulator-like stimulus effects [47]. Midazolam failed to substitute for 60% nitrous oxide when administered alone and when co-administered with midazolam failed to enhance the discriminative stimulus effects of nitrous oxide.

These findings suggest that nitrous oxide and midazolam do not share any discriminative stimulus properties. Likewise, gaboxadol, pentobarbital, and muscimol also only elicited vehicle-appropriate responding in nitrous oxide-trained mice.

4.2 Stimulus Effects of Inhalants: NMDA Receptors

Glutamate receptors are the primary excitatory receptors in the CNS. The NMDA receptor is one of the three subtypes of ionotropic glutamate receptors and is permeable to Ca^{2+}, Na^+, and K^+. Like the $GABA_A$ receptor, the NMDA receptor has a number of ligand binding domains [81, 82]. The channel is opened by the binding of glutamate in combination with the obligatory co-agonist glycine. The receptor can be blocked by antagonists acting through several mechanisms including those that act at the glutamate binding site, the glycine binding site, the polyamine site as well as by antagonists that bind within and block the channel itself.

There is a considerable body of literature demonstrating that a number of inhalants act as NMDA receptor antagonists in vitro and ex vivo. Benzene, xylene, ethylbenzene, and 1,1,1-trichloroethane [83] as well as isoflurane, sevoflurane, desflurane, and nitrous oxide inhibit NMDA-receptor function in recombinant receptors expressed in oocytes [65, 84–86]. Isoflurane and nitrous oxide also inhibit NMDA receptor activity in neuronal cultures and brain slices [87–91]. At the behavioral level toluene reduces the severity and lethality of NMDA-induced seizures [92] and administration of the NMDA glycine site co-agonist D-serine attenuates toluene-induced locomotor incoordination and memory impairment [93].

Drugs which attenuate NMDA receptor function by binding at the NMDA site within the channel and at the glycine co-agonist site can in many cases be differentiated from one another using drug discrimination [94–96]. The role of all three of these antagonist sites as contributors to the discriminative stimulus effects of toluene, 1,1,1-trichloroethane, trichloroethylene, isoflurane, and nitrous oxide has been fairly systematically explored. In mice trained to discriminate toluene vapor from vehicle the competitive NMDA receptor antagonist CGS-19755 produced a mean of 21% toluene-appropriate responding [51]. Likewise the uncompetitive channel blocker dizocilpine [(+)-MK-801] and the glycine-site antagonist L701,324 failed to substitute at greater than vehicle levels for toluene. These findings are consistent with previous data showing that toluene and xylene do not generalize in either C57BL/6 J or DBA/2 J mice trained to discriminate dizocilpine from vehicle [97]. However, both these studies are at odds with an experiment, which showed that 6,000 ppm toluene vapor produced a mean of 67% drug-lever selection in mice trained to discriminate 1.25 mg/kg of PCP from vehicle [74]. Dizocilpine is a highly selective NMDA receptor antagonist, whereas phencyclidine is less so. While there is no dispute that the discriminative stimulus effects of PCP are mediated by NMDA-receptor antagonism [98–100], PCP has been demonstrated to have greater downstream effects than does dizocilpine on other neurotransmitters including dopamine and acetylcholine [101–103]. This reduced selectivity appears to have implications in drug discrimination in that several studies have shown that PCP does not always fully generalize in dizocilpine-trained animals [97, 104, 105]. The results of the study by Bowen and colleagues may therefore be detecting some additional common stimulus component between PCP and toluene that is not present with dizocilpine, the most likely of which may be amphetamine-like dopamine receptor activity [106].

Neither dizocilpine nor PCP produced any appreciable drug-lever selection in mice trained to discriminate 1,1,1-trichloroethane vapor from air [43, 44]. Likewise in mice trained to discriminate either PCP from vehicle or dizocilpine from vehicle, 1,1,1-trichloroethane produced only a low level of partial substitution [74, 97]. Dizocilpine also failed to substitute for trichloroethylene vapor in mice [45]. The NMDA receptor glycine-site antagonist L701,324 produced only vehicle-lever selection in mice trained to discriminate either 1,1,1-trichloroethane [44] or trichloroethylene from vehicle [45].

Unlike hydrocarbon solvents, the volatile anesthetic isoflurane appears to possess a NMDA antagonist-like stimulus component. The competitive NMDA antagonist CGS-19755 produced full substitution, whereas the uncompetitive antagonist dizocilpine produced partial substitution and the NMDA glycine-site antagonist L701,324 produced no substitution [46] in mice trained to discriminate isoflurane vapor. These data are in agreement with prior findings that volatile inhalants inhibit binding of ^{3}H CGS-19755 as well as ^{3}H dizocilpine [107]. The complete substitution engendered by CGS-19755 and partial substitution by dizocilpine were, however, accompanied by substantial response-rate suppressing effects such that 4 of 8 mice in each test group were excluded from the lever selection data at the doses that produced the greatest percentage isoflurane-lever selection.

NMDA receptor antagonism also appears to play a significant role in producing the discriminative stimulus effects of nitrous oxide [47]. Neither CGS-19755 nor L701,324 generalized in mice trained to discriminate 60% nitrous oxide from oxygen. However, both dizocilpine and the low-affinity NMDA channel blocker memantine partially substituted for nitrous oxide. While the substitution of dizocilpine for nitrous oxide was not complete it does appear to be selective given that a low dose of dizocilpine was also capable of significantly enhancing the discriminative stimulus effects of nitrous oxide itself.

4.3   Common Stimulus Effects of Inhalants and Ethanol

The receptor mechanisms underlying the discriminative stimulus effects of ethanol have been discussed in detail in another chapter of the present work. Briefly, GABA$_A$ positive modulators and NMDA antagonist will robustly substitute for ethanol [23, 26, 108–110]. As the discriminative stimulus of inhalants appear to be mediated by one or both of these receptors it follows that inhalants should produce ethanol-like discriminative stimulus effects. 1,1,1-trichloroethane and toluene both produced concentration-dependent substitution in mice trained to discriminate 1 g/ kg i.p. ethanol from saline [111]. Consistent with the concept of asymmetric substitution of drug mixtures in animals trained to discriminate components of that mixture, ethanol only produced partial substitution in mice trained to discriminate 1,1,1-trichloroethane, toluene, trichloroethylene, or nitrous oxide from vehicle [16, 45–47, 51]. The volatile anesthetic isoflurane, which has discriminative stimulus effects similar to both $GABA_A$ positive modulators and NMDA antagonists [46] as well as several additional vapor anesthetics, all robustly substitute in ethanol-trained mice [111, 112]. These results may be the consequence of the stimulus mixture components of volatile anesthetics and ethanol being sufficiently similar in nature that they can fully mimic one another. However, ethanol only produces partial substitution in isoflurane trained-mice [46]; therefore, their discriminative stimulus properties do not appear to be completely symmetrical as would be predicted if the relative contribution of $GABA_A$ and NMDA receptors to the stimulus effects of ethanol and volatile anesthetics were identical.

4.4 Stimulus Effects of Inhalants: Other Receptor Targets

Although $GABA_A$ and NMDA receptors are the most strongly implicated in the actions of inhalants, they are by no means the only possible ion channel receptors through which the stimulus effects of inhalants may be transduced. Toluene, perchloroethylene [113], nitrous oxide [65, 114], isoflurane, sevoflurane, and halothane [115–118] have all been shown to interact with nicotinic acetylcholine receptors. Toluene, trichloroethylene, 1,1,1-trichloroethane, and volatile anesthetics also enhance glycine receptor function [58, 119]. Lastly, isoflurane, halothane, toluene, 1,1,1-trichloroethane, and trichloroethylene all enhance 5-HT$_3$ receptor function [120–122]. The role of these ion channel receptors in the stimulus effects of inhalants has received little attention. What has been established is that neither nicotine nor the uncompetitive nicotinic antagonist mecamylamine substitute for the stimulus effects of 1,1,1-trichloroethane vapor and pretreatment with either compound does not alter the 1,1,1-trichloroethane concentration-effect curve [43]. These results suggest that nicotinic acetylcholine receptors are not involved in the discriminative stimulus effects of 1,1,1-trichloroethane. This lack of nicotinic involvement may extend to nitrous oxide based on the lack of generalization of nicotine in nitrous oxide-trained mice, but this finding is tentative given the limited number of conditions examined [52]. It remains an open question as to whether nicotinic receptors may be critical to the stimulus effects of other inhalants. Finally, the 5-HT$_3$ antagonist MDL-72222 does not generalize in toluene vapor-trained nor trichloroethylene vapor-trained mice (unpublished observations).

In addition to ion channel receptors, there is also some evidence supporting the hypothesis that some g-protein receptors including opioid, dopamine, and serotonin receptors are targets of inhalants. Acute exposure to toluene and 1,1,1-

trichloroethane decreased DAMGO binding to mu opioid receptors in some brain regions [123]. The kappa opioid antagonist nor-binaltorphimine (nor-BNI) and the mixed mu agonist/antagonist β-chlornaltrexamine but not the delta opioid antagonist naltrindole attenuated nitrous oxide antinociception [124, 125] in rodents. However, in humans naloxone did not alter nitrous oxide-induced changes in pain perception [126, 127].

Again, relatively little work has been done examining the extent to which opioid receptors may be involved in the stimulus effects of inhalants. It has been demonstrated that the opioid antagonist naltrexone did not attenuate the discriminate stimulus effects of 1,1,1-trichloroethane [43] nor did naltrexone alter N_2O's subjective or cognitive impairing effects in human subjects [126, 127]. The mu opioid agonist morphine produced only vehicle-appropriate responding in mice trained to discriminate 1,1,1-trichloroethane from air [43] or 60% nitrous oxide from vehicle [47]. The delta opioid agonist SNC-80 failed to substitute in nitrous oxide-trained mice [47]. Lastly, the kappa opioid agonist U50,488 failed to substitute in animals trained to discriminate trichloroethylene [45], nitrous oxide [47], isoflurane (unpublished data), or toluene (unpublished data). Interestingly, in an earlier study nitrous oxide failed to substitute in morphine-trained rats, but did substitute in rats trained to discriminate the kappa opioid agonist ethylketocyclazocine [128]. However, it was noted that naltrexone failed to block the substitution of nitrous oxide for ethylketocyclazocine, and it was suggested that the results may have not been a consequence of an interaction with opioid receptors. Some opioids such as cyclazocine share stimulus effects with the uncompetitive NMDA antagonist PCP [129] which could be the underlying mechanism for this effect. However, the stimulus effects of ethylketocyclazocine have been repeatedly demonstrated to be opioid receptor mediated [130–132], therefore the mechanism through which ethylketocyclazocine and nitrous oxide may share stimulus effects is unclear.

Repeated treatment with toluene results in reductions in dopamine D2 as well as serotonin receptor binding [133] and also produces signs of serotonin syndrome in rats including head weaving, rigidity, and Straub tail [134]. Direct infusion of toluene into the ventral tegmental area (VTA) has been shown to increase both VTA and nucleus accumbens dopamine release [135]. A high dose of i.p. toluene administered once per day for 7 consecutive days also increased dopamine and serotonin levels in some brain regions [136], as did a single 8-h period of exposure to 1,000 ppm toluene vapor [137]. In another study a shorter treatment with toluene increased locomotor activity, but did not alter extracellular dopamine levels [138]. Isoflurane also has been shown to increase extracellular dopamine release as well as inhibit dopamine transporters in synaptosomes [139], but it does not alter dopamine D2 receptor ligand binding as measured by positron emission tomography (PET) in Rhesus monkeys [140]. In contrast, a second PET study showed that isoflurane appeared to enhance dopamine transporter inhibition produced by both cocaine and the dopamine reuptake inhibitor GBR 12909 [141].

These studies suggest that inhalant effects on catecholamine receptors appear to be more common following extended or chronic exposure, which may limit their role in acute discriminative stimulus effects. A limited number of drugs altering dopamine and 5-HT receptor function have been examined for their ability to elicit inhalant-like discriminative stimuli. Specifically, the 5-HT$_{1A}$ agonist 5-OH-DPAT failed to produce nitrous oxide-like stimulus effects in mice [47]. The mixed 5-HT agonist m-chlorphenylpiperazine (mCPP) which has stimulus effects likely mediated by 5-HT$_{2C}$ receptors [142–144] and has previously been shown to generalize to ethanol [145] also failed to substitute in mice trained to discriminate trichloroethylene [45], nitrous oxide [47], or isoflurane (unpublished observation) from air. A recent study failed to demonstrate that D-amphetamine had any nitrous oxide-like stimulus effects in mice [47]. However, an earlier study reported that toluene will very reliably elicit a high level of partial substitution in mice trained to discriminate D-amphetamine from vehicle [106]. These last results have yet to be replicated or extended to tests of dopaminergic agents in inhalant trained subjects, therefore the question as to whether dopamine receptor mechanisms are involved in the stimulus effects of toluene remains uncertain.

5   Cross-Substitution Studies Comparing Inhalants

The ability of drug discrimination to identify inhalants with similar stimulus effects may provide a means of rapidly classifying novel inhalants according to underlying pharmacological actions by comparing them to previously profiled reference inhalants. In order for this potential to be realized, inhalants with different pharmacological mechanisms should not cross-substitute for one another. This outcome appears likely if the pharmacological differences between two inhalants are sufficiently large. It is less certain whether cross-substitution results comparing inhalants to one another can detect more subtle differences in mechanism such as those discussed in the previous section between hydrocarbon solvents and isoflurane. This issue is not unique to inhalants but is also problematic in drug discrimination studies when comparing drugs that act upon the same receptor, but through different binding sites such as benzodiazepines and barbiturates [146].

Relatively few studies have been conducted examining the cross-substitution profiles of inhalants with one another. A summary of the results of these experiments is presented in Table 2. The most extensive inhalant cross-substitution study was conducted in mice trained to discriminate 12,000 ppm 1,1,1-trichloroethane vapor from air [15]. As previously noted, the stimulus effects of 1,1,1-trichloroethane appear to be most like those produced by positive GABA$_A$ modulators such as benzodiazepines and barbiturates [43, 44]. These GABA$_A$ positive modulator-like properties are shared by toluene [36, 51, 111] as well as trichloroethylene [45]. In 1,1,1-trichloroethane-trained mice both toluene and trichloroethylene produced complete substitution. A somewhat lower level of partial substitution was engendered by two additional aromatic hydrocarbons, ethylbenzene and o-xylene, as well as by the chlorinated hydrocarbon tetrachloroethylene. The volatile anesthetics isoflurane, desflurane, enflurane, and halothane were also tested in 1,1,1-trichloroethane-trained mice. Like 1,1,1-trichloroethane, isoflurane

Table 2 Maximal percentage drug-lever responding for volatile and gaseous compounds tested in subjects trained to discriminate different inhalants from vehicle

Test inhalant	Training inhalant				
	Toluene	1,1,1-Trichloroethane	Trichloroethylene	Isoflurane	Nitrous oxide
Aromatic hydrocarbons					
Toluene	–	100[b]	93[e]	95[d]	72[f]
Ethylbenzene	64[a]	62[b]			
O-xylene		74[b]			
Chlorinated hydrocarbons					
1,1,1-Trichloroethane		–	90[e]		44[f]
Trichloroethylene	81[b]		–	88[d]	
Perchloroethylene	70[b]		100[e]		
Volatile anesthetics					
Isoflurane	20[a]	50[b]	75[e]	–	39[f]
Desflurane		85[b]			
Enflurane		100[b]		100[d]	
Methoxyflurane			95[e]		47[f]
Halothane		100[b]		95[d]	
Other					
Nitrous oxide		15[c]		31[d]	
2-Butanol (odorant)		0[b]			3[f]

[a]Reference [32]
[b]Reference [15]
[c]Reference [44]
[d]Reference [46]
[e]Reference [45]
[f]Reference [16]

has GABA$_A$ positive modulator-like stimulus effects, but unlike 1,1,1-trichloroethane it also possesses NMDA antagonist-like stimulus properties [46]. This mixture-like stimulus profile of isoflurane is reminiscent of that produced by ethanol [75]. Ethanol produces intermediate levels of substitution in animals trained to discriminate GABA$_A$ positive modulators and this is likely due to overshadowing by the additional components of ethanol's stimulus [23, 147]. If the same concepts hold true for inhalants, it would be predicted that isoflurane should at best produce partial substitution in 1,1,1-trichloroethane-trained subjects. This was indeed the case as isoflurane resulted in a maximum of 50% drug-lever selection [15]. However, isoflurane appears to be an exception among volatile inhalants in this regard as desflurane, enflurane, and halothane all produced full substitution in 1,1,1-trichloroethane-trained mice [15, 43]. This may reflect actual differences in the pharmacology underlying the stimulus effects of volatile anesthetics, methodological factors, or inherent variability. No studies have yet been conducted using other volatile anesthetics as training stimuli to address this question. Finally, nitrous oxide produced only vehicle-appropriate responding in 1,1,1-trichloroethane-trained mice [15], which is consistent with data indicating that nitrous oxide's stimulus is not $GABA_A$ positive modulator-like [47].

In contrast to 1,1,1-trichloroethane, the stimulus effects of nitrous oxide have an uncompetitive NMDA antagonist-like component but no $GABA_A$ positive modulator-like properties [47]. In mice trained to discriminate 60% nitrous oxide from vehicle, 1,1,1-trichloroethane, isoflurane, and methoxyflurane all produced less than 50% nitrous oxide-lever selection [16]. The poor substitution produced by 1,1,1-trichloroethane is consistent with its lack of NMDA antagonist-like stimulus effects [44]. Likewise although isoflurane has NMDA antagonist-like stimulus effects they could have been overshadowed by its GABAergic stimulus component or failed to substitute due to the stimulus effects of isoflurane being more similar to competitive than uncompetitive NMDA antagonists [46]. Interestingly toluene vapor produced a higher level of partial substitution in nitrous oxide-trained mice than any of the other inhalants which were examined [47]. This outcome is inconsistent with what would have been predicted based on the lack of NMDA antagonist-like stimulus effects of toluene [51]. One possible explanation is that both nitrous oxide and toluene possess a common but as yet unidentified stimulus component. This speculative interpretation is somewhat strengthened by the inability of any of the receptor-selective probe compounds which have been tested in toluene-trained or nitrous oxide-trained mice to fully mimic the stimulus effects of either inhalant [47, 51]. Additional studies exploring the receptors underlying the stimulus effects of both toluene and nitrous oxide will be necessary to resolve this apparent inconsistency.

Lastly, cross-substitution of several inhalants has been examined in mice trained to discriminate 6,000 ppm isoflurane vapor from air [46]. As would be predicted the related volatile anesthetic enflurane as well as halothane fully substituted for isoflurane. As previously discussed the discriminative stimulus of isoflurane appears to be composed of a $GABA_A$ positive modulator-like as well as competitive NMDA antagonist-like effects. Consistent with the notion that each of the components of a stimulus mixture is perceived as independent elements [79], both toluene and trichloroethylene which have $GABA_A$ positive modulator-like stimulus effects fully substituted for isoflurane. In contrast, nitrous oxide substitutes poorly in isoflurane-trained mice [46]. This result may be the consequence of a relatively weak NMDA antagonist-like component in isoflurane's stimulus or the dissimilarities between the stimulus effects of competitive and uncompetitive NMDA antagonists [148–150].

6 Summary and Conclusions

The mechanisms underlying the pharmacological effects of inhalants are poorly understood, especially those properties which are most important in promoting their abuse. Our lack of knowledge is exacerbated by the fact that there are dozens of

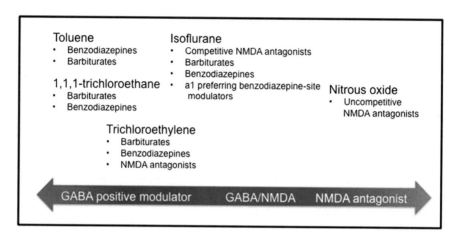

Fig. 1 Summary of greater than vehicle level cross-substitution produced by different classes of GABA_A positive modulators and NMDA antagonists for five training inhalants. Classes of cross-substitution drugs below each inhalant are listed in order of greatest to least cross-substitution. Toluene and 1,1,1-trichloroethane only had GABA_A positive modulator-like stimulus effects. Nitrous oxide had only NMDA antagonist-like stimulus effects. Isoflurane and trichloroethylene had mixed stimulus effects as depicted by their relative position on the X axis

different inhalants, the class is highly heterogeneous in form and structure, and many, indeed perhaps most inhalants interact with multiple receptor targets. At the present time some of the most powerful behavioral techniques (e.g., self-administration) that have proven invaluable to understanding the receptor systems involved in the abuse-related effects of other drugs have not been successfully adapted to study inhalants. Further, the toxicity of most inhalants precludes studies in humans closing off another important research strategy. Without some means of delineating the receptor systems involved in the abuse-related effects inhalants, development of pharmacological treatments to curb inhalant use and prevent relapse to inhalant abuse will continue to be seriously compromised. Drug discrimination is perhaps the most promising paradigm currently available for exploring the abuse-related effects of inhalants.

Figure 1 presents an overview of the cross-substitution results in mice trained to discriminate five different inhalants from their respective vehicles. The discriminative stimulus effects of the aromatic hydrocarbon solvent toluene, as well as the chlorinated hydrocarbons 1,1,1-trichloroethane and trichloroethylene, are mediated to a considerable extent by positive GABA_A modulatory effects, similar to those produced by barbiturates and classical benzodiazepines. The lack of substitution by the alpha 1 subunit preferring nonbenzodiazepine hypnotic zaleplon also supports the argument that GABA_A receptors composed of alpha 2, 3, or 5 subunits may mediate the discriminative stimulus effects of aromatic hydrocarbon solvents. There appears to be a more barbiturate-like stimulus component produced by the chlorinated hydrocarbons than by toluene, but this may be a consequence of alterations in relative selectivity produced by different training doses as opposed to more fundamental mechanistic differences. The stimulus effects of trichloroethylene may also have a NMDA antagonist-like component although the modest level of partial substitution produced by uncompetitive NMDA antagonists makes this conclusion more tentative. In contrast, the stimulus effects of isoflurane appear to be composed of both positive $GABA_A$ modulatory actions similar to barbiturates and alpha 1 subunit selective benzodiazepines, as well as NMDA antagonist-like effects most like those produced by competitive NMDA antagonists. Finally, the discriminative stimulus effects of nitrous oxide appear to be most similar to those of uncompetitive NMDA antagonists, although incomplete substitution suggests that other as yet unidentified mechanisms are probably also involved. Finally, despite considerable data showing that inhalants alter responses mediated by other receptors the available data is not supportive of other mechanisms as mediators of inhalants acute discriminative stimuli.

Taken as a whole it appears that drug discrimination can reveal differences in the underlying neurochemical actions of inhalants that would likely be indistinguishable using other techniques. The utility of drug discrimination as a means to categorize inhalants according to abuse-related pharmacological effects is therefore encouraging. The predicative power of this technique may be constrained to some degree by the inability of drug discrimination to consistently tease apart subtle differences in specific sites of action at the same receptor. Lastly, the limited evidence available at the present time suggests that cross-substitution studies comparing novel inhalants to a panel of well-characterized archetypal reference inhalants can provide some suggestions as to the underlying neurochemical mechanisms responsible for the stimulus effects of novel inhalants. While these data may be sufficient to tentatively categorize novel inhalants according to mechanism of action, cross-substitution studies are unlikely to be as informative or as definitive as experiments in which the inhalant of interest itself serves as the training drug.

References

1. Johnston LD, O'Malley PM, Miech RA, Bachman JG, Schulenberg JE (2014) Monitoring the future national results on adolescent drug use 1975–2014: overview of key findings on adolescent drug use. Institute for Social Research, The University of Michigan, Ann Arbor
2. Bruckner JV, Peterson RG (1981) Evaluation of toluene and acetone inhalant abuse. II. Model development and toxicology. Toxicol Appl Pharmacol 61:302–312
3. Hannigan JH, Bowen SE (2010) Reproductive toxicology and teratology of abused toluene. Syst Biol Reprod Med 56:184–200
4. Takagi M, Lubman DI, Yucel M (2011) Solvent-induced leukoencephalopathy: a disorder of adolescence? Subst Use Misuse 46(Suppl 1):95–98
5. Tormoehlen LM, Tekulve KJ, Nanagas KA (2014) Hydrocarbon toxicity: a review. Clin Toxicol (Phila) 52:479–489
6. Bowen SE, Batis JC, Paez-Martinez N, Cruz SL (2006) The last decade of solvent research in animal models of abuse: mechanistic and behavioral studies. Neurotoxicol Teratol 28:636–647

7. Cruz SL (2012) The latest evidence in the neuroscience of solvent misuse: an article written for service providers. Subst Use Misuse 46(Suppl 1):62–67
8. Villatoro JA, Cruz SL, Ortiz A, Medina-Mora ME (2011) Volatile substance misuse in Mexico: correlates and trends. Subst Use Misuse 46(Suppl 1):40–45
9. Eggertson L (2014) Opal fuel reduces gas-sniffing and suicides in Australia. CMAJ 186: E229–E230
10. Balster RL, Cruz SL, Howard MO, Dell CA, Cottler LB (2009) Classification of abused inhalants. Addiction 104:878–882
11. Glennon RA (1986) Discriminative stimulus properties of phenylisopropylamine derivatives. Drug Alcohol Depend 17:119–134
12. Glennon RA (1989) Stimulus properties of hallucinogenic phenalkylamines and related designer drugs: formulation of structure-activity relationships. NIDA Res Monogr 94:43–67
13. Balster RL (1998) Neural basis of inhalant abuse. Drug Alcohol Depend 51:207–214
14. Shelton KL, Balster RL (2011) Inhalant drug discrimination: methodology, literature review and future directions. In: Glennon RA, Young R (eds) Drug discrimination: applications to medicinal chemistry and drug studies, 1st edn. Wiley, Hoboken
15. Shelton KL (2009) Discriminative stimulus effects of inhaled 1,1,1-trichloroethane in mice: comparison to other hydrocarbon vapors and volatile anesthetics. Psychopharmacology (Berl) 203:431–440
16. Richardson KJ, Shelton KL (2014) Discriminative stimulus effects of nitrous oxide in mice: comparison with volatile hydrocarbons and vapor anesthetics. Behav Pharmacol 25:2–11
17. Wood RW (1978) Stimulus properties of inhaled substances. Environ Health Perspect 26:69–76
18. Wood RW, Rees DC, Laties VG (1983) Behavioral effects of toluene are modulated by stimulus control. Toxicol Appl Pharmacol 68:462–472
19. Rees DC, Knisely JS, Jordan S, Balster RL (1987) Discriminative stimulus properties of toluene in the mouse. Toxicol Appl Pharmacol 88:97–104
20. Knisely JS, Rees DC, Balster RL (1990) Discriminative stimulus properties of toluene in the rat. Neurotoxicol Teratol 12:129–133
21. Benignus VA (1981) Health effects of toluene: a review. Neurotoxicology 2:567–588
22. Baker LE, Pynnonen D, Poling A (2004) Influence of reinforcer type and route of administration on gamma-hydroxybutyrate discrimination in rats. Psychopharmacology (Berl) 174:220–227
23. Balster RL, Grech DM, Bobelis DJ (1992) Drug discrimination analysis of ethanol as an N-methyl-D-aspartate receptor antagonist. Eur J Pharmacol 222:39–42
24. Bienkowski P, Stefanski R, Kostowski W (1996) Competitive NMDA receptor antagonist, CGP 40116, substitutes for the discriminative stimulus effects of ethanol. Eur J Pharmacol 314:277–280
25. Grant KA, Colombo G (1993) Pharmacological analysis of the mixed discriminative stimulus effects of ethanol. Alcohol Alcohol Suppl 2:445–449
26. Shelton KL, Balster RL (1994) Ethanol drug discrimination in rats: substitution with GABA agonists and NMDA antagonists. Behav Pharmacol 5:441–451
27. Haney M, Hart C, Collins ED, Foltin RW (2005) Smoked cocaine discrimination in humans: effects of gabapentin. Drug Alcohol Depend 80:53–61
28. Katz JL, Sharpe LG, Jaffe JH, Shores EI, Witkin JM (1991) Discriminative stimulus effects of inhaled cocaine in squirrel monkeys. Psychopharmacology (Berl) 105:317–321
29. Wessinger WD, Martin BR, Balster RL (1985) Discriminative stimulus properties and brain distribution of phencyclidine in rats following administration by injection and smoke inhalation. Pharmacol Biochem Behav 23:607–612
30. Macenski MJ, Shelton KL (2001) Self-administered ethanol as a discriminative stimulus in rats. Drug Alcohol Depend 64:243–247
31. Shelton KL, Macenski MJ (1998) Discriminative stimulus effects of self-administered ethanol. Behav Pharmacol 9(4):329–336

32. Shelton KL (2007) Inhaled toluene vapor as a discriminative stimulus. Behav Pharmacol 18:219–229
33. Cometto-Muniz JE, Cain WS, Abraham MH, Gola JM (2001) Ocular and nasal trigeminal detection of butyl acetate and toluene presented singly and in mixtures. Toxicol Sci 63:233–244
34. Cometto-Muniz JE, Cain WS, Abraham MH, Gola JM (2002) Psychometric functions for the olfactory and trigeminal detectability of butyl acetate and toluene. J Appl Toxicol 22:25–30
35. Winters W, Devriese S, Eelen P, Veulemans H, Nemery B, van den Bergh O (2001) Symptom learning in response to odors in a single odor respiratory learning paradigm. Ann N Y Acad Sci 933:315–318
36. Rees DC, Coggeshall E, Balster RL (1985) Inhaled toluene produces pentobarbital-like discriminative stimulus effects in mice. Life Sci 37:1319–1325
37. DE Vry J, Slangen JL (1986) Effects of training dose on discrimination and cross-generalization of chlordiazepoxide, pentobarbital and ethanol in the rat. Psychopharmacology (Berl) 88:341–345
38. Stolerman IP, Naylor C, Elmer GI, Goldberg SR (1999) Discrimination and self-administration of nicotine by inbred strains of mice. Psychopharmacology (Berl) 141:297–306
39. York JL (1978) Efficacy of ethanol as a discriminative stimulus in ethanol-preferring and ethanol-nonpreferring rats. Experientia 34:224–225
40. Shelton KL, Slavova-Hernandez G (2009) Characterization of an inhaled toluene drug discrimination in mice: effect of exposure conditions and route of administration. Pharmacol Biochem Behav 92:614–620
41. Benignus VA, Muller KE, Graham JA, Barton CN (1984) Toluene levels in blood and brain of rats as a function of toluene level in inspired air. Environ Res 33:39–46
42. Lammers JH, van Asperen J, DE Groot D, Rijcken WR (2005) Behavioural effects and kinetics in brain in response to inhalation of constant or fluctuating toluene concentrations in the rat. Environ Toxicol Pharmacol 19:625–634
43. Shelton KL (2010) Pharmacological characterization of the discriminative stimulus of inhaled 1,1,1-trichloroethane. J Pharmacol Exp Ther 333:612–620
44. Shelton KL, Nicholson KL (2012) GABAA-positive modulator selective discriminative stimulus effects of 1,1,1-trichloroethane vapor. Drug Alcohol Depend 121:103–109
45. Shelton KL, Nicholson KL (2014) Pharmacological classification of the abuse-related discriminative stimulus effects of trichloroethylene vapor. J Drug Alcohol Res 3:235839
46. Shelton KL, Nicholson KL (2010) GABA(A) positive modulator and NMDA antagonist-like discriminative stimulus effects of isoflurane vapor in mice. Psychopharmacology (Berl) 212:559–569
47. Richardson KJ, Shelton KL (2015) N-methyl-D-aspartate receptor channel blocker-like discriminative stimulus effects of nitrous oxide gas. J Pharmacol Exp Ther 352:156–165
48. Urban BW, Bleckwenn M, Barann M (2006) Interactions of anesthetics with their targets: non-specific, specific or both? Pharmacol Ther 111:729–770
49. Duncan JR, Lawrence AJ (2013) Conventional concepts and new perspectives for understanding the addictive properties of inhalants. J Pharmacol Sci 122:237–243
50. Del Re AM, Dopico AM, Woodward JJ (2006) Effects of the abused inhalant toluene on ethanol-sensitive potassium channels expressed in oocytes. Brain Res 1087:75–82
51. Shelton KL, Nicholson KL (2013) Benzodiazepine-like discriminative stimulus effects of toluene vapor. Eur J Pharmacol 720:131–137
52. Richardson KJ (2014) Characterization of the discriminative stimulus effects of nitrous oxide. Doctoral dissertation, Virginia Commonwealth University, Richmond
53. Olsen RW, Sieghart W (2009) GABA A receptors: subtypes provide diversity of function and pharmacology. Neuropharmacology 56:141–148
54. Mohler H (2006) GABA(A) receptor diversity and pharmacology. Cell Tissue Res 326:505–516

55. Greiner AS, Larach DR (1989) The effect of benzodiazepine receptor antagonism by flumazenil on the MAC of halothane in the rat. Anesthesiology 70:644–648
56. Hapfelmeier G, Zieglgansberger W, Haseneder R, Schneck H, Kochs E (2000) Nitrous oxide and xenon increase the efficacy of GABA at recombinant mammalian GABA(A) receptors. Anesth Analg 91:1542–1549
57. Czech DA, Quock RM (1993) Nitrous oxide induces an anxiolytic-like effect in the conditioned defensive burying paradigm, which can be reversed with a benzodiazepine receptor blocker. Psychopharmacology (Berl) 113:211–216
58. Beckstead MJ, Weiner JL, Eger EI 2nd, Gong D. H, Mihic SJ (2000) Glycine and gamma-aminobutyric acid(A) receptor function is enhanced by inhaled drugs of abuse. Mol Pharmacol 57:1199–1205
59. Beckley JT, Woodward JJ (2011) The abused inhalant toluene differentially modulates excitatory and inhibitory synaptic transmission in deep-layer neurons of the medial prefrontal cortex. Neuropsychopharmacology 36:1531–1542
60. Maciver MB (2009) Abused inhalants enhance GABA-mediated synaptic inhibition. Neuropsychopharmacology 34:2296–2304
61. WILLIAMS JM, Stafford D, Steketee JD (2005) Effects of repeated inhalation of toluene on ionotropic GABA A and glutamate receptor subunit levels in rat brain. Neurochem Int 46:1–10
62. Grasshoff C, Drexler B, Rudolph U, Antkowiak B (2006) Anaesthetic drugs: linking molecular actions to clinical effects. Curr Pharm Des 12:3665–3679
63. Nagashima K, Zorumski CF, Izumi Y (2005) Nitrous oxide (laughing gas) facilitates excitability in rat hippocampal slices through gamma-aminobutyric acid A receptor-mediated disinhibition. Anesthesiology 102:230–234
64. Orii R, Ohashi Y, Halder S, Giombini M, Maze M, Fujinaga M (2003) GABAergic interneurons at supraspinal and spinal levels differentially modulate the antinociceptive effect of nitrous oxide in Fischer rats. Anesthesiology 98:1223–1230
65. Yamakura T, Harris RA (2000) Effects of gaseous anesthetics nitrous oxide and xenon on ligand-gated ion channels. Comparison with isoflurane and ethanol. Anesthesiology 93:1095–1101
66. Greenblatt EP, Meng X (2001) Divergence of volatile anesthetic effects in inhibitory neurotransmitter receptors. Anesthesiology 94:1026–1033
67. Harris BD, Moody EJ, Basile AS, Skolnick P (1994) Volatile anesthetics bidirectionally and stereospecifically modulate ligand binding to GABA receptors. Eur J Pharmacol 267:269–274
68. Jia F, Yue M, Chandra D, Homanics GE, Goldstein PA, Harrison NL (2008) Isoflurane is a potent modulator of extrasynaptic GABA(A) receptors in the thalamus. J Pharmacol Exp Ther 324:1127–1135
69. Rau V, Iyer SV, Oh I, Chandra D, Harrison N, Eger EI 2nd, Fanselow MS, Homanics GE, Sonner JM (2009) Gamma-aminobutyric acid type A receptor alpha 4 subunit knockout mice are resistant to the amnestic effect of isoflurane. Anesth Analg 109:1816–1822
70. Wood RW, Coleman JB, Schuler R, Cox C (1984) Anticonvulsant and antipunishment effects of toluene. J Pharmacol Exp Ther 230:407–412
71. Wiley JL, Bale AS, Balster RL (2003) Evaluation of toluene dependence and cross-sensitization to diazepam. Life Sci 72:3023–3033
72. Bowen SE, Wiley JL, Balster RL (1996) The effects of abused inhalants on mouse behavior in an elevated plus-maze. Eur J Pharmacol 312:131–136
73. Rago L, Kiivet RA, Harro J, Pold M (1988) Behavioral differences in an elevated plus-maze: correlation between anxiety and decreased number of GABA and benzodiazepine receptors in mouse cerebral cortex. Naunyn Schmiedebergs Arch Pharmacol 337:675–678
74. Bowen SE, Wiley JL, Jones HE, Balster RL (1999) Phencyclidine- and diazepam-like discriminative stimulus effects of inhalants in mice. Exp Clin Psychopharmacol 7:28–37

75. Grant KA (1999) Strategies for understanding the pharmacological effects of ethanol with drug discrimination procedures. Pharmacol Biochem Behav 64:261–267
76. Mariathasan EA, Garcha HS, Stolerman IP (1991) Discriminative stimulus effects of amphetamine and pentobarbitone separately and as mixtures in rats. Behav Pharmacol 2:405–415
77. Shoaib M, Baumann MH, Rothman RB, Goldberg SR, Schindler CW (1997) Behavioural and neurochemical characteristics of phentermine and fenfluramine administered separately and as a mixture in rats. Psychopharmacology (Berl) 131:296–306
78. Stolerman IP, Mariathasan EA, Garcha HS (1991) NIDA Res Monogr 116:277–306
79. Stolerman IP, Mariathasan EA, White JA, Olufsen KS (1999) Drug mixtures and ethanol as compound internal stimuli. Pharmacol Biochem Behav 64:221–228
80. Rees DC, Knisely JS, Balster RL, Jordan S, Breen TJ (1987) Pentobarbital-like discriminative stimulus properties of halothane, 1,1,1-trichloroethane, isoamyl nitrite, flurothyl and oxazepam in mice. J Pharmacol Exp Ther 241:507–515
81. Mcbain CJ, Mayer ML (1994) N-methyl-D-aspartic acid receptor structure and function. Physiol Rev 74:723–760
82. Wroblewski JT, Danysz W (1989) Modulation of glutamate receptors: molecular mechanisms and functional implications. Annu Rev Pharmacol Toxicol 29:441–474
83. Cruz SL, Balster RL, Woodward JJ (2000) Effects of volatile solvents on recombinant N-methyl-D-aspartate receptors expressed in Xenopus oocytes. Br J Pharmacol 131:1303–1308
84. Jevtovic-Todorovic V, Todorovic SM, Mennerick S, Powell S, Dikranian K, Benshoff N, Zorumski CF, Olney JW (1998) Nitrous oxide (laughing gas) is an NMDA antagonist, neuroprotectant and neurotoxin. Nat Med 4:460–463
85. Mennerick S, Jevtovic-Todorovic V, Todorovic SM, Shen W, Olney JW, Zorumski CF (1998) Effect of nitrous oxide on excitatory and inhibitory synaptic transmission in hippocampal cultures. J Neurosci 18:9716–9726
86. Ogata J, Shiraishi M, Namba T, Smothers CT, Woodward JJ, Harris RA (2006) Effects of anesthetics on mutant N-methyl-D-aspartate receptors expressed in Xenopus oocytes. J Pharmacol Exp Ther 318:434–443
87. Balon N, Dupenloup L, Blanc F, Weiss M, Rostain JC (2003) Nitrous oxide reverses the increase in striatal dopamine release produced by N-methyl-D-aspartate infusion in the substantia nigra pars compacta in rats. Neurosci Lett 343:147–149
88. DE Sousa SL, Dickinson R, Lieb WR, Franks NP (2000) Contrasting synaptic actions of the inhalational general anesthetics isoflurane and xenon. Anesthesiology 92:1055–1066
89. Keita H, Henzel-Rouelle D, Dupont H, Desmonts JM, Mantz J (1999) Halothane and isoflurane increase spontaneous but reduce the N-methyl-D-aspartate-evoked dopamine release in rat striatal slices: evidence for direct presynaptic effects. Anesthesiology 91:1788–1797
90. Ranft A, Kurz J, Deuringer M, Haseneder R, Dodt HU, Zieglgansberger W, Kochs E, Eder M, Hapfelmeier G (2004) Isoflurane modulates glutamatergic and GABAergic neurotransmission in the amygdala. Eur J Neurosci 20:1276–1280
91. Wakasugi M, Hirota K, Roth SH, Ito Y (1999) The effects of general anesthetics on excitatory and inhibitory synaptic transmission in area CA1 of the rat hippocampus in vitro. Anesth Analg 88:676–680
92. Cruz SL, Gauthereau MY, Camacho-Munoz C, Lopez-Rubalcava C, Balster RL (2003) Effects of inhaled toluene and 1,1,1-trichloroethane on seizures and death produced by N-methyl-D-aspartic acid in mice. Behav Brain Res 140:195–202
93. Lo PS, Wu CY, Sue HZ, Chen HH (2009) Acute neurobehavioral effects of toluene: involvement of dopamine and NMDA receptors. Toxicology 265:34–40
94. Balster RL (1991) Discriminative stimulus properties of phencyclidine and other NMDA antagonists. NIDA Res Monogr 116:163–180

95. Nicholson KL, Balster RL (2009) The discriminative stimulus effects of N-methyl-D-aspartate glycine-site ligands in NMDA antagonist-trained rats. Psychopharmacology (Berl) 203:441–451
96. Nicholson KL, Jones HE, Balster RL (1997) Evaluation of the reinforcing and discriminative stimulus effects of the N-methyl-D-aspartate competitive antagonist NPC 17742 in rhesus monkeys. Behav Pharmacol 8:396–407
97. Shelton KL, Balster RL (2004) Effects of abused inhalants and GABA-positive modulators in dizocilpine discriminating inbred mice. Pharmacol Biochem Behav 79:219–228
98. Balster RL (1989) Behavioral pharmacology of PCP, NMDA and sigma receptors. NIDA Res Monogr 95:270–274
99. Jackson A, Sanger DJ (1988) Is the discriminative stimulus produced by phencyclidine due to an interaction with N-methyl-D-aspartate receptors? Psychopharmacology (Berl) 96:87–92
100. Willetts J, Balster RL (1988) Role of N-methyl-d-aspartate receptor stimulation and antagonism in PCP discrimination in rats. In: Domino EF, Kamenka JM (eds) Sigma and phencyclidine-like compounds as molecular probes in biology. NPP Books, Ann Arbor, pp 397–406
101. Crosby MJ, Hanson JE, Fleckenstein AE, Hanson GR (2002) Phencyclidine increases vesicular dopamine uptake. Eur J Pharmacol 438:75–78
102. Nankai M, Klarica M, Fage D, Carter C (1998) The pharmacology of native N-methyl-D-aspartate receptor subtypes: different receptors control the release of different striatal and spinal transmitters. Prog Neuropsychopharmacol Biol Psychiatry 22:35–64
103. Snell LD, Yi SJ, Johnson KM (1988) Comparison of the effects of MK-801 and phencyclidine on catecholamine uptake and NMDA-induced norepinephrine release. Eur J Pharmacol 145:223–226
104. France CP, Moerschbaecher JM, Woods JH (1991) MK-801 and related compounds in monkeys: discriminative stimulus effects and effects on a conditional discrimination. J Pharmacol Exp Ther 257:727–734
105. Geter-Douglass B, Witkin JM (1997) Dizocilpine-like discriminative stimulus effects of competitive NMDA receptor antagonists in mice. Psychopharmacology (Berl) 133:43–50
106. Bowen SE (2006) Increases in amphetamine-like discriminative stimulus effects of the abused inhalant toluene in mice. Psychopharmacology (Berl) 186:517–524
107. Martin DC, Plagenhoef M, Abraham J, Dennison RL, Aronstam RS (1995) Volatile anesthetics and glutamate activation of N-methyl-D-aspartate receptors. Biochem Pharmacol 49:809–817
108. Barry H III, Krimmer EC (1978) Similarities and differences in discriminative stimulus effects of chlordiazepoxide, pentobarbital, ethanol and other sedatives. In: Colpaert FC, Rosecrans JA (eds) Stimulus properties of drugs: ten Years of progress. Elsevier, Amsterdam, pp 31–51
109. Butelman EE, Baron SS, Woods JJ (1993) Ethanol effects in pigeons trained to discriminate MK-801, PCP or CGS-19755. Behav Pharmacol 4:57–60
110. York JL, Bush R (1982) Studies on the discriminative stimulus properties of ethanol in squirrel monkeys. Psychopharmacology (Berl) 77:212–216
111. Rees DC, Knisely JS, Breen TJ, Balster RL (1987) Toluene, halothane, 1,1,1-trichloroethane and oxazepam produce ethanol-like discriminative stimulus effects in mice. J Pharmacol Exp Ther 243:931–937
112. Bowen SE, Balster RL (1997) Desflurane, enflurane, isoflurane and ether produce ethanol-like discriminative stimulus effects in mice. Pharmacol Biochem Behav 57:191–198
113. Bale AS, Meacham CA, Benignus VA, Bushnell PJ, Shafer TJ (2005) Volatile organic compounds inhibit human and rat neuronal nicotinic acetylcholine receptors expressed in Xenopus oocytes. Toxicol Appl Pharmacol 205:77–88
114. Suzuki T, Ueta K, Sugimoto M, Uchida I, Mashimo T (2003) Nitrous oxide and xenon inhibit the human (alpha 7)5 nicotinic acetylcholine receptor expressed in Xenopus oocyte. Anesth Analg 96:443–448

115. Dilger JP, Brett RS, Mody HI (1993) The effects of isoflurane on acetylcholine receptor channels.: 2. Currents elicited by rapid perfusion of acetylcholine. Mol Pharmacol 44:1056–1063

116. Mowrey DD, Liu Q, Bondarenko V, Chen Q, Seyoum E, Xu Y, Wu J, Tang P (2013) Insights into distinct modulation of alpha7 and alpha7beta2 nicotinic acetylcholine receptors by the volatile anesthetic isoflurane. J Biol Chem 288:35793–35800

117. Wachtel RE, Wegrzynowicz ES (1991) Mechanism of volatile anesthetic action on ion channels. Ann N Y Acad Sci 625:116–128

118. Yamashita M, Mori T, Nagata K, Yeh JZ, Narahashi T (2005) Isoflurane modulation of neuronal nicotinic acetylcholine receptors expressed in human embryonic kidney cells. Anesthesiology 102:76–84

119. Krasowski MD, Harrison NL (2000) The actions of ether, alcohol and alkane general anaesthetics on GABAA and glycine receptors and the effects of TM2 and TM3 mutations. Br J Pharmacol 129:731–743

120. Lopreato GF, Phelan R, Borghese CM, Beckstead MJ, Mihic SJ (2003) Inhaled drugs of abuse enhance serotonin-3 receptor function. Drug Alcohol Depend 70:11–15

121. Solt K, Stevens RJ, Davies PA, Raines DE (2005) General anesthetic-induced channel gating enhancement of 5-hydroxytryptamine type 3 receptors depends on receptor subunit composition. J Pharmacol Exp Ther 315:771–776

122. Suzuki T, Koyama H, Sugimoto M, Uchida I, Mashimo T (2002) The diverse actions of volatile and gaseous anesthetics on human-cloned 5-hydroxytryptamine 3 receptors expressed in Xenopus oocytes. Anesthesiology 96:699–704

123. Paez-Martinez N, Ambrosio E, Garcia-Lecumberri C, Rocha L, Montoya GL, Cruz SL (2008) Toluene and TCE decrease binding to mu-opioid receptors, but not to benzodiazepine and NMDA receptors in mouse brain. Ann N Y Acad Sci 1139:390–401

124. Emmanouil DE, Quock RM (2007) Advances in understanding the actions of nitrous oxide. Anesth Prog 54:9–18

125. Koyama T, Fukuda K (2010) Involvement of the kappa-opioid receptor in nitrous oxide-induced analgesia in mice. J Anesth 24:297–299

126. Zacny JP, Janiszewski D, Sadeghi P, Black ML (1999) Reinforcing, subjective, and psychomotor effects of sevoflurane and nitrous oxide in moderate-drinking healthy volunteers. Addiction 94:1817–1828

127. Zacny JP, Sparacino G, Hoffmann P, Martin R, Lichtor JL (1994) The subjective, behavioral and cognitive effects of subanesthetic concentrations of isoflurane and nitrous oxide in healthy volunteers. Psychopharmacology (Berl) 114:409–416

128. Hynes MD, Hymson DL (1984) Nitrous oxide generalizes to a discriminative stimulus produced by ethylketocyclazocine but not morphine. Eur J Pharmacol 105:155–159

129. Holtzman SG (1980) Phencyclidine-like discriminative effects of opioids in the rat. J Pharmacol Exp Ther 214:614–619

130. Hein DW, Young AM, Herling S, Woods JH (1981) Pharmacological analysis of the discriminative stimulus characteristics of ethylketazocine in the rhesus monkey. J Pharmacol Exp Ther 218:7–15

131. Herling S, Shannon HE (1982) Discriminative effects of ethylketazocine in the rat stereospecificity and antagonism by naloxone. Life Sci 31:2371–2374

132. Shearman GT, Herz A (1982) Discriminative stimulus properties of narcotic and non-narcotic drugs in rats trained to discriminate opiate kappa-receptor agonists. Psychopharmacology (Berl) 78:63–66

133. Celani MF, Fuxe K, Agnati LF, Andersson K, Hansson T, Gustafsson JA, Battistini N, Eneroth P (1983) Effects of subacute treatment with toluene on central monoamine receptors in the rat. Reduced affinity in [3H]5-hydroxytryptamine binding sites and in [3H]spiperone binding sites linked to dopamine receptors. Toxicol Lett 17:275–281

134. Castilla-Serna L, Barragan-Mejia MG, Rodriguez-Perez RA, Garcia Rillo A, Reyes-Vazquez C. (1993) Effects of acute and chronic toluene inhalation on behavior, monoamine

metabolism and specific binding (3H-serotonin and 3H-norepinephrine) of rat brain. Arch Med Res 24:169–176

135. Riegel AC, Zapata A, Shippenberg TS, French ED (2007) The abused inhalant toluene increases dopamine release in the nucleus accumbens by directly stimulating ventral tegmental area neurons. Neuropsychopharmacology 32:1558–1569

136. Riegel AC, Ali SF, Torinese S, French ED (2004) Repeated exposure to the abused inhalant toluene alters levels of neurotransmitters and generates peroxynitrite in nigrostriatal and mesolimbic nuclei in rat. Ann N Y Acad Sci 1025:543–551

137. Rea TM, Nash JF, Zabik JE, Born GS, Kessler WV (1984) Effects of toluene inhalation on brain biogenic amines in the rat. Toxicology 31:143–150

138. Kondo H, Huang J, Ichihara G, Kamijima M, Saito I, Shibata E, Ono Y, Hisanaga N, Takeuchi Y, Nakahara D (1995) Toluene induces behavioral activation without affecting striatal dopamine metabolism in the rat: behavioral and microdialysis studies. Pharmacol Biochem Behav 51:97–101

139. El-Maghrabi EA, Eckenhoff RG (1993) Inhibition of dopamine transport in rat brain synaptosomes by volatile anesthetics. Anesthesiology 78:750–756

140. Nader MA, Grant KA, Gage HD, Ehrenkaufer RL, Kaplan JR, Mach RH (1999) PET imaging of dopamine D2 receptors with [18F]fluoroclebopride in monkeys: effects of isoflurane- and ketamine-induced anesthesia. Neuropsychopharmacology 21:589–596

141. Tsukada H, Nishiyama S, Kakiuchi T, Ohba H, Sato K, Harada N, Nakanishi S (1999) Isoflurane anesthesia enhances the inhibitory effects of cocaine and GBR12909 on dopamine transporter: PET studies in combination with microdialysis in the monkey brain. Brain Res 849:85–96

142. Callahan PM, Cunningham KA (1994) Involvement of 5-HT2C receptors in mediating the discriminative stimulus properties of m-chlorophenylpiperazine (mCPP). Eur J Pharmacol 257:27–38

143. Fiorella D, Helsley S, Rabin RA, Winter JC (1995) 5-HT2C receptor-mediated phosphoinositide turnover and the stimulus effects of m-chlorophenylpiperazine. Psychopharmacology (Berl) 122:237–243

144. Fiorella D, Rabin RA, Winter JC (1995) The role of the 5-HT2A and 5-HT2C receptors in the stimulus effects of m-chlorophenylpiperazine. Psychopharmacology (Berl) 119:222–230

145. Grant KA, Colombo G, Gatto GJ (1997) Characterization of the ethanol-like discriminative stimulus effects of 5-HT receptor agonists as a function of ethanol training dose. Psychopharmacology (Berl) 133:133–141

146. Ator NA, Kautz MA (2000) Differentiating benzodiazepine- and barbiturate-like discriminative stimulus effects of lorazepam, diazepam, pentobarbital, imidazenil and zaleplon in two- versus three-lever procedures. Behav Pharmacol 11:1–14

147. York JL (1978) A comparison of the discriminative stimulus effects of ethanol, barbital, and phenobarbital in rats. Psychopharmacology (Berl) 60:19–23

148. Gold LH, Balster RL (1993) Effects of NMDA receptor antagonists in squirrel monkeys trained to discriminate the competitive NMDA receptor antagonist NPC 12626 from saline. Eur J Pharmacol 230:285–292

149. Tricklebank MD, Singh L, Oles RJ, Preston C, Iversen SD (1989) The behavioural effects of MK-801: a comparison with antagonists acting non-competitively and competitively at the NMDA receptor. Eur J Pharmacol 167:127–135

150. Witkin JM, Steele TD, Sharpe LG (1997) Effects of strychnine-insensitive glycine receptor ligands in rats discriminating dizocilpine or phencyclidine from saline. J Pharmacol Exp Ther 280:46–52

The Discriminative Stimulus Properties of Hallucinogenic and Dissociative Anesthetic Drugs

Tomohisa Mori and Tsutomu Suzuki

Abstract The subjective effects of drugs are related to the kinds of feelings they produce, such as euphoria or dysphoria. One of the methods that can be used to study these effects is the drug discrimination procedure. Many researchers have been trying to elucidate the mechanisms that underlie the discriminative stimulus properties of abused drugs (e.g., alcohol, psychostimulants, and opioids). Over the past two decades, patterns of drug abuse have changed, so that club/recreational drugs such as phencyclidine (PCP), 3,4-methylenedioxymethamphetamine (MDMA), ketamine, and cannabinoid, which induce perceptual distortions, like hallucinations, are now more commonly abused, especially in younger generations. In particular, the abuse of designer drugs, which aim to mimic the subjective effects of psychostimulants (e.g., MDMA or amphetamines), has been problematic. However, the mechanisms of the discriminative stimulus effects of hallucinogenic and dissociative anesthetic drugs are not yet fully clear. This chapter focuses on recent findings regarding hallucinogenic and dissociative anesthetic drug-induced discriminative stimulus properties in animals.

Keywords Discriminative stimulus properties • Hallucinogens • Psychedelics • Serotonin • Sigma-1 receptor

Contents

T. Mori and T. Suzuki (✉)
Department of Toxicology, Hoshi University School of Pharmacy and Pharmaceutical Sciences; Hoshi, Japan
e-mail: suzuki@hoshi.ac.jp

© Springer International Publishing Switzerland 2016
Curr Topics Behav Neurosci (2018) 39: 141–152
DOI 10.1007/7854_2016_29
Published Online: 1 September 2016

1   Introduction

The most important determinant of a substance's abuse potential is the nature of the subjective effects that are produced by the drug's influence on the central nervous system. Alcohol, psychostimulants – like methamphetamine and cocaine – and opioids – such as morphine and heroin – produce a drug state that includes feelings referred to as euphoria. Hallucinogens and dissociative anesthetics have also been misused or abused mainly for recreational drugs. With regard to the relationship between drug-induced subjective effects and abuse potential, animal models have been developed to study the components of action of abused drugs that bear on their subjective effects in humans. One method that has considerable potential in this regard is the drug discrimination procedure, which has been used to study the mechanisms that underlie the discriminative stimulus properties of abused drugs, and the similarities among the discriminative stimulus properties of abused drugs.

Use of the club drugs 3,4-methylenedioxymethamphetamine (MDMA), including new psychoactive substances, lysergic acid diethylamide (LSD), became popular in the past few decades. Phencyclidine (PCP) and ketamine, which induce perceptual distortions (e.g., hallucinations, illusions) and disordered thinking (e.g., paranoia), are classified as dissociative anesthetic drugs. *Salvia divinorum* contains salvinorin A, which is a selective k-opioid receptor agonist and has dissociative effects, has been misused [1, 2]. On the other hand, it has been proposed that hallucinogenic effects mediated by sigma-1 receptors [3] are closely related to NMDA receptors or serotonin receptors [4, 5]. Even though these hallucinogenic drugs sometimes induce psychotomimetic effects, which are closely related to bad trips and dysphoria in humans, they have been abused for at least two decades. Interestingly, these hallucinogenic/psychedelic drugs induce both rewarding and aversive effects, depending on the details of conditioning as measured by conditioned place preference procedures in animals. While the discriminative stimulus properties of a hallucinogenic drug may be responsible for or be related to its rewarding or aversive effects, it is not yet clear exactly how the discriminative stimulus properties of hallucinogenic drugs influence for their reinforcing or aversive effects [6, 7].

Hallucinogenic drugs can be divided into distinct classes according to their chemical structures and pharmacological actions. Since the discriminative stimulus properties of a hallucinogenic drug are believed to be mediated by receptor mechanisms thought to be important for hallucinogenic effects, these drugs might substitute for the discriminative stimulus properties of other drugs (e.g., the non-hallucinogenic compound lisuride at least partially substitutes for the discriminative stimulus properties of LSD, which are mediated by the activation of serotonergic $5\text{-}HT_{1A}$ and $5\text{-}HT_2$ receptors) [8–10]. In most cases, each type of hallucinogenic drug exerts distinct discriminative stimulus properties. Thus, the discriminative stimulus properties of a hallucinogenic drug depend on its hallucinogenic effects and/or mechanisms of action. Several recent reports have provided new insight into the mechanisms of the discriminative stimulus properties of hallucinogenic drugs. The present chapter focuses on the mechanism(s) of the discriminative stimulus of hallucinogenic/psychotomimetic drugs. Furthermore, the possible relationship between the discriminative stimulus properties of hallucinogenic drugs and their reinforcing or aversive effects in animals was also investigated.

2 Discriminative Stimulus Effects of 5-HT-Related Compounds

MDMA and LSD (and related compounds, such as the hallucinogenic derivatives of phenethylamine and tryptamine) are known to regulate serotonergic systems to induce hallucinogenic effects. MDMA mainly releases serotonin from nerve terminals, and to a lesser extent dopamine, and, thereby, produces an enhanced mood with increased well-being or dysphoria and perceptual changes (in addition to hallucinations, illusions, and disordered thinking) in humans. Additionally, a history of MDMA use may influence the subsequent vulnerability to the use and abuse of MDMA in humans. In rodents, a large and growing body of evidence suggests that MDMA can induce hyperlocomotion and reinforcing/rewarding, aversive and discriminative stimulus properties [11].

The serotonin receptor superfamily consists of 14 subtypes that have been classified based on gene structure, amino acid sequence homology, and intracellular signaling cascades, and at least seven families of serotonin receptors ($5\text{-}HT_1$, $5\text{-}HT_2$, $5\text{-}HT_3$, $5\text{-}HT_4$, $5\text{-}HT_6$, and $5\text{-}HT_7$) have been identified. Serotonin $5\text{-}HT_2$ and $5\text{-}HT_{1A}$ receptor agonists have opposite behavioral effects; however, activation of these receptors has a synergistic action on the locomotor activity induced by MDMA [12]. The synthetic tryptamine hallucinogen N,N-dipropyltryptamine partially to fully substitutes for the discriminative stimulus properties of hallucinogens like LSD, psilocybin, and MDMA, and LSD produces MDMA-like discriminative stimulus properties in rats [13], indicating that these 5-HT-related compounds show similar discriminative stimulus properties. $5\text{-}HT_{1A}$ receptor agonists exert MDMA-like discriminative stimulus properties; whereas, a $5\text{-}HT_{1A}$ receptor antagonist partially antagonizes the discriminative stimulus properties of MDMA in rats [14]. The activation of $5\text{-}HT_{1A}$ receptors elicits the stimulus properties of the tryptaminergic hallucinogen 5-MeO-DMT [15], indicating that the agonist actions of $5\text{-}HT_{1A}$ receptors play a role in the discriminative stimulus properties of serotonin-related hallucinogenic drugs. On the other hand, it has been clearly demonstrated that the activation of $5\text{-}HT_2$ receptors plays a significant role in the discriminative stimulus properties of LSD [15]. The discriminative stimulus properties of MDMA and LSD are more potently attenuated by 5-HT$_2$ receptor antagonists than by 5-HT$_{1A}$ receptor antagonists in rats [7]. The perceptual changes, emotional excitation, and adverse responses induced by MDMA are reduced by 5-HT$_2$ receptor antagonists in humans [16]. A more recent study showed that serotonin 5-HT$_2$ receptors are crucial for the reinforcing effects induced by MDMA [17]. These results indicate that the activation of 5-HT$_2$ receptor is an essential element of the discriminative stimulus properties and subjective effects of serotonin-related hallucinogenic drugs, which are closely related to their reinforcing and/or aversive effects, and that a 5-HT$_{1A}$-mediated component may have facilitatory functions [7].

It is well known that psychostimulants increase not only dopamine levels in the synaptic cleft of the terminals of the dopaminergic system, but also serotonin and noradrenaline levels. In humans, both methamphetamine and MDMA induce an increase in wakefulness and euphoria [18, 19], and it is difficult to discriminate between them based on their subjective effects in humans [20]. Thus, MDMA and other psychostimulants generally produce similar subjective effects in humans. Previous animal studies have shown that while cocaine does not substitute for the discriminative stimulus properties of MDMA, MDMA substitutes for the discriminative stimulus properties of cocaine [21]. Amphetamine partially substitutes for the discriminative stimulus properties of MDMA [22]. In contrast, MDMA does not substitute for the discriminative stimulus properties of methamphetamine [7]. In cross-substitution tests, MDMA and methylphenidate do not cross-substitute for each other in rats that have been trained to discriminate between MDMA or methylphenidate and saline [7], indicating that the discriminative stimulus properties of MDMA are distinctly different from those of other psychostimulants in rats. As mentioned above, the serotonergic system plays an important role in the discriminative stimulus properties of MDMA. However, a high dose of MDMA increases the release of dopamine, and may substitute for the discriminative stimulus properties of psychostimulants. Interestingly, recent research may provide an answer. Amphetamine substitutes for the discriminative stimulus properties of MDMA in rats that have been trained to discriminate between a high dose, but not a low dose, of MDMA and saline [23]. The discriminative stimulus properties of MDMA depend on the training doses (dopamine vs. 5-HT); lower doses of MDMA enhance serotonin, whereas higher doses of MDMA is required to enhance the dopamine release. In the case of humans, high dose of MDMA was associated with more drug-related problems [24], MDMA is frequently taken in combination with other substances to boost its effects [25]. Therefore, subjective effects of MDMA in humans are mainly mediated by the activation of serotonergic systems in the case of regular use. On the other hand, MDMA increases "negative" mood; whereas, methamphetamine enhances only "positive" mood in humans [20]. In fact, activation of dopaminergic system is partly involved in the euphoric effects of MDMA in humans [26]. In contrast, MDMA-induced perceptual changes and emotional excitation are mediated by serotonergic system [27]. Thus, MDMA and other psychostimulants, like methamphetamine, exert some overlapping and divergent effects. Particularly, serotonin-related subjective changes may explain why MDMA and other serotonin-related drugs are used recreationally.

3 Discriminative Stimulus Effects of PCP and k-Opioid Receptor Agonist

Ketamine and PCP, which are noncompetitive N-methyl-D-aspartate (NMDA) receptor antagonists that induce a dissociative anesthetic effect, produce psychotomimetic effects, such as nightmares, hallucinations, and delusions. Noncompetitive NMDA receptor antagonists, such as PCP and MK-801, but not the noncompetitive NMDA receptor antagonist 3-(2-carboxypiperazin-4-yl)propyl-1-phosphonic acid, partially substituted for the discriminative stimulus properties of the barbiturate pentobarbital [28]; whereas, noncompetitive NMDA receptor antagonists, but not competitive NMDA receptor agonists, substituted for the selective k-opioid receptor agonist 2-(3,4-dichlorophenyl)-N-methyl-N-[(1R,2R)-2-pyrrolidin-1-ylcyclohexyl]acetamide (U-50,488H) [29]. These findings suggest that the discriminative stimulus properties of competitive and noncompetitive NMDA receptor antagonists are different from each other. Further, the spectrum of behaviors induced by competitive and noncompetitive NMDA-receptor antagonists is totally different: PCP and MK-801 induce potent hyperlocomotion with ataxia, which might be related to the induction of the psychotomimetic effects of these drugs [30], whereas competitive NMDA receptor antagonists induce sedation. Since PCP, like ketamine, is not selective for NMDA receptors (i.e., PCP and ketamine can regulate the dopaminergic and serotonergic systems and sigma-1 receptor function), it is likely that several components might be involved in the cue of the discriminative stimulus properties of NMDA receptor antagonist in animals; thus, representing a "compound" or "complex" discriminative cue.

k-opioid receptors are widely distributed in regions in the brain that are closely related to rewarding effects, aversive effects, mood and cognitive functions, such as the ventral tegmental area, substantia nigra, nucleus accumbens, striatum, amygdala, locus coeruleus, hypothalamus, and dorsal raphe nucleus in human and rat brains, and are also located in the spinal cord and peripheral tissues [28], which suggests that k-opioid receptor ligands may regulate many functions in the brain. Previous studies have shown that k-opioid receptor agonists exert antinociceptive effects without producing robust reinforcing or rewarding effects. Further, k-opioid receptor agonists exert antinociceptive effects without producing robust reinforcing/rewarding effects. On the other hand, the k-opioid receptor agonist spiradoline causes sedation and dysphoria but no euphoria [31], whereas enadoline induces feelings of depersonalization in humans [32]. Furthermore, Salvinorin A also produces strong dissociative effects and memory impairment, which only partially overlap with classic hallucinogen effects [1]. Therefore, k-opioid receptor agonists produce hallucinogenic effects and dysphoria [31, 29]. Most k-opioid receptor agonists, including salvinorin A, but not the μ-opioid receptor agonists morphine or fentanyl or the δ-opioid receptor agonist SNC80, can substitute for the discriminative stimulus properties of the prototypic k-opioid receptor agonists U50,488H and U69593 [29, 33, 34]. These previous findings indicate that the cue of the discriminative stimulus properties of k-opioid receptor agonists is not shared by the discriminative stimulus properties of other opioid receptor agonists, and closely linked to dysphoric (aversive) effects.

PCP and MK-801 substitute for the discriminative stimulus properties of U50,488H [29]. Furthermore, the discriminative stimulus properties of U50,488H, the substitution of PCP for the discriminative stimulus properties of U50,488H, and the discriminative stimulus properties of ketamine were significantly blocked by the sigma-1 receptor antagonist NE-100 ([35, 36]; for an overview of sigma-1 receptors, see next section). On the other hand, sigma-1 receptor agonists such as (+)-pentazocine and SKF10,047 completely substituted for the discriminative stimulus properties of U50,488H [35], indicating that the discriminative stimulus properties of k-opioid receptor agonists and the k-opioid receptor agonist-like discriminative stimulus properties of noncompetitive NMDA receptor antagonists are at least in part mediated by sigma-1 receptors. It should be noted here that partial substitution of fluvoxamine, which has sigma-1 receptor agonistic action [37], for the discriminative stimulus properties of MDMA was completely suppressed by NE-100. Thus, a sigma-1 receptor agonist, k-opioid receptor agonist, and noncompetitive NMDA receptor antagonist-related cue may be related to psychotomimetic-like discriminative stimulus properties.

4 Hallucination and Sigma-1 Receptors

The sigma-1 receptor agonist SKF10,047 produces hallucinogenic/psychotomimetic effects. U50,488H-induced aversive effects, which are related to its psychotomimetic potential, are completely suppressed by sigma-1 receptor antagonist [35]. Further, it was believed that the hallucinogenic effects of PCP were mediated by sigma-1 receptors. Sigma-1 receptors are specifically localized at the interface between endoplasmic reticulum (ER) and mitochondria, the so-called mitochondria-associated ER membrane (MAM) inside the ER, and regulate Ca^{2+} signaling by stabilizing 1,4,5-triphosphate (IP_3) receptors as an ER chaperone protein [38]. The activity of sigma-1 receptors could be reciprocally inhibited by an association with binding immunoglobulin protein (BiP) through the formation of a sigma-1 receptor-BiP complex. Sigma-1 receptor agonists binding to sigma-1 receptors could exhibit chaperone activity by breaking the tether of the sigma-1 receptor-BiP complex [39], and enhance the Ca^{2+} through IP_3 receptors [40]. On the other hand, a sigma-1 receptor agonist may cause a translocation of sigma-1 receptor from the MAM to the plasma membrane where the sigma-1 receptor may bind to receptors (D_1 or NMDA-receptor) or ion channels (e.g., Kv1.2 channel)

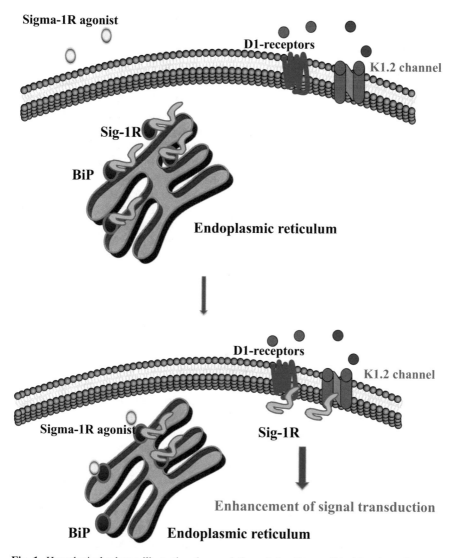

Fig. 1 Hypothetical scheme illustrating the regulation of signaling mediated by sigma-1 receptors. Sigma-1 receptors at the mitochondrion-associated endoplasmic reticulum (ER) function as ligand-activated molecular chaperones. Sig-1R agonists cause the dissociation of Sig-1Rs from another ER chaperone, binding immunoglobulin protein (BiP), allowing translocation of Sig-1Rs from ER to G-protein couples receptors and/or channels to regulate their signal transduction

that are regulating the signaling [41–43]. Recently, the endogenous hallucinogenic amine N,N-dimethyltryptamine (DMT) was shown to be an endogenous sigma-1 receptor ligand, and DMT and sigma-1 receptor agonists were shown to induce the dissociation of sigma-1 receptors from the sigma-1 receptor-BiP complex [3]. As noted above, the sigma-1 receptor antagonist NE-100 significantly attenuated the discriminative stimulus properties of U-50,488H and the U-50,488H-like discriminative effects of PCP. However, the mechanism that underlies the involvement of sigma-1 receptors in the discriminative stimulus properties of U50,488H and the U50,488H-like discriminative stimulus properties of PCP remains unclear. One possibility is that k-opioid receptor agonists [34] as well as PCP [44] can activate extracellular signal-regulated kinase (ERK), and this activation of ERK induces the up-regulation of sigma-1 receptors [45]. Sigma-1 receptors translocated from the ER to the cellular membrane by sigma-1 receptor agonists negatively or positively regulate Src kinase, dopamine D_1 receptors, neurotropic tyrosine kinase receptor type 2 (TrkB), NMDA receptors, and $Kv_{1,2}$ channels [43, 46] (see Fig. 1). Such intracellular events might be involved in the psychotomimetic-like discriminative stimulus properties. Taken together, these results suggest that k-opioid receptor agonists and noncompetitive NMDA receptor antagonists may regulate endogenous sigma-1 receptor systems by regulating DMT, which induces a hallucinogenic effect. Therefore, the release of DMT by k-opioid receptor agonists and noncompetitive NMDA receptor antagonists should be addressed in future research.

5   Conclusion

Serotonin-related compounds and noncompetitive NMDA receptor antagonists/k-opioid receptor agonists induce hallucinations in humans and discriminative properties and reinforcing and aversive effects in animals. Previous studies have indicated that the activation of $5-HT_2$ receptors plays a role in the discriminative stimulus properties of U50,488H, PCP, MDMA, and LSD in animals [15, 47]. Even though these hallucinogenic drugs induce similar behavioral phenotypes in some cases, each type of drug exerts different discriminative stimulus properties by regulating different receptors and signals. LSD and MDMA do not substitute for the discriminative stimulus properties of PCP in rats [15]. The discriminative stimulus properties of PCP were diminished by combination with LSD or MDMA in rats, presumably due to masking effects. A recent study showed that MDMA can regulate the endogenous k-opioid system mediated by the activation of $5-HT_2$ receptors [48]. Therefore, it is possible that the hallucinogenic effects of U50,488H, PCP, MDMA, and LSD are mediated, at least in part, through the activation of $5-HT_2$ receptors followed by sigma-1 receptors. While these drugs share some similarities in their mechanism of action, they differ with regard to the cue of their discriminative stimulus properties. On the other hand, tetrahydrocannabinol induced more robust cognitive impairment than MDMA, and their co-administration did not exacerbate the effects of either drug alone on cognitive function. However, the co-administration of tetrahydrocannabinol with MDMA increased subjective drug effects and drug strength compared with MDMA alone, which may explain the widespread use of this combination [49]. MDMA did not induce cannabinoid-like discriminative stimulus properties in rats [50]. These results suggest that cannabinoid receptor agonist has distinct discriminative stimulus properties compared to its serotonergic-related effects. It should be noted here that humans can recognize hallucinogenic as a subjective effects induced by drugs. Nobody knows that animals could recognize whether they are having a hallucination or hallucinogenic drug-induced discriminative stimulus properties are related to hallucinogenic state, however, hallucinogenic and dissociative anesthetic drugs induce abnormal behaviors (e.g., head weaving, head-twitching, and ataxia). Furthermore, little is known about the specific regions that may part in the discriminative stimulus effects of hallucinogenic and dissociative anesthetic drugs. Such future findings may give us a better understanding of the underlying mechanisms of the discriminative stimulus effects of hallucinogenic and dissociative anesthetic drugs.

In conclusion, most hallucinogenic/psychotomimetic drugs induce distinct discriminative stimulus properties in animals, which may be related to their reinforcing or aversive effects. It is well known that most hallucinogenic drugs induce euphoria as well as dysphoria in humans depending on the situation. Thus, the discriminative stimulus properties of hallucinogens provide a reliable tool for investigating the subjective effects in humans. The discriminative stimulus properties of hallucinogenic drugs can be classified based on the underlying mechanism by which they exert their effects, such as whether they are mediated by 5-HT_2/ sigma-1 (even though these receptors might be cross-linked). Based on previous results, the mechanisms of the discriminative stimulus properties of hallucinogenic drugs are related, at least partially, to their aversive effects. Interestingly, we recently interviewed 10 ex-polydrug abusers who were undergoing rehabilitation and asked them about the difference between the subjective effects of methamphetamine and hallucinogens, such as MDMA and cannabinoid. All of them stated that the subjective effects of MDMA and cannabinoid are totally different from those of methamphetamine, and there is no relapse for MDMA or cannabinoid, unlike in the case of methamphetamine. It is unclear how hallucinogenic effects may induce aversive and reinforcing effects accompanied by subjective effects/ discriminative stimulus. MDMA was not a potent reinforcer in a self-administration study [6]; the ex-polydrug abusers mentioned above stated that they just enjoy the hallucination. Further research should address these points.

Acknowledgement This work was supported in part by grants for Research on Regulatory Science of Pharmaceuticals and Medical Devices from the Ministry of Health, Labour and Welfare, Japan (MHLW) to TS and/or TM, and by JSPS KAKENHI Grant Number 15 K07977.

References

1. MacLean KA, Johnson MW, Reissig CJ, Prisinzano TE, Griffiths RR (2013) Dose-related effects of salvinorin A in humans: dissociative, hallucinogenic, and memory effects. Psychopharmacology (Berl) 226:381–392
2. Sami M, Piggott K, Coysh C, Fialho A (2015) Psychosis, psychedelic substance misuse and head injury: a case report and 23 year follow-up. Brain Inj 29:1383–1386

3. Su TP, Hayashi T, Vaupel DB (2009) When the endogenous hallucinogenic trance amine N, N-dimethyltryptamine meets the sigma-1 receptor. Sci Signal 2:pe12

4. Cozzi NV, Gopalakrishnan A, Anderson LL, Feih JT, Shulgin AT, Daley PF, Ruoho AE (2009) Dimethyltryptamine and other hallucinogenic tryptamines exhibit substrate behavior at the serotonin uptake transporter and the vesicle monoamine transporter. J Neural Transm (Vienna) 116:1591–1599

5. Pabba M, Wong AY, Ahlskog N, Hristova E, Biscaro D, Nassrallah W, Ngsee JK, Snyder M, Beique JC, Bergeron R (2014) NMDA receptors are upregulated and trafficked to the plasma membrane after sigma-1 receptor activation in the rat hippocampus. J Neurosci 34:11325–11338

6. De La Garza R II, Fabrizio KR, Gupta A (2007) Relevance of rodent models of intravenous MDMA self-administration to human MDMA consumption patterns. Psychopharmacology (Berl) 189:425–434

7. Mori T, Uzawa N, Kazawa H, Watanabe H, Mochizuki A, Shibasaki M, Yoshizawa K, Higashiyama K, Suzuki T (2014) Differential substitution for the discriminative stimulus effects of 3,4-methylenedioxymethamphetamine and methylphenidate in rats. J Pharmacol Exp Ther 350(2):403–11

8. Fiorella D, Rabin RA, Winter JC (1995) Role of 5-HT2A and 5-HT2C receptors in the stimulus effects of hallucinogenic drugs. II: reassessment of LSD false positives. Psychopharmacology (Berl) 121:357–363

9. Marona-Lewicka D, Kurrasch-Orbaugh DM, Selken JR, Cumbay MG, Lisnicchia JG, Nichols DE (2002) Re-evaluation of lisuride pharmacology: 5-hydroxytryptamine 1A receptor-mediated behavioral effects overlap its other properties in rats. Psychopharmacology (Berl) 164:93–107

10. Nielsen EB (1985) Discriminative stimulus properties of lysergic acid diethylamide in the monkey. J Pharmacol Exp Ther 234:244–249

11. Cole JC, Sumnall HR (2003) The pre-clinical behavioural pharmacology of 3,4-methylenedioxymethamphetamine (MDMA). Neurosci Biobehav Rev 27:199–217

12. Mori T, Ito S, Kuwaki T, Yanagisawa M, Sakurai T, Sawaguchi T (2010) Monoaminergic neuronal changes in orexin-deficient mice. Neuropharmacology 58:826–832

13. Fantegrossi WE, Reissig CJ, Katz EB, Yarosh HL, Rice KC, Winter JC (2008) Hallucinogen-like effects of N, N-dipropyltryptamine (DPT): possible mediation by serotonin 5-HT1A and 5-HT2A receptors in rodents. Pharmacol Biochem Behav 88(3):358–65

14. Glennon RA, Young R (2000) MDMA stimulus generalization to the 5-HT(1A) serotonin agonist 8-hydroxy-2-(di-n-propylamino)tetralin. Pharmacol Biochem Behav 66:483–488

15. Winter JC (2009) Hallucinogens as discriminative stimuli in animals: LSD, phenethylamines, and tryptamines. Psychopharmacology (Berl) 203:251–263

16. van Wel JH, Kuypers KP, Theunissen EL, Bosker WM, Bakker K, Ramaekers JG (2012) Effects of acute MDMA intoxication on mood and impulsivity: role of the 5-HT2 and 5-HT1 receptors. PLoS One 7, e40187

17. Orejarena MJ, Lanfumey L, Maldonado R, Robledo P (2011) Involvement of 5-HT2A receptors in MDMA reinforcement and cue-induced reinstatement of MDMA-seeking behaviour. Int J Neuropsychopharmacol 14:927–940

18. Cami J, Farré M, Mas M, Roset PN, Poudevida S, Mas A, San L, de la Torre R (2000) Human pharmacology of 3,4-methylenedioxymethamphetamine ("ecstasy"): psychomotor performance and subjective effects. J Clin Psychopharmacol 20:455–466

19. Tancer M, Johanson CE (2003) Reinforcing, subjective, and physiological effects of MDMA in humans: a comparison with D-amphetamine and mCPP. Drug Alcohol Depend 72:33–44

20. Kirkpatrick MG, Gunderson EW, Perez AY, Haney M, Foltin RW, Hart CL (2012) A direct comparison of the behavioral and physiological effects of methamphetamine and 3,4-methylenedioxymethamphetamine (MDMA) in humans. Psychopharmacology (Berl) 219:109–122

21. Bondareva T, Wesołowska A, Dukat M, Lee M, Young R, Glennon RA (2005) S(+)- and R(−) N-methyl-1-(3,4-methylenedioxyphenyl)-2-aminopropane (MDMA) as discriminative stimuli: effect of cocaine. Pharmacol Biochem Behav 82:531–538

22. Goodwin AK, Pynnonen DM, Baker LE (2003) Serotonergic-dopaminergic mediation of MDMA's discriminative stimulus effects in a three-choice discrimination. Pharmacol Biochem Behav 74:987–995
23. Harper DN, Langen AL, Schenk S (2014) A 3-lever discrimination procedure reveals differences in the subjective effects of low and high doses of MDMA. Pharmacol Biochem Behav 116:9–15
24. Parrott AC (2005) Chronic tolerance to recreational MDMA (3,4-methylenedioxymethamphetamine) or Ecstasy. J Psychopharmacol 19:71–83
25. Mohamed WM, Ben Hamida S, Cassel JC, de Vasconcelos AP, de Jones BC (2011) MDMA: interactions with other psychoactive drugs. Pharmacol Biochem Behav 99(4):759–74
26. Liechti ME, Vollenweider FX (2000) Acute psychological and physiological effects of MDMA ("Ecstasy") after haloperidol pretreatment in healthy humans. Eur Neuropsychopharmacol 10:289–295
27. Liechti ME, Saur MR, Gamma A, Hell D, Vollenweider FX (2000) Psychological and physiological effects of MDMA ("Ecstasy") after pretreatment with the 5-HT(2) antagonist ketanserin in healthy humans. Neuropsychopharmacology 23:396–404
28. Willetts J, Balster RL (1989) Pentobarbital-like discriminative stimulus effects of N-methyl-D-aspartate antagonists. J Pharmacol Exp Ther 249:438–443
29. Mori T, Nomura M, Yoshizawa K, Nagase H, Sawaguchi T, Narita M et al (2006) Generalization of NMDA-receptor agonists U-50,488H, but not TRK-820 in rats. J Pharmacol Sci 100:157–161
30. Mori T, Baba J, Ichimaru Y, Suzuki T (2000) Effects of rolipram, a selective inhibitor of phosphodiesterase 4, on hyperlocomotion induced by several abused drugs in mice. Jpn J Pharmacol 83:113–118
31. Rimoy GH, Wright DM, Bhaskar NK, Rubin PC (1994) The cardiovascular and central nervous system effects in the human of U-62066E. A selective opioid receptor agonist. Eur J Clinic Pharmacol 46:203–207
32. Walsh SL, Strain EC, Abreu ME, Bigelow GE (2001) Enadoline, a selective kappa opioid agonist: comparison with butorphanol and hydromorphone in humans. Psychopharmacology (Berl) 157:151–162
33. Baker LE, Panos JJ, Killinger BA, Peet MM, Bell LM, Haliw LA et al (2009) Comparison of the discriminative stimulus effects of salvinorin A and its derivatives to U69,593 and U50,488 in rats. Psychopharmacology (Berl) 203:203–211
34. Yoshizawa K, Narita M, Saeki M, Narita M, Isotani K, Horiuchi H et al (2011) Activation of extracellular signal-regulated kinase is critical for the discriminative stimulus effects induced by U-50,488H. Synapse 65:1052–1061
35. Mori T, Yoshizawa K, Nomura M, Isotani K, Torigoe K, Tsukiyama Y et al (2012) Sigma-1 receptor function is critical for both the discriminative stimulus and aversive effects of the kappa-opioid receptor agonist U-50,488H. Addict Biol 17:717–724
36. Narita M, Yoshizawa K, Aoki K, Takagi M, Miyatake M, Suzuki T (2001) A putative sigma1 receptor antagonist NE-100 attenuates the discriminative stimulus effects of ketamine in rats. Addict Biol 6:373–376
37. Narita N, Hashimoto K, Tomitaka S, Minabe Y (1996) Interactions of selective serotonin reuptake inhibitors with subtypes of sigma receptors in rat brain. Eur J Pharmacol 20 (307):117–119
38. Hayashi T, Su TP (2007) Sigma-1 receptor chaperones at the ER-mitochondrion interface regulate Ca(2+) signaling and cell survival. Cell 131:596–610
39. Fujimoto M, Hayashi T, Urfer R, Mita S, Su TP (2012) Sigma-1 receptor chaperones regulate the secretion of brain-derived neurotropic factor. Synapse 66:630–639
40. Hayashi T, Su TP (2001) Regulating ankyrin dynamics: roles of sigma-1 receptors. Proc Natl Acad Sci U S A 98:491–496
41. Buch S, Yao H, Guo M, Mori T, Su TP, Wang J (2011) Cocaine and HIV-1 interplay: molecular mechanisms of action and addiction. J Neuroimmune Pharmacol 6:503–515

42. Hayashi T, Tsai SY, Mori T, Fujimoto M, Su TP (2011) Targeting ligand-operated chaperone sigma-1 receptors in the treatment of neuropsychiatric disorders. Expert Opin Ther Targets 15:557–577
43. Kourrich S, Hayashi T, Chuang JY, Tsai SY, Su TP, Bonci A (2013) Dynamic interaction between sigma-1 receptor and Kv1.2 shapes neuronal and behavioral responses to cocaine. Cell 152:236–247
44. Kyosseva SV, Owens SM, Elbein AD, Karson CN (2001) Differential and region-specific activation of mitogen-activated protein kinases following chronic administration of phency-clidine in rat brain. Neuropsychopharmacology 24:267–277
45. Cormaci G, Mori T, Hayashi T, Su TP (2007) Protein kinase A activation down-regulates, whereas extracellular signal-regulated kinase activation up-regulates sigma-1 receptors in B-104 cells: implication for neuroplasticity. J Pharmacol Exp Ther 320:202–210
46. Kourrich S, Su TP, Fujimoto M, Bonci A (2012) The sigma-1 receptor: roles in neuronal plasticity and disease. Trends Neurosci 35:762–771
47. Bronson ME, Lin YP, Burchett K, Picker MJ, Dykstra LA (1993) Serotonin involvement in the discriminative stimulus effects of kappa opioids in pigeons. Psychopharmacology (Berl) 111:69–77
48. Di Benedetto M, Bastias Candia Sdel C, D'Addario C, Porticella EE, Cavina C, Candeletti S et al (2011) Regulation of opioid gene expression in the rat brainstem by 3,4-methylenedioxymethamphetamine (MDMA): role of serotonin and involvement of CREB and ERK cascade. Naunyn Schmiedebergs Arch Pharmacol 383:169–178
49. Dumont GJ, van Hasselt JG, de Kam M, van Gerven JM, Touw DJ, Buitelaar JK et al (2011) Acute psychomotor, memory and subjective effects of MDMA and THC co-administration over time in healthy volunteers. J Psychopharmacol 25:478–489
50. Barrett RL, Wiley JL, Balster RL, Martin BR (1995) Pharmacological specificity of delta 9-tetrahydrocannabinol discrimination in rats. Psychopharmacology (Berl) 118:419–424

Discriminative Stimulus Properties of Phytocannabinoids, Endocannabinoids, and Synthetic Cannabinoids

Jenny L. Wiley, R. Allen Owens, and Aron H. Lichtman

Abstract Psychoactive cannabinoids from the marijuana plant (phytocannabinoids), from the body (endocannabinoids), and from the research lab (synthetic cannabinoids) produce their discriminative stimulus effects by stimulation of CB_1 receptors in the brain. Early discrimination work with phytocannabinoids confirmed that Δ^9-tetrahydrocannabinol (Δ^9-THC) is the primary psychoactive constituent of the marijuana plant, with more recent work focusing on characterization of the contribution of the major endocannabinoids, anandamide and 2-arachidonoylglycerol (2-AG), to Δ^9-THC-like internal states. Collectively, these latter studies suggest that endogenous increases in both anandamide and 2-AG seem to be optimal for mimicking Δ^9-THC's discriminative stimulus effects, although suprathreshold concentrations of anandamide also appear to be Δ^9-THC-like in discrimination assays. Recently, increased abuse of synthetic cannabinoids (e.g., "fake marijuana") has spurred discrimination studies to inform regulatory authorities by predicting which of the many synthetic compounds on the illicit market are most likely to share Δ^9-THC's abuse liability. In the absence of a reliable model of cannabinoid self-administration (specifically, Δ^9-THC self-administration), cannabinoid discrimination represents the most validated and pharmacologically selective animal model of an abuse-related property of cannabinoids – i.e., marijuana's subjective effects. The influx of recent papers in which cannabinoid discrimination is highlighted attests to its continued relevance as a valuable method for scientific study of cannabinoid use and abuse.

J.L. Wiley (✉)
RTI International, 3040 Cornwallis Road, Research Triangle Park, NC 27709, USA
e-mail: jwiley@rti.org

R.A. Owens and A.H. Lichtman
Department of Pharmacology and Toxicology, Virginia Commonwealth University, Box 980613, Richmond, VA, USA

© Springer International Publishing Switzerland 2016
Curr Topics Behav Neurosci (2018) 39: 153–174
DOI 10.1007/7854_2016_24
Published Online: 8 June 2016

Keywords 2-Arachidonoylglycerol • Anandamide • Cannabinoids •
Endocannabinoids • FAAH inhibitors • MAGL inhibitors • Synthetic
cannabinoids • Δ^9-THC

Contents

1   Introduction

Cannabinoids are chemicals derived primarily from three sources: plants of the
Cannabis genus (phytocannabinoids), the body (endocannabinoids), and laboratories (synthetic cannabinoids). Despite their disparate origins and different structural
templates (Fig. 1), psychoactive cannabinoids bind to and activate cannabinoid CB_1
receptors, which are found in largest concentrations in the brain [1]. This mechanism underlies their ability to serve as discriminative stimuli [2]. Many cannabinoids also activate cannabinoid CB_2 receptors [3], which are primarily, but not
exclusively [4, 5], located in the periphery [6]. The sections below take a critical
look at preclinical cannabinoid discrimination research, in which agonists from
each cannabinoid source and cannabinoid antagonists were trained as discriminative stimuli. A concluding section discusses the translational implications of this
research.

2   Phytocannabinoids

In the 1960s, Mechoulam and colleagues [7] identified Δ^9-tetrahydrocannabinol
(Δ^9-THC) as the primary psychoactive constituent of *Cannabis sativa*. Although
Δ^9-THC is largely responsible for the subjective "high" experienced by users, the
marijuana plant contains many other psychoactive and inactive cannabinoid substances, including cannabinol, cannabidiol, and Δ^8-THC [8, 9]. Major
non-cannabinoid constituents of the plant include terpenoids [10]. The influence

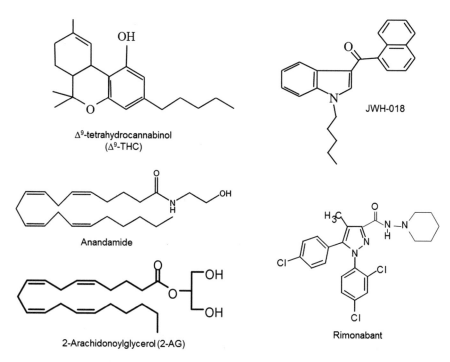

Fig. 1 Chemical structures of cannabinoids from each of the three sources: phytocannabinoids (Δ^9-THC), synthetic cannabinoids (agonist: JWH-018; antagonist: rimonabant), and endocannabinoids (anandamide and 2-AG)

of interaction(s) of these cannabinoid and noncannabinoid constituents on the pharmacological effects of Δ^9-THC has not been completely determined. Given the recent loosening of legal restrictions on use of marijuana for medicinal or recreational purposes, this area is ripe for further research.

To date, preclinical drug discrimination research with phytocannabinoids has focused almost exclusively on Δ^9-THC as a discriminative stimulus, with only a few studies reporting use of other phytocannabinoids as training drugs (e.g., [11]). Δ^9-THC's ability to serve as a discriminative stimulus has been demonstrated in rats [12], gerbils [13], pigeons [14], mice [15, 16], nonhuman primates [17], and humans [18]. Psychoactivity rests in the (−)-isomer, as (+)-Δ^9-THC does not substitute for Δ^9-THC [19]. Early studies showed that some phytocannabinoids, including Δ^8-THC, $\Delta^{9,11}$-THC, and cannabinol, substituted in Δ^9-THC-trained animals [20–22], as did Δ^9-THC's 11-hydroxy metabolites [20]. In contrast, cannabidiol, a non-psychoactive phytocannabinoid that binds to CB_1 receptors only with high micromolar affinity [23], did not generalize to Δ^9-THC [20, 24]. Later research showed that phytocannabinoid potency for producing Δ^9-THC-like discriminative stimulus effects was associated with binding affinity for CB_1 receptors in the brain [25], suggesting CB_1 receptor mediation of Δ^9-THC's discriminative stimulus effects. This hypothesis received additional support from the finding that the prototypic CB_1 receptor antagonist rimonabant shifted the Δ^9-THC substitution dose-effect curve in Δ^9-THC-trained rats to the right [2].

Δ^9-THC's discriminative stimulus effects exhibit pharmacological selectivity for other classes of psychoactive cannabinoid agonists [17, 26, 27]. -Non-cannabinoid drugs generally fail to substitute [28]. Further, Δ^9-THC discrimination is considered a reliable animal model of marijuana intoxication [29]. As such, this model has been used to explore the physiological underpinnings of this intoxication [30] and to screen for marijuana-like abuse liability [31], as reviewed below (see Sects. 3 and 4, respectively).

3   Endocannabinoids

Discovery of CB_1 and CB_2 cannabinoid receptors in the late 1980s/early 1990s resulted in efforts to identify the endogenous substance(s) that activated these receptors, a drive that led eventually to discovery and characterization of the endocannabinoid system. In the brain, this system is one of the several lipid signaling systems and is comprised of the cannabinoid receptors, their signaling pathways, two predominant endogenous ligands, and synthetic and metabolic pathways for these endocannabinoids. To date, drug discrimination research has focused on examination of the endogenous ligands, anandamide [32] and 2-arachidonoylglycerol (2-AG) [33], and their respective primary metabolic enzymes, fatty acid amide hydrolase (FAAH) [34] and monoacylglycerol lipase (MAGL) [35].

3.1   *Anandamide and Anandamide Analogs*

Shortly after anandamide's discovery, investigators attempted to train rats to discriminate anandamide, an effort that failed [30, 36]. Subsequent evaluation of anandamide substitution in animals trained to discriminate Δ^9-THC or CP55,940 was inconclusive, with some studies reporting substitution [36, 37] and others reporting failure to substitute [38–40]. Similarly, systemic injection with 2-AG also failed to substitute for Δ^9-THC in a Δ^9-THC discrimination in rodents [40]. Failure of these efforts to establish a profile of the discriminative stimulus effects of endocannabinoids was attributed to their rapid metabolism [35, 41–43]. For anandamide, this hypothesis received tentative support through the finding that anandamide substituted for Δ^9-THC in Δ^9-THC-trained mice when co-administered with the nonselective amidase inhibitor phenylmethyl sulfonyl fluoride [16]. However, since reliable and selective tools to inhibit endocannabinoid metabolism were not yet available when endocannabinoids were initially discovered, cannabinoid chemists turned to synthesis of metabolically stable analogs.

These chemicals allowed examination of structure–activity relationships with the goal of determining how endocannabinoids interacted with the same (CB_1) receptor as Δ^9-THC despite notable differences in their chemical structures [44, 45]. In addition, potential physiological roles of endocannabinoids were explored through behavioral observations following administration of the analogs to living animals, including evaluation of their discriminative stimulus effects. For example, studies reported that the methylated anandamide analogs, R-(+)-methanandamide, 2-methylarachidonyl-2'-fluoroethylamide (O-875), and arachidonylcyclopropylamide, fully substituted for Δ^9-THC in Δ^9-THC-trained rhesus monkeys [37, 46]. In Δ^9-THC-trained rodents, methylated anandamide analogs with a methyl group at C-1 of the ethanolamide constituent or at C-3 of the arachidonyl group also produced the highest degree of substitution, with minimal generalization of analogs with other types of substitutions, including saturation of the arachidonyl constituent, substitution for the ethanolamide constituent or for the terminal hydroxyl [38, 39, 47]. When it occurred, substitution in rats was sometimes accompanied by decreases in overall responding [39], suggesting overlap of the stimulus effects of these compounds with Δ^9-THC. Differences were also suggested by the finding that methanandamide substituted fully in rats trained to discriminate 3 mg/kg Δ^9-THC from vehicle [48, 49], but substituted only partially or not at all in rats trained to discriminate 5.6 or 30 mg/kg Δ^9-THC [48–50]. This association of higher training doses with greater specificity in drug discrimination procedures has been noted previously for other classes of drugs [51].

To delve further into the degree to which anandamide and Δ^9-THC share cannabimimetic discriminative stimulus effects, selected anandamide analogs that substituted for Δ^9-THC in Δ^9-THC-trained animals were used as training drugs in discrimination procedures. Unlike anandamide itself, these analogs showed a reasonable degree of discriminability, and studies in rodents reported successful acquisition of discriminations for methanandamide, O-1812, and AM-1346 [50, 52–55]. In rats, Δ^9-THC also substituted for the anandamide analog used as a training drug [52, 54, 55]. Anandamide engendered greater substitution for methanandamide than it did for Δ^9-THC in rodents trained to discriminate methanandamide or Δ^9-THC from vehicle, particularly when the interval between injection and testing was short (3 min) [53] or the methanandamide training dose was high (70 mg/kg) [50]. Despite these apparent similarities between the discriminative stimulus effects of Δ^9-THC and anandamide analogs, differences also emerged. For example, Δ^9-THC did not occasion responding on the methanandamide-associated lever at a high methanandamide training dose (70 mg/kg) in mice [50]. In addition, rimonabant antagonism of Δ^9-THC's discriminative stimulus effects in rats was surmountable, whereas its antagonism of methanandamide's discriminative stimulus effects (with 10 mg/kg training dose) was not [53]. These results suggest that Δ^9-THC and rimonabant are competitive for a binding site, but that methanandamide and rimonabant interact noncompetitively. Consistent with these other differences, the discriminative stimulus effects of the 70 mg/kg training dose of methanandamide in mice were not altered by rimonabant [50], suggesting that a non-CB_1 receptor mechanism may contribute to discrimination of high doses of methanandamide. Whether these differences are related to differences in training dose and/or species has not been determined; however, much of the recent cannabinoid discrimination research in rodents has been conducted in mice, primarily because of the facility with which this species is subject to genetic manipulation. For the most part, research with analogs of endocannabinoids in wildtype rodents has been abandoned in favor of studies with endocannabinoid metabolic enzyme inhibitors and in transgenic mice.

3.2   Endocannabinoid Discrimination in Transgenic Mice

As knowledge about the endocannabinoid system grew and the metabolic pathways for anandamide and 2-AG were delineated [35, 41, 42], transgenic mice became available and provided new opportunities to examine the potential contribution of endocannabinoid mechanisms to cannabinoid discrimination. The inability of CB_1 knockout mice to acquire a Δ^9-THC discrimination reinforced the hypothesis that the discriminative stimulus effects of psychoactive cannabinoids were mediated via activation of the CB_1 receptor [56], an idea that also received support from findings that CB_1 (but not CB_2) receptor antagonists blocked the discriminative stimulus effects of cannabinoids [57]. In mice devoid of either the catabolic enzyme FAAH or MAGL, brain levels of anandamide or 2-AG were elevated, respectively [58, 59]. $FAAH^{(-/-)}$ and $MAGL^{(-/-)}$ mice also exhibit distinct phenotypes that implicate endocannabinoid involvement in a number of physiological and behavioral processes, including pain [60, 61], seizures [62], learning and memory [63–65], and energy metabolism [66, 67].

To date, only $FAAH^{(-/-)}$ mice have been used in drug discrimination procedures. Whereas discriminability of anandamide has not been demonstrated in wildtype mice, it is readily discriminated in $FAAH^{(-/-)}$ mice in a two-lever milk-reinforced procedure [68] and in a water t-maze procedure [69]. Further, substitution of Δ^9-THC, but not the fatty acid amide oleamide, was observed in these mice. $FAAH^{(-/-)}$ mice have also been trained to discriminate Δ^9-THC in both procedures, with full cross-substitution of anandamide in these mice, but not in $FAAH^{(+/+)}$ mice tested in parallel experiments [69, 70]. While O-1812 substituted in both $FAAH^{(-/-)}$ and $FAAH^{(+/+)}$ mice, it was more potent in $FAAH^{(-/-)}$ mice [70], suggesting that the higher phenotypic brain levels of anandamide in these mice may have enhanced the cannabimimetic potency of this anandamide analog. Interestingly, co-administration of the nonspecific amidase inhibitor phenylmethyl sulfonyl fluoride, which would be predicted to increase levels of fatty acid amides such as anandamide, also enhanced the potency of Δ^9-THC and CP55,940 in Δ^9-THC discrimination in rats [71]. Rimonabant attenuation of Δ^9-THC and anandamide substitution in $FAAH^{(-/-)}$ mice trained to discriminate each drug suggests CB_1 receptor activation as a shared mechanism underlying their discriminative stimulus effects [68–70].

3.3 FAAH and MAGL Inhibitors

Recent synthesis of selective and dual FAAH and MAGL inhibitors which prevent the rapid hydrolysis of anandamide and/or 2-AG have opened up new opportunities for investigation of the psychoactive effects of endocannabinoids, particularly when combined with results from use of transgenic mice. These enzyme inhibitors produce substantial increases in brain levels of anandamide and/or 2-AG in mice [72–77]. Several selective and dual FAAH and MAGL inhibitors have been evaluated in drug discrimination alone and in combination with exogenously administered doses of anandamide or 2-AG.

Alone, selective FAAH inhibitors, URB-597 and PF-3845, and anandamide transport inhibitors, AM-404 and UCM-707, did not substitute in rodents trained to discriminate Δ^9-THC [40, 47, 78]. URB-597 also did not substitute for Δ^9-THC in Δ^9-THC-trained rhesus monkeys [79]. In combination with exogenously administered anandamide, however, URB-597 potentiated substitution of anandamide for Δ^9-THC such that full dose-dependent and rimonabant-reversible substitution was observed in rats and rhesus monkey [47, 79]. In mice, PF-3845, but not URB-597, also enhanced anandamide's efficacy in producing Δ^9-THC-like discriminative stimulus effects (maximum $= 64\%$ Δ^9-THC-lever responding), although full substitution still was not achieved [40]. Together, these results suggest that FAAH inhibition alone (with associated increases in endogenous anandamide) is not sufficient to engender Δ^9-THC-like discriminative stimulus effects without an additional influx of exogenous anandamide, suggesting that endogenous anandamide levels would not be high enough to produce Δ^9-THC-like subjective effects in humans.

As described throughout this section, most drug discrimination research with endocannabinoids has concentrated on anandamide, with less attention being given to 2-AG. The single study [40] in which 2-AG was evaluated in mice trained to discriminate Δ^9-THC reported that 2-AG did not substitute when administered alone or when co-administered with a dose of the nonselective MAGL inhibitor, N-arachidonylmaleimide [80]. A previous study had found that this dose enhanced other cannabimimetic effects of 2-AG in mice [81]. In recent years, pharmacological and genetic tools that selectively alter brain 2-AG levels have allowed emphasis to shift to examination of 2-AG's role in cannabinoid discriminative stimulus effects. While MAGL$^{(-/-)}$ mice have not yet been evaluated in a drug discrimination procedure, several MAGL inhibitors have been tested. Alone, the MAGL inhibitor JZL184 partially substituted for Δ^9-THC in wildtype mice or rats trained to discriminate Δ^9-THC from vehicle [40, 69, 70, 82], although one study reported that Δ^9-THC-trained mice responded almost solely on the vehicle-associated lever following JZL184 administration [78].

Results for other selective MAGL inhibitors have been mixed. Whereas KML29 did not substitute for Δ^9-THC in Δ^9-THC-trained wildtype mice [83], MJN110 produced full dose-dependent substitution in CP55,940-trained mice [84]. JZL184 also fully substituted for CP55,940 in these mice whereas it had only partially substituted for Δ^9-THC in previous Δ^9-THC discrimination studies, raising the possibility that efficacy of the cannabinoid used as the training drug might affect endocannabinoid generalization profiles (CP55,940 and Δ^9-THC as full and partial CB_1 receptor agonists, respectively). Rimonabant attenuation of substitution by MJN110 and JZL184 confirmed a role for CB_1 receptor mediation in the CP55,940-like discriminative stimulus effects of these MAGL inhibitors [84].

The pattern of inconsistent substitution of MAGL inhibitors alone in cannabinoid discrimination procedures spurred investigators to examine the effects of dual inhibition of FAAH and MAGL. Dual inhibition has been achieved in one of two ways: (1) administration of a MAGL inhibitor to $FAAH^{(-/-)}$ mice and (2) administration of dual FAAH and MAGL inhibitor(s) to wildtype mice. Using the first method, studies have reported that JZL184 fully substituted for Δ^9-THC in FAAH $^{(-/-)}$ mice trained to discriminate Δ^9-THC in a food-reinforced procedure [70, 82] and partially substituted for Δ^9-THC in a water maze discrimination procedure [69]. In $FAAH^{(-/-)}$ mice trained to discriminate anandamide, a similar pattern was observed, with full substitution of the MAGL inhibitor KML29 for anandamide in a food-reinforced procedure [83] and partial substitution of JZL184 for anandamide in a water maze procedure [69]. In wildtype mice trained to discriminate Δ^9-THC, full substitution was observed with combined administration of the FAAH inhibitor PF-3845 and the MAGL inhibitor JZL184, whereas the combination of another FAAH inhibitor URB-597 and JZL184 produced responding primarily on the vehicle-associated lever [78]. This latter effect may be species-specific or related to differences in the Δ^9-THC training dose because increased responding on the Δ^9-THC-associated lever following combined administration of URB-597 and JZL184 has been reported in rats trained to discriminate Δ^9-THC from vehicle [40]. Consistent substitution for Δ^9-THC in wildtype mice trained to discriminate Δ^9-THC from vehicle has been reported for dual FAAH/MAGL inhibitors, JZL195 and SA-57 [70, 78, 82]. Substitution of both compounds was attenuated by rimonabant, suggesting CB_1 receptor mediation.

More recently, SA-57 was trained as a discriminative stimulus in wildtype mice [85]. This observation is the first reported instance of an endocannabinoid metabolic enzyme inhibitor serving as a discriminative stimulus and provides future opportunities to investigate directly the psychoactive effects of endogenous cannabinoids. In addition to demonstrating dose- and time-dependent discriminative stimulus effects, the results showed cross-substitution for CP55,940 and SA-57, with CP55,940 producing full substitution in mice trained to discriminate SA-57 and SA-57 producing full substitution in mice trained to discriminate CP55,940. SA-57's discriminative stimulus effects were blocked by rimonabant, but not by SR144528, suggesting that CB_1, but not CB_2, receptors played a role in this novel discrimination. Interestingly, SA-57 also substituted for anandamide in $FAAH^{(-/-)}$ mice. Since these mice lack FAAH, these results suggest that the increased 2-AG produced by the compound's inhibition of MAGL shares discriminative stimulus effects with anandamide. The implications of this finding for functioning of the endocannabinoid system have not yet been fully delineated.

3.4 Endocannabinoid Discrimination Summary

In summary, anandamide and 2-AG interact in a complex manner to induce cannabimimetic discriminative stimulus effects. While anandamide has been reported to substitute for Δ^9-THC, it does so only upon exogenous administration, most often with concomitant inhibition of its metabolism by FAAH. FAAH inhibitors do not induce Δ^9-THC-like discriminative stimulus effects when administered alone. In contrast, endogenous increase in 2-AG concentration via administration of an MAGL inhibitor appears more effective in promoting discriminative stimulus effects similar to those produced by Δ^9-THC and CP55,940, although variability across compounds has been noted. Further, endogenous increases in both anandamide and 2-AG seem to be optimal for mimicking Δ^9-THC's discriminative stimulus effects. Given the relatively recent availability of $MAGL^{(-/-)}$ mice and selective pharmacological tools, additional insights into the ways in which the endocannabinoid system contributes to subjective states that resemble those produced by marijuana intoxication are likely to be forthcoming as research continues in this area.

4 Synthetic Cannabinoids

Synthetic cannabinoids are a class of novel psychoactive substances that were originally developed as research chemicals to probe cannabinoid receptors and to search for compounds with potential therapeutic use. In the early 2000s, they were diverted and started to appear on drug abuse monitoring sites in products labeled "Spice" or "herbal incense" [86]. As reviewed previously [87], these compounds produce Δ^9-THC-like intoxication in humans [88, 89] and engender Δ^9-THC-like discriminative stimulus effects in rodents and nonhuman primates [27, 90]. Binding and other pharmacological properties of these compounds have been reviewed elsewhere [91, 92]. Collectively, preclinical data on these compounds have been used to support drug policy decisions in the USA, including classification of synthetic cannabinoids as schedule 1.

Previous studies have shown that CP55,940 (a bicyclic cannabinoid) and WIN55,212-2 (an aminoalkylindole) dose-dependently substituted and cross-substituted for Δ^9-THC in rodents and nonhuman primates [15, 30]. Replacement of the morpholinoethyl group of WIN55,212-2 with a pentyl chain resulted in JWH-018 (1-pentyl-3-1-naphthoylindole), the first synthetic cannabinoid identified in a Spice product. Early studies demonstrated that JWH-018 and other indole- and pyrrole-derived synthetic cannabinoids exhibit orderly structure–activity relationships in binding assays and in a battery of pharmacological tests in mice, with good correlations between CB_1 receptor binding affinities and potencies for centrally mediated cannabinoid effects [90, 93, 94]. To the extent that these compounds were tested in discrimination procedures, they also were shown to dose-dependently substitute for psychoactive cannabinoids (Δ^9-THC, CP55,940, or methanandamide) in rats and rhesus monkeys [54, 90, 95–98], again with potencies that were consistent with their CB_1 receptor affinities. Further, substitution of JWH-018 and JWH-073 occurred following inhalation in mice trained to discriminate intraperitoneal Δ^9-THC from vehicle [99]. For some naphthoylindoles, duration of their Δ^9-THC-like discriminative stimulus effects appeared to differ from that of Δ^9-THC itself [95, 96].

JWH-018 and other structural variants of this compound comprised the first major wave of synthetic cannabinoids to be diverted for abuse. As these compounds were systematically banned, other compounds based on different structural templates quickly took their place. For example, substitution of a phenylacetyl group for the naphthoyl substituent of JWH-018 resulted in a series of phenylacetylindoles [100, 101]. Several of these compounds (JWH-203, JWH-204, JWH-205, JWH-250) have high CB_1 receptor affinity and were shown to substitute for Δ^9-THC in rodents trained to discriminate Δ^9-THC from vehicle [16, 95]; however, another compound (JWH-202) that had low CB_1 receptor affinity did not substitute [16]. XLR-11 and UR-144 are two additional compounds that have appeared on the illicit market within the last few years [102]. These tetramethylcyclopropyl ketone indoles were derived from a series described by Abbott Laboratories [103, 104] and both of these compounds produced dose-dependent substitution for Δ^9-THC in mice [31] and in rats [105]. Further, their Δ^9-THC-like discriminative stimulus effects were attenuated by co-administration of rimonabant [31], suggesting CB_1 receptor mediation.

While XLR-11 and UR-144 still occasionally appear in confiscated samples in the USA, indazole cannabinoids (e.g., AB-CHMINACA, AB-PINACA, AB-FUBINACA) have largely replaced naphthoylindoles and their derivatives as the most common compounds identified in recent samples [106–108]. As would be predicted by their high affinity for CB_1 receptors, AB-CHMINACA, AB-PINACA, and AB-FUBINACA fully substituted for Δ^9-THC in rodents trained to discriminate Δ^9-THC from vehicle [105, 109], as did the quinolinyl carboxylates, PB-22 (1-pentyl-1H-indole-3-carboxylic acid 8-quinolinyl ester) and 5F-PB-22, and the amantane-derived indole AKB-48 [105]. In contrast, the benzimidazole FUBIMINA only partially substituted for Δ^9-THC in mouse drug discrimination, which is consistent with its modest CB_1 receptor affinity [109].

As described above, synthetic cannabinoids that diverge structurally from the prototypic JWH-018 in a number of ways are Δ^9-THC-like in rodent Δ^9-THC discrimination procedures, with potencies for substitution corresponding closely to their CB_1 receptor binding affinities. In recent studies, cross-substitution between Δ^9-THC and synthetic cannabinoids has been examined. Results of two studies showed that JWH-018 served as a discriminative stimulus in rats and rhesus monkeys and that Δ^9-THC and other psychoactive cannabinoids (e.g., JWH-073) substituted for JWH-018 whereas non-cannabinoids (e.g., benzodiazepines) did not [110, 111]. JWH-018 discrimination also served as the basis for a study that examined the discriminative stimulus effects of open-ring degradants of the tetramethylcyclopropyl ketones (XLR-11, UR-144, and A834735). The formation of these open-ring degradants are associated with repeated exposure of the parent compounds to high heat (B.F. Thomas, unpublished data), suggesting that the chemicals contained in purchased "herbal incense" products may not be identical to the chemicals the user inhales during smoking or vaping. Δ^9-THC and the open-ring degradants of the three tetramethylcyclopropyl ketones, but not a carboxy degradant of PB-22, were shown to produce full dose-dependent substitution in mice trained to discriminate JWH-018 from vehicle (J.L. Wiley and B.F. Thomas, unpublished data). Further, the open-ring degradants exhibited greater potencies and efficacies than the respective tetramethylcyclopropyl ketones from which they were derived. These findings emphasize that consideration of the actual exposure profile of the user is crucial in estimating in vivo potencies for chemicals that are inhaled after combustion or volatilization.

In an elegant study examining the relationship between efficacy and potency in a cannabinoid discrimination paradigm, Järbe et al. [112] reported successful acquisition of discriminations based upon a range of training doses of AM-5983 [(1-((1-methylpiperidin-2-yl)methyl)-1H-indol-3-yl)(naphthalen-1-yl)methanone], an indole-derived synthetic cannabinoid with high affinity and high efficacy at CB_1 receptors. Results showed that Δ^9-THC, methanandamide, WIN55,212-2, and the R- and S-isomers of AM-5983 dose-dependently substituted for racemic AM-5983. While generalization dose-effect curves for all cannabinoids showed rightward shifts as the AM-5983 training dose increased, potencies for the partial agonists Δ^9-THC and methanandamide exhibited greater enhancement than did those for the full agonists WIN55,212-2 and AM-5983. These results suggest that cannabinoid discrimination is sensitive to efficacy differences among compounds as well as differences in their potencies.

In summary, most of the discrimination research on abused synthetic cannabinoids has found that the discriminative stimulus effects of these compounds are Δ^9-THC-like in animal models, with potencies that correlate strongly with their affinities for the CB_1 receptor. Rimonabant blockade further supports CB_1 receptor mediation of the Δ^9-THC-like discriminative stimulus effects of synthetic cannabinoids [31]. To the extent tested, cross-substitution between synthetic cannabinoids and Δ^9-THC also has been demonstrated. Together, these findings are consistent with anecdotal reports that synthetic cannabinoids produce a marijuana-like intoxication [89] and continue to support the use of cannabinoid discrimination to predict abuse liability of these compounds.

5 Cannabinoid Antagonist Discrimination

Only a handful of studies have examined the discriminative stimulus effects of the CB_1 receptor antagonist/inverse agonist rimonabant. After early studies failed to establish a discrimination with rimonabant using food reinforcement [113, 114], Järbe et al. [115, 116] reported successful training of a rimonabant discrimination in rats using a taste aversion paradigm. In this paradigm, injection with the toxin lithium chloride was systematically paired with rimonabant administration to induce discriminated aversion to a drinking solution. The toxin was not injected prior to vehicle training sessions; hence, the absence of the rimonabant cue served as a safety signal to the thirsty rats that the solution was safe to drink. Acquisition was demonstrated by consistently low levels of drinking in the presence of rimonabant and high levels of drinking in its absence (as compared to rats that received the same schedule of rimonabant and vehicle injections, but did not receive lithium chloride). Results of substitution tests in the discriminating rats showed that rimonabant and its diarylpyrazole analog AM-251 substituted whereas CB_2 receptor antagonists, SR144528 and AM630, did not [115]. Δ^9-THC also failed to substitute for rimonabant when administered alone; however, when administered in combination with rimonabant, it attenuated rimonabant's suppression of drinking, suggesting opposing actions at the CB_1 receptor [115, 116]. Hence, rimonabant appears to be discriminable, but only under modified experimental conditions.

Use of a taste aversion procedure is one way to train discrimination of a drug with low discriminability. In the case of an antagonist, a second method is to train discrimination on a baseline of chronic agonist administration. Using this method, McMahon and colleagues embarked upon a series of studies in which they successfully trained rimonabant discrimination in a shock avoidance paradigm in monkeys who were being administered ongoing daily doses of Δ^9-THC [117, 118]. Attempts to train the discrimination in monkeys that were not receiving daily doses of Δ^9-THC failed, suggesting that chronic Δ^9-THC, and the accompanying dependence, was a necessary requisite for acquisition of rimonabant discrimination. Systematic examination of the discrimination revealed that discontinuation of the daily Δ^9-THC injection produced greater responding on the rimonabant-associated lever as well as overt behaviors that were similar to those observed during cannabinoid withdrawal in other species [119]. Administration of another CB_1 receptor antagonist AM-251 also increased responding on the rimonabant-associated lever [117]. In contrast, supplemental administration of Δ^9-THC, anandamide, CP55,940 or WIN55,212-2 prior to the rimonabant training dose produced responding primarily on the vehicle-associated lever [79, 117]. Similarly, administration of non-cannabinoid drugs failed to substitute for rimonabant. Together, these results support the hypothesis that rimonabant discrimination represents a useful model with which to investigate cannabinoid dependence.

Although rimonabant does not appear to be readily trainable as a discriminative stimulus using traditional procedures in naïve animals, discrimination with a rimonabant analog O-6629 was established [120]. O-6629 is one of a series of rimonabant analogs in which various substituents have been substituted for the 3-substituent of its pyrazole core. Unlike rimonabant, however, this set of analogs produces a battery of in vivo cannabinoid effects in inbred and CB_1 knockout mice [121]. In addition, their potencies in these tests are not strongly correlated with their binding affinities for the CB_1 receptor. Results of the O-6629 discrimination study showed that O-6629 produced dose-dependent substitution for the training dose as did another 3-substituent analog O-6658 [120]. In contrast, neither rimonabant nor

Δ^9-THC substituted for O-6629. Further, O-6629 did not substitute for Δ^9-THC-trained mice and did not alter the discriminative stimulus effects of Δ^9-THC when administered in combination with the Δ^9-THC training dose. These results suggest that this set of 3-substituent rimonabant analogs represents a novel class of cannabinoids with an unknown mechanism of action.

6 Translational Aspects of Cannabinoid Discrimination

Development of a reliable model of Δ^9-THC self-administration has proven difficult [122]. To date, Δ^9-THC self-administration has not been demonstrated in rodents and has been shown in squirrel monkeys in only a single lab [123]. Consequently, cannabinoid discrimination represents the most validated and pharmacologically selective animal model of an abuse-related property of cannabinoids – i.e., marijuana's subjective effects [29]. As such, its translational implications are several. First, the increased loosening of regulations surrounding the medicinal and recreational use of phytocannabinoids in the USA and other western countries has focused attention on separation and identification of the many chemicals contained in marijuana and determination of their pharmacological effects, alone and in combination. For example, several reports have suggested that non-psychoactive constituents of marijuana (e.g., cannabidiol) may contribute to the nature of its subjective and other abuse-related effects [24, 124, 125]. In addition, therapeutic potential for various non-Δ^9-THC constituents within the plant has been proposed [126–128]. Results from cannabinoid discrimination studies have been and will continue to be helpful in characterizing interactive effects among phytocannabinoids and in determining whether a constituent proposed for medicinal use is likely to have Δ^9-THC-like effects, which may be considered aversive by inexperienced users in a medical context [129, 130]. Other potential cannabinoid medications based upon manipulation of endocannabinoid synthesis, metabolism (e.g., FAAH and MAGL inhibitors), or transport also may be screened for Δ^9-THC-like psychoactivity through the use of carefully designed Δ^9-THC discrimination studies. Synthesis of selective CB_2 agonists and peripherally restricted CB_1 receptor agonists and antagonists are additional foci of drug development efforts [131, 132], for which drug discrimination studies may predict whether lead candidates may possess unintended Δ^9-THC-like subjective effects. For cannabinoids synthesized for more nebulous purposes (e.g., "Spice," "herbal incense"), discrimination is a crucial tool in prediction of abuse liability and provision of scientific data required for classifying these compounds as schedule 1 [31, 105, 109]. Use of selective CB_1 and CB_2 antagonists in the context of discrimination of cannabinoid agonists has confirmed CB_1 receptor mediation of the discriminative stimulus effects of Δ^9-THC and psychoactive synthetic cannabinoids whereas rimonabant discrimination has been useful for examination of factors related to cannabinoid dependence. Finally, cannabinoid discrimination provides a pharmacologically selective method for examination of underlying function of endocannabinoid system in Δ^9-THC-like intoxication.

7 Conclusions

Cannabinoid discrimination has a long history, stretching from the 1970s when the discriminative stimulus effects of Δ^9-THC initially were established to its current multi-purpose uses. Over the course of this period, drug discrimination has been used to characterize phytocannabinoids, endocannabinoids, synthetic cannabinoids, and cannabinoid antagonists. The influx of recent papers in which cannabinoid discrimination is highlighted attests to its continued relevance as a valuable method for scientific study of cannabinoid use and abuse.

Acknowledgments Preparation of this review was supported in part by National Institute on Drug Abuse grants DA-03672 and DA-026449; NIDA had no further role in the writing of the review or in the decision to submit the paper for publication.

References

1. Herkenham M, Lynn AB, Johnson MR, Melvin LS, de Costa BR, Rice KC (1991) Characterization and localization of cannabinoid receptors in rat brain: a quantitative in vitro autoradiographic study. J Neurosci 11:563–583
2. Wiley JL, Lowe JA, Balster RL, Martin BR (1995) Antagonism of the discriminative stimulus effects of delta 9-tetrahydrocannabinol in rats and rhesus monkeys. J Pharmacol Exp Ther 275:1–6
3. Matsuda LA, Lolait SJ, Brownstein MJ, Young AC, Bonner TI (1990) Structure of a cannabinoid receptor and functional expression of the cloned cDNA. Nature 346:561–564
4. Van Sickle MD, Duncan M, Kingsley PJ, Mouihate A, Urbani P, Mackie K et al (2005) Identification and functional characterization of brainstem cannabinoid CB2 receptors. Science 310:329–332
5. Xi ZX, Peng XQ, Li X, Song R, Zhang HY, Liu QR et al (2011) Brain cannabinoid CB2 receptors modulate cocaine's actions in mice. Nat Neurosci 14:1160–1166
6. Galiegue S, Mary S, Marchand J, Dussossoy D, Carriere D, Carayon P et al (1995) Expression of central and peripheral cannabinoid receptors in human immune tissues and leukocyte subpopulations. Eur J Biochem 232:54–61
7. Gaoni Y, Mechoulam R (1964) Isolation, structure, and partial synthesis of an active constituent of hashish. J Am Chem Soc 86:1646–1647
8. Mechoulam R, Braun P, Gaoni Y (1972) Synthesis of Δ^1-tetrahydrocannabinol and related cannabinoids. J Am Chem Soc 94:6159–6165
9. Radwan MM, ElSohly MA, El-Alfy AT, Ahmed SA, Slade D, Husni AS et al (2015) Isolation and pharmacological evaluation of minor cannabinoids from high-potency Cannabis sativa. J Nat Prod 78:1271–1276
10. Russo EB (2011) Taming THC: potential cannabis synergy and phytocannabinoid-terpenoid entourage effects. Br J Pharmacol 163:1344–1364

11. Järbe TU, Henriksson BG (1974) Discriminative response control produced with hashish, tetrahydrocannabinols (delta-8-THC and delta-9-THC), and other drugs. Psychopharmacologia (Berl) 40:1–16

12. Järbe TUC, McMillan DE (1979) Discriminative stimulus properties of tetrahydrocannabinols and related drugs in rats and pigeons. Neuropharmacology 18:1023–1024

13. Järbe TU, Johansson JO, Henriksson BG (1975) Delta9-tetrahydrocannabinol and pentobarbital as discriminative cues in the Mongolian Gerbil (Meriones unguiculatus). Pharmacol Biochem Behav 3:403–410

14. Henriksson BG, Johansson JO, Järbe TU (1975) Delta 9-tetrahydrocannabinol produced discrimination in pigeons. Pharmacol Biochem Behav 3:771–774

15. McMahon LR, Ginsburg BC, Lamb RJ (2008) Cannabinoid agonists differentially substitute for the discriminative stimulus effects of Delta(9)-tetrahydrocannabinol in C57BL/6J mice. Psychopharmacology (Berl) 198:487–495

16. Vann RE, Warner JA, Bushell K, Huffman JW, Martin BR, Wiley JL (2009) Discriminative stimulus properties of Delta9-tetrahydrocannabinol (THC) in C57Bl/6J mice. Eur J Pharmacol 615:102–107

17. Gold LH, Balster RL, Barrett RL, Britt DT, Martin BR (1992) A comparison of the discriminative stimulus properties of delta 9-tetrahydrocannabinol and CP 55,940 in rats and rhesus monkeys. J Pharmacol Exp Ther 262:479–486

18. Lile JA, Kelly TH, Pinsky DJ, Hays LR (2009) Substitution profile of Delta(9)-tetrahydrocannabinol, triazolam, hydromorphone, and methylphenidate in humans discriminating Delta(9)-tetrahydrocannabinol. Psychopharmacology (Berl) 203:241–250

19. Järbe TUC, Swedberg MDB, Mechoulam R (1981) A repeated tests procedure to assess onset and duration of the cue properties of (-)Δ^9-THC, (-)Δ^8-THC-DMH and (+)Δ^8-THC. Psychopharmacology (Berl) 75:152–157

20. Järbe TUC, McMillan DE (1980) Δ^9-THC as a discriminative stimulus in rats and pigeons: generalization to THC metabolites and SP-111. Psychopharmacology (Berl) 71:281–289

21. Semjonow A, Binder M (1985) Generalization of the discriminative stimulus properties of Δ9-THC to Δ9(11)-THC in rats. Psychopharmacology (Berl) 85:178–183

22. Wiley JL, Barrett RL, Britt DT, Balster RL, Martin BR (1993) Discriminative stimulus effects of delta 9-tetrahydrocannabinol and delta 9-11-tetrahydrocannabinol in rats and rhesus monkeys. Neuropharmacology 32:359–365

23. Thomas BF, Gilliam AF, Burch DF, Roche MJ, Seltzman HH (1998) Comparative receptor binding analyses of cannabinoid agonists and antagonists. J Pharmacol Exp Ther 285:285–292

24. Vann RE, Gamage TF, Warner JA, Marshall EM, Taylor NL, Martin BR et al (2008) Divergent effects of cannabidiol on the discriminative stimulus and place conditioning effects of Delta(9)-tetrahydrocannabinol. Drug Alcohol Depend 94:191–198

25. Compton DR, Rice KC, De Costa BR, Razdan RK, Melvin LS, Johnson MR et al (1993) Cannabinoid structure-activity relationships: correlation of receptor binding and in vivo activities. J Pharmacol Exp Ther 265:218–226

26. Compton DR, Gold LH, Ward SJ, Balster RL, Martin BR (1992) Aminoalkylindole analogs: cannabimimetic activity of a class of compounds structurally distinct from delta 9-tetrahydrocannabinol. J Pharmacol Exp Ther 263:1118–1126

27. Wiley JL, Huffman JW, Balster RL, Martin BR (1995) Pharmacological specificity of the discriminative stimulus effects of delta 9-tetrahydrocannabinol in rhesus monkeys. Drug Alcohol Depend 40:81–86

28. Barrett RL, Wiley JL, Balster RL, Martin BR (1995) Pharmacological specificity of delta 9-tetrahydrocannabinol discrimination in rats. Psychopharmacology (Berl) 118:419–424

29. Balster RL, Prescott WR (1992) Δ^9-Tetrahydrocannabinol discrimination in rats as a model for cannabis intoxication. Neurosci Biobehav Rev 16:55–62

30. Wiley JL (1999) Cannabis: discrimination of "internal bliss"? Pharmacol Biochem Behav 64:257–260

31. Wiley JL, Marusich JA, Lefever TW, Grabenauer M, Moore KN, Thomas BF (2013) Cannabinoids in disguise: delta-9-Tetrahydrocannabinol-like effects of tetramethylcyclopropyl ketone indoles. Neuropharmacology 75:145–154

32. Devane WA, Hanus L, Breuer A, Pertwee RG, Stevenson LA, Griffin G et al (1992) Isolation and structure of a brain constituent that binds to the cannabinoid receptor. Science 258:1946–1949

33. Sugiura T, Waku K (2000) 2-arachidonoylglycerol: a possible multifunctional lipid mediator in the nervous and immune systems. Ann N Y Acad Sci 905:344–346

34. Cravatt BF, Giang DK, Mayfield SP, Boger DL, Lerner RA, Gilula NB (1996) Molecular characterization of an enzyme that degrades neuromodulatory fatty-acid amides. Nature 384:83–87

35. Dinh TP, Carpenter D, Leslie FM, Freund TF, Katona I, Sensi SL et al (2002) Brain monoglyceride lipase participating in endocannabinoid inactivation. Proc Natl Acad Sci U S A 99:10819–10824

36. Wiley JL, Balster RL, Martin BR (1995) Discriminative stimulus effects of anandamide in rats. Eur J Pharmacol 276:49–54

37. McMahon L (2009) Apparent affinity estimates of rimonabant in combination with ananda-mide and chemical analogs of anandamide in rhesus monkeys discriminating Δ9-tetrahydro-cannabinol. Psychopharmacology 203:219–228

38. Burkey RT, Nation JR (1997) (R)-methanandamide, but not anandamide, substitutes for delta-9-THC in a drug-discrimination procedure. Exp Clin Psychopharmacol 5:195–202

39. Wiley JL, Ryan WJ, Razdan RK, Martin BR (1998) Evaluation of cannabimimetic effects of structural analogs of anandamide in rats. Eur J Pharmacol 355:113–118

40. Wiley JL, Walentiny DM, Wright MJ Jr, Beardsley PM, Burston JJ, Poklis JL et al (2014) Endocannabinoid contribution to Delta(9)-tetrahydrocannabinol discrimination in rodents. Eur J Pharmacol 737:97–105

41. Boger DL, Fecik RA, Patterson JE, Miyauchi H, Patricelli MP, Cravatt BF (2000) Fatty acid amide hydrolase substrate specificity. Bioorg Med Chem Lett 10:2613–2616

42. Cravatt BF, Lichtman AH (2002) The enzymatic inactivation of the fatty acid amide class of signaling lipids. Chem Phys Lipids 121:135–148

43. Deutsch DG, Chin SA (1993) Enzymatic synthesis and degradation of anandamide, a cannabinoid receptor agonist. Biochem Pharmacol 46:791–796

44. Adams IB, Compton DR, Martin BR (1998) Assessment of anandamide interaction with the cannabinoid brain receptor: SR 141716A antagonism studies in mice and autoradiographic analysis of receptor binding in rat brain. J Pharmacol Exp Ther 284:1209–1217

45. Adams IB, Ryan W, Singer M, Thomas BF, Compton DR, Razdan RK et al (1995) Evaluation of cannabinoid receptor binding and in vivo activities for anandamide analogs. J Pharmacol Exp Ther 273:1172–1181

46. Wiley JL, Golden KM, Ryan WJ, Balster RL, Razdan RK, Martin BR (1997) Evaluation of cannabimimetic discriminative stimulus effects of anandamide and methylated fluoroanandamide in rhesus monkeys. Pharmacol Biochem Behav 58:1139–1143

47. Solinas M, Tanda G, Justinova Z, Wertheim CE, Yasar S, Piomelli D et al (2007) The endogenous cannabinoid anandamide produces delta-9-tetrahydrocannabinol-like discriminative and neurochemical effects that are enhanced by inhibition of fatty acid amide hydrolase but not by inhibition of anandamide transport. J Pharmacol Exp Ther 321:370–380

48. Järbe T, Lamb R, Lin S, Makriyannis A (2000) Delta9-THC training dose as a determinant for (R)-methanandamide generalization in rats: a systematic replication. Behav Pharmacol 11:81–86

49. Järbe TU, Lamb RJ, Makriyannis A, Lin S, Goutopoulos A (1998) Delta-9-THC training dose as a determinant for (R)-methanandamide generalization in rats. Psychopharmacology (Berl) 140:519–522

50. Wiley JL, Matthew Walentiny D, Vann RE, Baskfield CY (2011) Dissimilar cannabinoid substitution patterns in mice trained to discriminate Delta(9)-tetrahydrocannabinol or methanandamide from vehicle. Behav Pharmacol 22:480–488

51. Mansbach RS, Balster RL (1991) Pharmacological specificity of the phencyclidine discriminative stimulus in rats. Pharmacol Biochem Behav 39:971–975

52. Järbe T, Li C, Liu Q, Makriyannis A (2009) Discriminative stimulus functions in rats of AM1346, a high-affinity CB1R selective anandamide analog. Psychopharmacology 203:229–239

53. Järbe TU, Lamb RJ, Lin S, Makriyannis A (2001) (R)-methanandamide and delta 9-THC as discriminative stimuli in rats: tests with the cannabinoid antagonist SR-141716 and the endogenous ligand anandamide. Psychopharmacology (Berl) 156:369–380

54. Järbe TU, Li C, Vadivel SK, Makriyannis A (2010) Discriminative stimulus functions of methanandamide and delta(9)-THC in rats: tests with aminoalkylindoles (WIN55,212-2 and AM678) and ethanol. Psychopharmacology (Berl) 208:87–98

55. Wiley JL, LaVecchia KL, Karp NE, Kulasegram S, Mahadevan A, Razdan RK et al (2004) A comparison of the discriminative stimulus effects of Delta(9)-tetrahydrocannabinol and O-1812, a potent and metabolically stable anandamide analog, in rats. Exp Clin Psychopharmacol 12:173–179

56. Baskfield CY, Martin BR, Wiley JL (2004) Differential effects of delta9-tetrahydrocannabinol and methanandamide in CB1 knockout and wild-type mice. J Pharmacol Exp Ther 309:86–91

57. Järbe TU, Liu Q, Makriyannis A (2006) Antagonism of discriminative stimulus effects of delta(9)-THC and (R)-methanandamide in rats. Psychopharmacology (Berl) 184:36–45

58. Chanda PK, Gao Y, Mark L, Btesh J, Strassle BW, Lu P et al (2010) Monoacylglycerol lipase activity is a critical modulator of the tone and integrity of the endocannabinoid system. Mol Pharmacol 78:996–1003

59. Cravatt BF, Demarest K, Patricelli MP, Bracey MH, Giang DK, Martin BR et al (2001) Supersensitivity to anandamide and enhanced endogenous cannabinoid signaling in mice lacking fatty acid amide hydrolase. Proc Natl Acad Sci U S A 98:9371–9376

60. Lichtman AH, Shelton CC, Advani T, Cravatt BF (2004) Mice lacking fatty acid amide hydrolase exhibit a cannabinoid receptor-mediated phenotypic hypoalgesia. Pain 109:319–327

61. Petrenko AB, Yamazaki M, Sakimura K, Kano M, Baba H (2014) Augmented tonic pain-related behavior in knockout mice lacking monoacylglycerol lipase, a major degrading enzyme for the endocannabinoid 2-arachidonoylglycerol. Behav Brain Res 271:51–58

62. Clement AB, Hawkins EG, Lichtman AH, Cravatt BF (2003) Increased seizure susceptibility and proconvulsant activity of anandamide in mice lacking fatty acid amide hydrolase. J Neurosci 23:3916–3923

63. Kishimoto Y, Cagniard B, Yamazaki M, Nakayama J, Sakimura K, Kirino Y et al (2015) Task-specific enhancement of hippocampus-dependent learning in mice deficient in monoacylglycerol lipase, the major hydrolyzing enzyme of the endocannabinoid 2-arachidonoylglycerol. Front Behav Neurosci 9:134

64. Pan B, Wang W, Zhong P, Blankman JL, Cravatt BF, Liu QS (2011) Alterations of endocannabinoid signaling, synaptic plasticity, learning, and memory in monoacylglycerol lipase knock-out mice. J Neurosci 31:13420–13430

65. Varvel SA, Wise LE, Niyuhire F, Cravatt BF, Lichtman AH (2007) Inhibition of fatty-acid amide hydrolase accelerates acquisition and extinction rates in a spatial memory task. Neuropsychopharmacology 32:1032–1041

66. Taschler U, Radner FP, Heier C, Schreiber R, Schweiger M, Schoiswohl G et al (2011) Monoglyceride lipase deficiency in mice impairs lipolysis and attenuates diet-induced insulin resistance. J Biol Chem 286:17467–17477

67. Tourino C, Oveisi F, Lockney J, Piomelli D, Maldonado R (2010) FAAH deficiency promotes energy storage and enhances the motivation for food. Int J Obes (Lond) 34:557–568

68. Walentiny DM, Gamage TF, Warner JA, Nguyen TK, Grainger DB, Wiley JL et al (2011) The endogenous cannabinoid anandamide shares discriminative stimulus effects with Delta (9)-tetrahydrocannabinol in fatty acid amide hydrolase knockout mice. Eur J Pharmacol 656:63–67

69. Wiley JL, Lefever TW, Pulley NS, Marusich JA, Cravatt BF, Lichtman AH (2016). Just add water: cannabinoid discrimination in a water t-maze with FAAH(-/-) and FAAH(+/+) mice. Behav Pharmacol (in press)

70. Walentiny DM, Vann RE, Wiley JL (2015) Phenotypic assessment of THC discriminative stimulus properties in fatty acid amide hydrolase knockout and wildtype mice. Neuropharmacology 93:237–242

71. Vann RE, Walentiny DM, Burston JJ, Tobey KM, Gamage TF, Wiley JL (2012) Enhancement of the behavioral effects of endogenous and exogenous cannabinoid agonists by phenylmethyl sulfonyl fluoride. Neuropharmacology 62:1019–1027

72. Ahn K, Johnson DS, Mileni M, Beidler D, Long JZ, McKinney MK et al (2009) Discovery and characterization of a highly selective FAAH inhibitor that reduces inflammatory pain. Chem Biol 16:411–420

73. Chang JW, Niphakis MJ, Lum KM, Cognetta AB 3rd, Wang C, Matthews ML et al (2012) Highly selective inhibitors of monoacylglycerol lipase bearing a reactive group that is bioisosteric with endocannabinoid substrates. Chem Biol 19:579–588

74. Fegley D, Gaetani S, Duranti A, Tontini A, Mor M, Tarzia G et al (2005) Characterization of the fatty acid amide hydrolase inhibitor cyclohexyl carbamic acid 3′-carbamoyl-biphenyl-3-yl ester (URB597): effects on anandamide and oleoylethanolamide deactivation. J Pharmacol Exp Ther 313:352–358

75. Long JZ, Nomura DK, Cravatt BF (2009) Characterization of monoacylglycerol lipase inhibition reveals differences in central and peripheral endocannabinoid metabolism. Chem Biol 16:744–753

76. Niphakis MJ, Johnson DS, Ballard TE, Stiff C, Cravatt BF (2012) O-hydroxyacetamide carbamates as a highly potent and selective class of endocannabinoid hydrolase inhibitors. ACS Chem Neurosci 3:418–426

77. Niphakis MJ, Cognetta AB 3rd, Chang JW, Buczynski MW, Parsons LH, Byrne F et al (2013) Evaluation of NHS carbamates as a potent and selective class of endocannabinoid hydrolase inhibitors. ACS Chem Neurosci 4:1322–1332

78. Hruba L, Seillier A, Zaki A, Cravatt BF, Lichtman AH, Giuffrida A et al (2015) Simultaneous inhibition of fatty acid amide hydrolase and monoacylglycerol lipase shares discriminative stimulus effects with delta9-tetrahydrocannabinol in mice. J Pharmacol Exp Ther 353:261–268

79. Stewart JL, McMahon LR (2011) The fatty acid amide hydrolase inhibitor URB 597: interactions with anandamide in rhesus monkeys. Br J Pharmacol 164:655–666

80. Matuszak N, Muccioli GG, Labar G, Lambert DM (2009) Synthesis and in vitro evaluation of N-substituted maleimide derivatives as selective monoglyceride lipase inhibitors. J Med Chem 52:7410–7420

81. Burston JJ, Sim-Selley LJ, Harloe JP, Mahadevan A, Razdan RK, Selley DE et al (2008) N-arachidonyl maleimide potentiates the pharmacological and biochemical effects of the endocannabinoid 2-arachidonylglycerol through inhibition of monoacylglycerol lipase. J Pharmacol Exp Ther 327:546–553

82. Long JZ, Nomura DK, Vann RE, Walentiny DM, Booker L, Jin X et al (2009) Dual blockade of FAAH and MAGL identifies behavioral processes regulated by endocannabinoid crosstalk in vivo. Proc Natl Acad Sci U S A 106:20270–20275

83. Ignatowska-Jankowska BM, Ghosh S, Crowe MS, Kinsey SG, Niphakis MJ, Abdullah RA et al (2014) In vivo characterization of the highly selective monoacylglycerol lipase inhibitor KML29: antinociceptive activity without cannabimimetic side effects. Br J Pharmacol 171:1392–1407

84. Ignatowska-Jankowska B, Wilkerson JL, Mustafa M, Abdullah R, Niphakis M, Wiley JL et al (2015) Selective monoacylglycerol lipase inhibitors: antinociceptive versus cannabimimetic effects in mice. J Pharmacol Exp Ther 353:424–432

85. Owens RA, Ignatowska-Jankowska B, Mustafa M, Beardsley PM, Wiley JL, Niphakis MJ, et al (2016) The dual FAAH and MAGL inhibitor SA-57 elicits CB1 receptor mediated stimulus effects in mice. J Pharmacol Exp Ther (submitted)

86. Wiley JL, Marusich JA, Huffman JW, Balster RL, Thomas BF (2011b) Hijacking of basic research: the case of synthetic cannabinoids. RTI Press, Research Triangle Park

87. Jarbe TU, Gifford RS (2014) "Herbal incense": designer drug blends as cannabimimetics and their assessment by drug discrimination and other in vivo bioassays. Life Sci 97:64–71

88. Auwarter V, Dresen S, Weinmann W, Muller M, Putz M, Ferreiros N (2009) 'Spice' and other herbal blends: harmless incense or cannabinoid designer drugs? J Mass Spectrom 44:832–837

89. Vardakou I, Pistos C, Spiliopoulou C (2010) Spice drugs as a new trend: mode of action, identification and legislation. Toxicol Lett 197:157–162

90. Wiley JL, Compton DR, Dai D, Lainton JA, Phillips M, Huffman JW et al (1998) Structure-activity relationships of indole- and pyrrole-derived cannabinoids. J Pharmacol Exp Ther 285:995–1004

91. Manera C, Tuccinardi T, Martinelli A (2008) Indoles and related compounds as cannabinoid ligands. Mini Rev Med Chem 8:370–387

92. Wiley JL, Marusich JA, Huffman JW (2014) Moving around the molecule: relationship between chemical structure and in vivo activity of synthetic cannabinoids. Life Sci 97:55–63

93. Huffman JW, Dai D, Martin BR, Compton DR (1994) Design, synthesis and pharmacology of cannabimimetic indoles. Bioorg Med Chem Lett 4:563–566

94. Lainton JAH, Huffman JW, Martin BR, Compton DR (1995) 1-Alkyl-3-(1-naphthoyl)pyr-roles: a new cannabinoid class. Tetrahedron Lett 36:1401–1404

95. Gatch MB, Forster MJ (2014) Delta9-tetrahydrocannabinol-like discriminative stimulus effects of compounds commonly found in K2/Spice. Behav Pharmacol 25:750–757

96. Ginsburg BC, Schulze DR, Hruba L, McMahon LR (2012) JWH-018 and JWH-073: delta 9-tetrahydrocannabinol-like discriminative stimulus effects in monkeys. J Pharmacol Exp Ther 340:37–45

97. Hruba L, Ginsburg B, McMahon LR (2012) Apparent inverse relationship between cannabinoid agonist efficacy and tolerance/cross-tolerance produced by {Delta}9-tetrahydrocannabinol treatment in rhesus monkeys. J Pharmacol Exp Ther 342:843–849

98. Järbe TU, Deng H, Vadivel SK, Makriyannis A (2011) Cannabinergic aminoalkylindoles, including AM678 = JWH018 found in 'Spice', examined using drug (Delta9-tetrahydrocannabinol) discrimination for rats. Behav Pharmacol 22:498–507

99. Marshell R, Kearney-Ramos T, Brents LK, Hyatt WS, Tai S, Prather PL et al (2014) In vivo effects of synthetic cannabinoids JWH-018 and JWH-073 and phytocannabinoid Δ9-THC in mice: inhalation versus intraperitoneal injection. Pharmacol Biochem Behav 124:40–47

100. Huffman JW, Szklennik PV, Almond A, Bushell K, Selley DE, He H et al (2005) 1-Pentyl-3-phenylacetylindoles, a new class of cannabimimetic indoles. Bioorg Med Chem Lett 15:4110–4113

101. Wiley JL, Marusich JA, Martin BR, Huffman JW (2012) 1-Pentyl-3-phenylacetylindoles and JWH-018 share in vivo cannabinoid profiles in mice. Drug Alcohol Depend 123:148–153

102. Uchiyama N, Kawamura M, Kikura-Hanajiri R, Goda Y (2013) URB-754: a new class of designer drug and 12 synthetic cannabinoids detected in illegal products. Forensic Sci Int 227:21–32

103. Frost JM, Dart MJ, Tietje KR, Garrison TR, Grayson GK, Daza AV et al (2010) Indol-3-ylcycloalkyl ketones: effects of N1 substituted indole side chain variations on CB(2) cannabinoid receptor activity. J Med Chem 53:295–315

104. Frost JM, Dart MJ, Tietje KR, Garrison TR, Grayson GK, Daza AV et al (2008) Indol-3-yl-tetramethylcyclopropyl ketones: effects of indole ring substitution on CB2 cannabinoid receptor activity. J Med Chem 51:1904–1912

105. Gatch MB, Forster MJ (2015) Delta9-tetrahydrocannabinol-like effects of novel synthetic cannabinoids found on the gray market. Behav Pharmacol 26:460–468

106. Karinen R, Tuv SS, Oiestad EL, Vindenes V (2015) Concentrations of APINACA, 5F-APINACA, UR-144 and its degradant product in blood samples from six impaired drivers compared to previous reported concentrations of other synthetic cannabinoids. Forensic Sci Int 246:98–103

107. Shevyrin V, Melkozerov V, Nevero A, Eltsov O, Baranovsky A, Shafran Y (2014) Synthetic cannabinoids as designer drugs: new representatives of indol-3-carboxylates series and indazole-3-carboxylates as novel group of cannabinoids. Identification and analytical data. Forensic Sci Int 244:263–275

108. Uchiyama N, Shimokawa Y, Kikura-Hanajiri R, Demizu Y, Goda Y, Hakamatsuka T (2015) A synthetic cannabinoid FDU-NNEI, two 2-indazole isomers of synthetic cannabinoids AB-CHMINACA and NNEI indazole analog (MN-18), a phenethylamine derivative -OH-EDMA, and a cathinone derivative dimethoxy-alpha-PHP, newly identified in illegal products. Forensic Toxicol 33:244–259

109. Wiley JL, Marusich JA, Lefever TW, Antonazzo KR, Wallgren MT, Cortes RA et al (2015) AB-CHMINACA, AB-PINACA, and FUBIMINA: affinity and potency of novel synthetic cannabinoids in producing delta-9-tetrahydrocannabinol-like effects in mice. J Pharmacol Exp Ther 354:328–339

110. Rodriguez JS, McMahon LR (2014) JWH-018 in rhesus monkeys: differential antagonism of discriminative stimulus, rate-decreasing, and hypothermic effects. Eur J Pharmacol 740:151–159

111. Wiley JL, Marusich JA, Lefever TW, Cortes RA (2014) Cross-substitution of delta-9-tetrahydrocannabinol and JWH-018 in drug discrimination in rats. Pharmacol Biochem Behav 124:123–128

112. Jarbe TU, LeMay BJ, Halikhedkar A, Wood J, Vadivel SK, Zvonok A et al (2014) Differentiation between low- and high-efficacy CB1 receptor agonists using a drug discrimination protocol for rats. Psychopharmacology (Berl) 231:489–500

113. Mansbach RS, Rovetti CC, Winston EN, Lowe JA III (1996) Effects of the cannabinoid CB1 receptor antagonist SR141716A on the behavior of pigeons and rats. Psychopharmacology 124:315–322

114. Pério A, Rinaldi-Carmona M, Maruani J, Barth F, Le Fur G, Soubrie P (1996) Central mediation of the cannabinoid cue: activity of a selective CB1 antagonist, SR 141716A. Behav Pharmacol 7:65–71

115. Järbe TU, Li C, Vadivel SK, Makriyannis A (2008) Discriminative stimulus effects of the cannabinoid CB1 receptor antagonist rimonabant in rats. Psychopharmacology (Berl) 198:467–478

116. Järbe TU, Harris MY, Li C, Liu Q, Makriyannis A (2004) Discriminative stimulus effects in rats of SR-141716 (rimonabant), a cannabinoid CB1 receptor antagonist. Psychopharmacology (Berl) 177:35–45

117. McMahon LR (2006) Discriminative stimulus effects of the cannabinoid CB1 antagonist SR 141716A in rhesus monkeys pretreated with Delta9-tetrahydrocannabinol. Psychopharmacology (Berl) 188:306–314

118. McMahon LR, France CP (2003) Discriminative stimulus effects of the cannabinoid antagonist, SR 141716A, in delta-9-tetrahydrocannabinol-treated rhesus monkeys. Exp Clin Psychopharmacol 11:286–293

119. Stewart JL, McMahon LR (2010) Rimonabant-induced Delta9-tetrahydrocannabinol withdrawal in rhesus monkeys: discriminative stimulus effects and other withdrawal signs. J Pharmacol Exp Ther 334:347–356

120. Walentiny DM, Vann RE, Mahadevan A, Kottani R, Gujjar R, Wiley JL (2013) Novel 3-substituted rimonabant analogues lack Delta(9) -tetrahydrocannabinol-like abuse-related behavioural effects in mice. Br J Pharmacol 169:10–20

121. Wiley JL, Selley DE, Wang P, Kottani R, Gadthula S, Mahadeven A (2012) 3-Substituted pyrazole analogs of the cannabinoid type 1 (CB1) receptor antagonist rimonabant: cannabinoid agonist-like effects in mice via non-CB1, non-CB2 mechanism. J Pharmacol Exp Ther 340:433–444

122. Lefever TW, Marusich JA, Antonazzo KR, Wiley JL (2014) Evaluation of WIN 55,212-2 self-administration in rats as a potential cannabinoid abuse liability model. Pharmacol Biochem Behav 118:30–35

123. Justinova Z, Tanda G, Redhi GH, Goldberg SR (2003) Self-administration of delta9-tetrahydrocannabinol (THC) by drug naive squirrel monkeys. Psychopharmacology (Berl) 169:135–140

124. Hayakawa K, Mishima K, Hazekawa M, Sano K, Irie K, Orito K et al (2008) Cannabidiol potentiates pharmacological effects of delta(9)-tetrahydrocannabinol via CB(1) receptor-dependent mechanism. Brain Res 1188:157–164

125. Pertwee RG (2008) The diverse CB1 and CB2 receptor pharmacology of three plant cannabinoids: delta9-tetrahydrocannabinol, cannabidiol and delta9-tetrahydrocannabivarin. Br J Pharmacol 153:199–215

126. Hill AJ, Williams CM, Whalley BJ, Stephens GJ (2012) Phytocannabinoids as novel therapeutic agents in CNS disorders. Pharmacol Ther 133:79–97

127. Wilkinson JD, Whalley BJ, Baker D, Pryce G, Constanti A, Gibbons S et al (2003) Medicinal cannabis: is delta9-tetrahydrocannabinol necessary for all its effects? J Pharm Pharmacol 55:1687–1694

128. Zuardi AW, Hallak JE, Dursun SM, Morais SL, Sanches RF, Musty RE et al (2006) Cannabidiol monotherapy for treatment-resistant schizophrenia. J Psychopharmacol 20:683–686

129. Calhoun SR, Galloway GP, Smith DE (1998) Abuse potential of dronabinol (Marinol). J Psychoactive Drugs 30:187–196

130. Plasse TF, Gorter RW, Krasnow SH, Lane M, Shepard KV, Wadleigh RG (1991) Recent clinical experience with dronabinol. Pharmacol Biochem Behav 40:695–700

131. Horvath TL (2006) The unfolding cannabinoid story on energy homeostasis: central or peripheral site of action? Int J Obes (Lond) 30(Suppl 1):S30–S32

132. Pertwee RG (2006) The pharmacology of cannabinoid receptors and their ligands: an overview. Int J Obes (Lond) 30(Suppl 1):S13–S18

Discriminative Stimulus Properties of Opioid Ligands: Progress and Future Directions

Eduardo R. Butelman and Mary Jeanne Kreek

Abstract Opioid receptors (MOP-r, KOP-r, DOP-r, as well as NOP-r) and their endogenous neuropeptide agonist systems are involved in diverse neurobiological and behavioral functions, in health and disease. These functions include pain and analgesia, addictions, and psychiatric diseases (e.g., depression-, anxiety-like, and stress-related disorders). Drug discrimination assays have been used to characterize the behavioral pharmacology of ligands with affinity at MOP-r, KOP-r, or DOP-r (and to a lesser extent NOP-r). Therefore, drug discrimination studies with opioid ligands have an important continuing role in translational investigations of diseases that are affected by these neurobiological targets and their pharmacotherapy.

Keywords Addiction • Analgesia • Drug discrimination • Opioid • Prescription opioids

Contents

Ligands with affinity at the mu opioid receptor (MOP-r), kappa opioid receptor (KOP-r), or delta receptor (DOP-r) are among the most widely studied pharmacological classes studied in drug discrimination assays, largely preclinically, but also

E.R. Butelman (✉) and M.J. Kreek
Laboratory on the Biology of Addictive Diseases, The Rockefeller University, 1230 York Avenue, Box 171, New York, NY 10065, USA
e-mail: butelme@rockefeller.edu

© Springer International Publishing Switzerland 2016
Curr Topics Behav Neurosci (2018) 39: 175–192
DOI 10.1007/7854_2016_9
Published Online: 25 May 2016

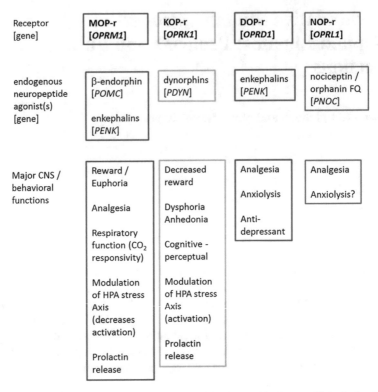

Receptor [gene]	MOP-r [OPRM1]	KOP-r [OPRK1]	DOP-r [OPRD1]	NOP-r [OPRL1]
endogenous neuropeptide agonist(s) [gene]	β-endorphin [POMC] enkephalins [PENK]	dynorphins [PDYN]	enkephalins [PENK]	nociceptin / orphanin FQ [PNOC]
Major CNS / behavioral functions	Reward / Euphoria Analgesia Respiratory function (CO$_2$ responsivity) Modulation of HPA stress Axis (decreases activation) Prolactin release	Decreased reward Dysphoria Anhedonia Cognitive - perceptual Modulation of HPA stress Axis (activation) Prolactin release	Analgesia Anxiolysis Anti-depressant	Analgesia Anxiolysis?

Fig. 1 Summary of opioid receptor and neuropeptide genes, and basic neurobiological and major behavioral functions. Opioid receptors (MOP-r, KOP-r, DOP-r, and NOP-r) are 7-transmembrane domain $G_{i/o}$-coupled receptors. *CNS* central nervous system, *HPA* hypothalamic-pituitary-adrenal

clinically. Ligands with affinity at nociceptin/orphanin FQ (NOP-r), another more recently discovered member of the opioid receptor family, have been studied to a lesser extent (Fig. 1). Opioid receptors and their endogenous neuropeptide agonist systems, as well as exogenous ligands, are involved in major clinical conditions, including pain and analgesia, addictions, and psychiatric diseases (e.g., depression-like, anxiety-like states, and stress-related disorders) (Fig. 1). Therefore, drug discrimination studies with opioid ligands have an important continuing role in translational investigations of diseases that are affected by these neurobiological targets and their pharmacotherapy.

Drug discrimination (as described in detail in chapters one and two in this volume) can be briefly described as a set of operant (or instrumental) techniques, by which a subject learns to emit a specific behavioral response in the presence of a particular drug stimulus (e.g., a specific dose of an opioid ligand) and a different behavioral response in the absence of the drug stimulus (e.g., when vehicle is administered). In humans, the behavioral response can be a verbal identifier (e.g., "Drug A" vs. "Drug B") or a response on a physical manipulandum. Non-operant approaches to drug discrimination have also been used, but will not be discussed extensively here. Drug discriminations are generally learnt by repeated pairings of a particular drug stimulus with a particular reward contingency (e.g., reward by responding on a particular lever in the presence of a particular drug stimulus).

This article will focus on major fields of in vivo and behavioral opioid pharmacology studied with drug discrimination assays. The citations are not exhaustive, and we selected representative references, for brevity. Overall, several laboratories have contributed to this rich field, for methodology, pharmacological analysis/drug development, and also for the study of basic behavioral and neurobiological processes (e.g., MOP-r agonist pharmacology, tolerance, dependence, and withdrawal).

Brief Summary of Drug Discrimination Studies with Opioid Ligands

Opioid ligands (initially MOP-r agonists, such as morphine or fentanyl) figure prominently in the evolution of drug discrimination as a tool in neuroscience and pharmacology research [1]. Thus, early studies with MOP-r agonists were used to explore and develop drug discrimination techniques, and to understand the in vivo pharmacology of MOP-r and KOP-r ligands [2–5]. For example, early studies determined the pharmacological specificity of the discriminative stimulus effects of standard MOP-r agonists such as fentanyl, and their relative potency, which was positively correlated with other major in vivo effects (e.g., analgesia) [3]. Other early drug discrimination studies directly examined the then-emerging field of multiple opioid receptors [6], for example, showing that rhesus monkeys trained to discriminate standard MOP-r agonists did not generalize compounds with KOP-r agonist effects (e.g., ethylketazocine) and vice versa [4, 7]. This differentiation in discriminative stimulus effects is consistent with divergent neurobiological functions of MOP-r and KOP-r systems in CNS (see Fig. 1).

Major uses of drug discrimination assays in the opioid field include: (a) Dose- and time-dependence of discriminative stimulus effects (e.g., allowing potency and time course comparisons between ligands in the same class) [8, 9]; (b) differentiation of drug classes (e.g., MOP-r, KOP-r, or DOP-r agonists versus each other [10–13]; (c) selective antagonism, including quantitative approaches such as in vivo apparent pA_2 analysis [14–16]; and (d) characterization of novel or atypical ligands that may have an opioid-receptor mediated effect [17, 18].

Influence of the Training Dose

An important feature of drug discrimination assays in the opioid field is the influence of the magnitude of the training dose of a compound, on pharmacological specificity and identity of the interoceptive drug stimulus [19–21]. Thus, it appears that the higher the training dose of a given ligand (e.g., a MOP-r agonist), the greater the pharmacological specificity of the interoceptive drug stimulus. There are a considerable number of possibilities that could explain the aforementioned phenomenon [20, 22], including the experimental conditions under study, and differential pharmacodynamic effects at the same receptor site. For example, ligands with partial agonist MOP-r effects (e.g., nalbuphine) can be generalized by rats trained to discriminate relatively low dose of a high efficacy MOP-r agonist (e.g., fentanyl), but not to a higher training dose thereof, due to a sub-maximal "plateau" in the signaling caused by partial agonists (which can be directly characterized with in vitro signaling assays) [21, 23–25].

Table 1 Examples of exogenous opioid receptor agonists and antagonists

	MOP-r	KOP-r	DOP-r	NOP-r
Agonists	Fentanyl morphine methadone oxycodone (*and other* *prescription opioids*)	U50,488 U69,593 Salvinorin A	SNC80 BW373U86	Ro 64-6198 SCH 221,510
Antagonists	CTAP (peptidic) Naltrexone (*relatively* *low doses*) β-funaltrexamine or clocinnamox (*irrevers-* *ible/functionally* *irreversible*)	Nor-BNI or JDtic (*very-* *long-lasting*) Naltrexone (*larger doses* *than those required to* *block MOP-r agonist* *effects*). LY645,6302 (CERC-501)	Naltrindole	J-113,397

Cross-Species and Translational Aspects

Important early contributions in opioid drug discrimination were made in rodents, pigeons, and non-human primates (see below). Drug discrimination is also one of the behavioral pharmacology methodologies that can be applied in a translational manner in humans and experimental animals, including studies with opioid ligands [26, 27]. One current obstacle to the expansion of such translational studies is the paucity (for practical, safety, and regulatory reasons) of clinically available pharmacological agents with selective KOP-r, DOP-r, or NOP-r effects. For example, most drug discrimination studies in humans exploring KOP-r function have had to employ mixed opioids (e.g., pentazocine, nalbuphine, or butorphanol) [28], which exhibit intermediate efficacy at both MOP-r and KOP-r [23, 29, 30]. By contrast, several selective MOP-r agonists can be studied in appropriately designed clinical studies (see Table 1, for selected examples). Of methodological interest, studies have also shown that specific verbal training instructions in humans can also affect the selectivity of opioid drug discrimination assays [31].

1 Mu-Opioid Receptor (MOP-r) Systems and Drug Discrimination

MOP-r systems are involved in many brain and behavioral functions, including pain, analgesia, motivation, reward and addictions, and neuroendocrine function (e.g., in the hypothalamic-pituitary-adrenal [HPA] stress axis, as well as in the prolactin release) [32]. MOP-r receptors are present in spinal and supra-spinal areas mediating the aforementioned functions. MOP-r agonists have major clinical importance as analgesics for moderate to severe pain [33], but are also drugs of abuse, and currently result in substantial mortality, mainly caused by respiratory depression [34, 35].

MOP-r agonists cause increases in activation of dopaminergic pathways, and this may be of special relevance to their euphoric and reinforcing effects, and abuse potential [36, 37]. The prototypical MOP-r agonists morphine and fentanyl were among the earliest opioid compounds to be studied in drug discrimination assays [38–40].

Generalization Across the Discriminative Effects of MOP-r Agonists

Drug discrimination techniques have contributed considerably to our understanding of the behavioral pharmacology of these important compounds. For example, different MOP-r agonists tend to share discriminative stimulus effects, and thus are "cross-generalized" by subjects trained to discriminate a specific MOP-r agonist, across a variety of conditions and species. As an early example, when fentanyl was trained as a discriminative stimulus in rats, other structurally diverse centrally penetrating MOP-r agonists (e.g., methadone and morphine) were generalized [38]. In a different example, rhesus monkeys trained to discriminate heroin generalized to its active metabolites including morphine and morphine-6-glucuronide, and structurally diverse MOP-r agonists, such as fentanyl and methadone [41]. This profile has allowed the use of drug discrimination assays to investigate abuse potential of novel MOP-r compounds or formulations. Thus, it can be postulated that novel compounds that do not generalize to abused compounds such as classic MOP-r agonists may present lesser abuse liability.

Pharmacological Specificity of MOP-r Agonist Discriminative Stimulus

One of the strengths of drug discrimination as an in vivo assay is that can exhibit prominent pharmacological specificity, in that compounds that have similar pharmacodynamic effects tend to produce similar discriminative stimuli (see above). Thus, if an MOP-r agonist is the training compound in a drug discrimination assay, other MOP-r agonists are generalized, whereas compounds from other pharmacological classes (such as KOP-r or DOP-r agonists, or compounds acting at other systems) are typically not generalized [14, 41–43]. This feature is especially useful for the study of novel compounds at doses below those that may produce overt behavioral effects, or for classes of compounds that may share some overt behavioral effects (such as antinociception or locomotor effects), but can be differentiated by their discriminative effects in animals and potentially in humans. One major clinically relevant example of such pharmacological specificity is the finding that mixed opioid ligands (such as pentazocine) can be differentiated from a selective MOP-r agonist (e.g., hydromorphone) in a three-way discrimination, that is, pentazocine vs hydromorphone vs. saline. A likely reason for such a profile is that mixed opioids such as pentazocine also have KOP-r mediated effects, in addition to intermediate efficacy at MOP-r receptors [26, 44]. The discriminative stimulus effects of such "mixed opioids" (including widely used compounds such as buprenorphine) vary across species. This is possibly due to differential receptor selectivity or differential signaling efficacy across species [30, 45–47].

Differential Pharmacodynamic Efficacy of MOP-r Ligands as Detected in Drug Discrimination Studies (High Efficacy Agonists Versus Partial Agonists)

Under specific conditions, drug discrimination studies can also be used to differentiate pharmacodynamic (or signaling) efficacy of MOP-r ligands. As mentioned above, when a relatively high dose of the high efficacy MOP-r agonist fentanyl was trained as discriminative stimulus in rats, other high efficacy agonists such as methadone could be generalized; whereas, compounds with MOP-r partial agonist effects (e.g., nalbuphine) were not fully generalized, presumably because they were unable to produce a neurobiological signal of the required intensity [21].

A second type of approach that has been used to detect differences in pharmacodynamic efficacy in drug discrimination assays includes the use of ligands that cause irreversible or functionally irreversible MOP-r antagonism, such as β-funaltrexamine or clocinnamox (see Table 1), which cause a decrease in MOP-r B_{max} [16, 48, 49]. As in other in vivo assays (such as antinociception), discriminative effects of partial agonists such as morphine were more sensitive to such a decrease in available MOP-r populations, than higher efficacy agonists such as fentanyl [16, 50, 51]. See also below for further discussion of tolerance to discriminative effects of opioids.

Overall, these differential discriminative patterns of partial MOP-r agonists versus higher efficacy agonists can potentially inform preclinical, clinical, and regulatory investigators to the relative profile and abuse potential of novel compounds, including compounds with effects partially mediated through non-opioid receptor mechanisms [18, 52–55].

Drug Discrimination Studies of MOP-r Agonist Tolerance

Repeated exposure to MOP-r agonists is known to result in tolerance, that is, a decrease in the observed effect of a given dose (often quantified as a rightward shift in agonist dose-effect curves). Tolerance can be observed clinically both in analgesia and drug addiction settings, and is of relevance to acute MOP-r toxicity (i.e., respiratory depression) in persons with different amounts of MOP-r exposure [33, 56]. Tolerance per se, in the context of appropriate medical control of pain, is not indicative of abuse or addiction. Neurobiological mechanisms of tolerance may differ across in vitro and in vivo endpoints, and are not fully understood, despite intensive study [57]. Tolerance has thus been detected after repeated MOP-r agonist exposure in experimental animals, in some, but not all drug discrimination studies [58–61]. For example, in rats trained to discriminate morphine from vehicle, dose-effect curves for morphine and other MOP-r agonists were shifted to the right (i.e., a decrease in potency observed due to tolerance) due to chronic morphine exposure (14–18 days). In this study [61], training was suspended during the chronic morphine exposure period, to minimize the risk of "re-training" to functionally decreasing doses of the training drug [61]. Differential methodological factors in training and testing (including the aforementioned suspension of training during chronic MOP-r agonist exposure) may underlie these apparent differences.

Drug Discrimination Studies of MOP-r Agonist Dependence and Withdrawal

Other neurobiological and behavioral hallmarks of chronic MOP-r agonist administration in humans and experimental animals are dependence and withdrawal. For example, repeated administration of MOP-r agonists (and of several other types of compounds, not necessarily all abused) results in neurobiological and behavioral adaptations that can be discerned upon drug discontinuation as a "withdrawal syndrome." The classic MOP-r agonist withdrawal syndrome is aversive (with a variety of subjective effects, including anxiogenesis) and includes autonomic/sympathetic over-activation, piloerection, tremors, diarrhea, and hypothalamic-pituitary-adrenal (HPA) hormonal axis activation [56, 62–64]. Several neurobiological and molecular mechanisms of MOP-r agonist dependence and withdrawal have been investigated [65–67]. Avoidance and escape from MOP-r withdrawal can be studied as processes underlying negative reinforcement in experimental animals [68, 69]. Similarly to tolerance above, the presence of dependence or withdrawal in the context of appropriate medical care for pain is not alone indicative of abuse or addiction.

Drug discrimination studies have also been used to characterize internal stimuli of withdrawal, in that animals that receive chronic MOP-r agonists (e.g., morphine) can be trained to respond differentially when drug is acutely discontinued, or when there is precipitated short-term withdrawal caused by a relatively small dose of an opioid antagonist (e.g., naloxone or naltrexone) [70]. For example, in rhesus monkeys chronically treated with morphine, a low dose of the opioid antagonist naltrexone can be trained as a discriminative stimulus, and short-term morphine discontinuation can be generalized to the naltrexone stimulus [70, 71]. Like other opioid discriminative stimuli, this endpoint can be sensitive and repeatable, and can be examined under conditions that do not cause robust overt withdrawal signs. Some drug discrimination studies also have investigated the phenomenon of "acute withdrawal," in which a single relatively large dose of MOP-r agonist is rapidly followed by treatment with an antagonist such as naloxone or naltrexone. This acute withdrawal may share some behavioral and neurobiological similarities to classic withdrawal mechanisms observed after chronic MOP-r agonist exposure (mentioned above) [72].

Overall, MOP-r antagonists alone (in the absence of MOP-r agonist exposure, see above) have low potency or effectiveness as discriminative stimuli. This may be an indication that in unperturbed subjects, major MOP-r systems have relatively low basal "tone" (i.e., limited endogenous agonist occupancy or receptor signaling). One exception may be the hypothalamic-pituitary-adrenal (HPA) axis, in that compounds such as naloxone and naltrexone can acutely cause increases in levels of stress hormones ACTH and cortisol, in human and non-human primates, in the absence of chronic MOP-r agonist exposure [73, 74].

Of interest, compounds used clinically to decrease certain withdrawal signs (such as the adrenergic α_2-agonist clonidine) do not robustly or dose-dependently block the postulated "withdrawal" discriminative stimulus, even though they partially block some subjective, overt and autonomic effects, in humans and non-human primates [70, 75, 76]. This is an illustration of drug discrimination as a practical tool to examine clinically important interoceptive experiences of withdrawal, which may be dissociated from overt signs, and may be of relevance to processes of continued addiction and relapse.

Relationship of Discriminative Effects of MOP-r Agonists to Other Behavioral, Physiological and Neurobiological Effects, and Abuse Potential

There is typically a positive correlation between the potency (e.g., ED_{50} values) of centrally penetrating MOP-r agonists in causing discriminative stimuli and in causing clinically relevant in vivo effects, including antinociception/analgesia, respiratory depression, and reward-related effects [77, 78]. Generally, operant drug discrimination tends to occur at smaller doses than some of the aforementioned endpoints (when studied within a species and route of administration). Thus, discriminative stimulus effects may be a useful predictive biomarker for these other effects of relevance to preclinical and clinical evaluations, including for abuse potential.

Generalization of novel compounds to the discriminative effects of known drugs of abuse (e.g., MOP-r agonists) is indicative of abuse potential, of relevance during the drug development process, and to regulatory evaluation [55, 79, 80]. In a related manner, compounds with opioid receptor components of action such as the analgesic tramadol can be evaluated for cross-generalization to standard MOP-r agonists (e.g., the prescription opioid hydromorphone) [18].

2 Kappa Opioid Receptor (KOP-r) Systems and Drug Discrimination

KOP-r receptors and their endogenous high efficacy agonist ligands (the dynorphins) are also widely distributed in areas in the CNS mediating motivated behaviors and euphoria/dysphoria (e.g., caudate–putamen and nucleus accumbens), learning, memory, and emotional processing (e.g., hippocampus and amygdala), neuroendocrine function (e.g., hypothalamus), as well as several cortical areas [81–83]. Exogenous KOP-r agonists tend to cause aversion and dysphoria, and also sedation and psychotomimetic or hallucinogenic effects, as investigated in preclinical and clinical studies [12, 84–86]. Of interest, KOP-r agonists tend to decrease synaptic dopamine overflow, an effect opposite to that of MOP-r agonists and other abused compounds, including cocaine and ethanol [36, 37, 87]. Plasticity in KOP-r/dynorphin systems (typically upregulation) has been detected experimentally after exposure to different drugs of abuse or diverse stresses; this appears to be a mechanism underlying escalation of drug taking, and relapse-like, depressant-like, or anxiety-like behaviors [88–92].

Selective KOP-r agonists or antagonists have been examined in several clinical studies [93–95], but have not been examined in formal drug discrimination studies in humans, to our knowledge. Of interest, salvinorin A, a diterpene derived from the plant *Salvia divinorum*, is a selective high efficacy KOP-r agonist and there has been a recent expansion in its non-medical use (see below) [96]. As alluded to above, clinically used compounds such as buprenorphine, nalbuphine, and butorphanol have considerable affinity at both MOP-r and KOP-r, and have intermediate signaling efficacy, which can vary according to experimental situation [29, 30].

Specificity of Discriminative Effects of KOP-r Agonists
Selective exogenous KOP-r agonists produce discriminative stimulus effects that are differentiated from those of MOP-r or DOP-r agonists, as studied in different non-human species [12, 14, 97]. Selective KOP-r agonists, be they synthetic or plant-derived (such as salvinorin A, from the plant *Salvia divinorum*, the focus of considerable non-medical use) produce dissociative, psychotomimetic, or

hallucinogenic effects in humans [84, 85]. However, KOP-r agonist-induced discriminative effects are distinct from those of pharmacologically unrelated compounds that produce dissociative or hallucinogenic effects in humans, including 5HT2A agonists such as LSD or psilocybin, or NMDA antagonists such as ketamine [98, 99]. Of translational relevance, human subjects who received salvinorin A have also reported that the interoceptive or experiential/subjective effects of this KOP-r ligand may differ from those of classic hallucinogens, for example, by causing more intense dissociative and somatic effects [100, 101]. This further illustrates the degree of in vivo pharmacological selectivity that can be afforded by drug discrimination assays, since such behavioral or experiential/subjective effects (dysphoria, dissociation, hallucinations, or psychotomimetic effects) may be ultimately produced by ligands acting directly on different receptor systems and neuronal pathways [102]. This suggests a continuing potential contribution of drug discrimination studies to more mechanism-based analysis of such experiential/ subjective effects [103, 104]. The relative potency of centrally penetrating KOP-r agonists in drug discrimination assays is also positively correlated with their potency in several other behavioral assays, including clinically undesirable effects such as sedation, and also with the translational neuroendocrine biomarker assay of prolactin levels [12, 93, 105].

3   Delta Opioid Receptor (DOP-r) Systems and Drug Discrimination

Peptidic and non-peptidic DOP-r agonists produce characteristic discriminative effects that can be differentiated from those of MOP-r and KOP-r agonists, typically by a lack of cross-generalization (also, peptidic DOP-r agonists have been typically administered by the i.c.v route, to bypass their exclusion by the blood–brain barrier) [10, 14, 106]. As an example, rhesus monkeys trained to discriminate the DOP-r agonist SNC80 exhibited at most nondose-dependent partial generalization when administered MOP-r or KOP-r agonists (morphine and U50,488 respectively) [10]. Likewise, selective antagonism studies also confirm the pharmacological specificity of these stimuli. For example, the DOP-r selective antagonist naltrindole can potently block the discriminative effects of the synthetic DOP-r agonists SNC80 or BW373U86 [10, 14]. These studies further illustrate the pharmacological specificity of drug discrimination assays, and their use in the characterization of novel ligands, including those that may produce subtle unconditioned behavioral effects.

Nociceptin/Orphanin Systems and Drug Discrimination
The NOP-r receptor system (and its endogenous agonist, nociception/orphanin FQ) have some genetic, functional and neuroanatomical homology to MOP-r, KOP-r, and DOP-r systems [107]. Certain clinically used ligands, such as buprenorphine, do have NOP-r mediated effects, although it is unknown to what extent these are important to their clinical profile [108]. NOP-r ligands have been investigated for different

pharmacotherapeutic indications, especially analgesia [109–111]. Based on the limited number of available studies in experimental animals [112, 113], NOP-r agonists also produce characteristic discriminative stimulus effects, distinct from those of MOP-r, KOP-r, or DOP-r agonists. Thus, in rats trained to discriminate the NOP-r agonist Ro 64-6198, morphine produced a maximum of 40% drug-appropriate responding, and KOP-r and DOP-r agonists each produced less than 25% drug-appropriate responding (mean values) [112]. In the same study, Ro 64-6198 only produced ≤25% morphine appropriate responding in a separate group of rats. This suggests that NOP-r agonists do not share interoceptive effects of standard agonists at MOP-r, KOP-r, or DOP-r.

Please see Table 1 for a summary of major agonists and antagonists that can be potentially useful for drug discrimination studies examining the above opioid receptor systems.

4 Current Trends and Potential Future Directions

As previously noted, drug discrimination assays with opioid ligands continue to have an important role in behavioral pharmacology, including studies of abuse potential in novel drugs. Thus, novel opioid ligands with discriminative stimulus effects different from those of MOP-r agonists such as heroin or abused prescription opioids could be considered to have decreased likelihood of abuse potential.

"Biased" Ligands
The pharmacological specificity of drug discrimination assays can also be used to investigate timely mechanistic questions in the larger opioid neurobiology field. For example, it is unknown to date whether opioid ligands acting at the same receptor, but with different downstream signaling "bias" (e.g., at G-protein, adenylyl cyclase, β-arrestin, and other downstream pathways) would have differential discriminative stimuli [114, 115]. For example, it would be of interest to determine whether MOP-r ligands with differential "bias" also have differential discriminative effects, as this may be relevant to their profile of desirable and undesirable effects in the clinic (e.g., analgesia vs. abuse potential).

Drug Discrimination as a Behavioral "Readout" in Studies of Opioid Neuro-biology, and as Dimensional Variables
Drug discrimination assays, due to their relative robustness, repeatability, and quantitative nature, are suitable for relatively low "n" studies. Such "behavioral readouts" can be useful in parallel with other techniques, including PET or fMRI neuroimaging, to provide a measure of the interoceptive qualities associated with a particular neuroimaging signal (e.g., receptor occupancy by a particular ligand, acting at MOP-r, KOP-r, or DOP-r). The pharmacological specificity of discriminative stimuli (mentioned above) would also be an asset in such biomarker studies. A relatively small number of studies have investigated the neuroanatomical site (s) of opioid-receptor mediated discriminative effects [116–118], and this may be potentially investigated further with neuroimaging approaches.

The potential of discriminative effects of opioids to be used as "dimensional" variables for the study of mental health, pain/analgesia, and specific addictions can also be considered [119]. For example, drug discrimination studies can be designed comparatively for humans and experimental animals, allowing translational mechanistic studies. In addition, drug discrimination dose-effect curves can be examined as dimensional markers for underlying neurobiological and pharmacodynamic mechanisms.

Sex Differences
Sex differences in discriminative effects of opioids have been studied primarily in rodent models [120, 121]. Some of the sex differences reported include a greater potency of the discriminative effects of the MOP-r agonist morphine in females vs. male rats, whereas the converse was observed for the discriminative effects of the KOP-r agonist U69,593 [120, 122]. Further studies on sex differences in discriminative effects of specific opioid ligands are of importance, for both basic and clinical science [123]. For example, a mechanistic investigation of the aforementioned sex differences in MOP-r induced discriminative effects may be translationally relevant to differential profiles of abuse of MOP-r agonists in women and men [124, 125].

Genetics, Epigenetics, and Clinical Status in Drug Discrimination Studies
Opioid receptor systems (and their endogenous ligands) exhibit clinically relevant genetic polymorphisms (e.g., at *OPRM1*, *OPRK1*, and *OPRD1*, as well as *POMC*, *PENK1*, and *PDYN*), and are also affected by epigenetics and environmental history [126–130]. Opioid receptor and neuropeptide systems are altered (for example, at the mRNA or protein level), by exposure to stress or to drugs of abuse, of relevance to diverse mental health conditions (see reviews) [89, 131]. Therefore, discriminative stimulus effects of opioid ligands could be hypothesized to differ based on such genetic, epigenetic, and stress/environmental factors, and this could be studied in appropriate animal models, including transgenic constructs, as well as in specific clinical populations. Overall, drug discrimination has the potential to remain a powerful methodology for modern neurobiological, neuropharmacological, and translational studies.

Acknowledgements Experimental work by the authors was supported by the NIH-NIDA grants DA011113 and DA017369 (ERB), and DA05130, which are gratefully acknowledged.

References

1. Colpaert FC (1999) Drug discrimination in neurobiology. Pharmacol Biochem Behav 64:337–345
2. Colpaert FC, Niemegeers CJ, Janssen PA (1975) The narcotic cue: evidence for the specificity of the stimulus properties of narcotic drugs. Arch Int Pharmacodyn Ther 218:268–276
3. Colpaert FC, Niemegeers CJ, Janssen PA (1976) The narcotic discriminative stimulus complex: relation to analgesic activity. J Pharm Pharmacol 28:183–187

4. Hein DW, Young AM, Herling S, Woods JH (1981) Pharmacological analysis of the discriminative stimulus characteristics of ethylketazocine in the rhesus monkey. J Pharmacol Exp Ther 218:7–15
5. Schaefer GJ, Holtzman SG (1978) Discriminative effects of cyclazocine in the squirrel monkey. J Pharmacol Exp Ther 205:291–301
6. Martin WR, Eades CG, Thompson JA, Huppler RE, Gilbert PE (1976) The effects of morphine- and nalorphine-like drugs in the nondependent and morphine-dependent chronic spinal dog. J Pharmacol Exp Ther 197:517–532
7. Woods JH, Young AM, Herling S (1982) Classification of narcotics on the basis of their reinforcing, discriminative, and antagonist effects in rhesus monkeys. Fed Proc 41:221–227
8. Butelman ER, Ko MC, Traynor JR, Vivian JA, Kreek MJ, Woods JH (2001) GR89,696: a potent kappa-opioid agonist with subtype selectivity in rhesus monkeys. J Pharmacol Exp Ther 298:1049–1059
9. France CP, de Costa BR, Jacobson AE, Rice KC, Woods JH (1990) Apparent affinity of opioid antagonists in morphine-treated rhesus monkeys discriminating between saline and naltrexone. J Pharmacol Exp Ther 252:600–604
10. Brandt MR, Negus SS, Mello NK, Furness MS, Zhang X, Rice KC (1999) Discriminative stimulus effects of the nonpeptidic delta-opioid agonist SNC80 in rhesus monkeys. J Pharmacol Exp Ther 290:1157–1164
11. Carey GJ, Bergman J (2001) Enadoline discrimination in squirrel monkeys: effects of opioid agonists and antagonists. J Pharmacol Exp Ther 297:215–223
12. Dykstra LA, Gmerek DE, Winger G, Woods JH (1987) Kappa opioids in rhesus monkeys. I. Diuresis, sedation, analgesia and discriminative stimulus effects. J Pharmacol Exp Ther 242:413–420
13. Mori T, Yoshizawa K, Ueno T, Nishiwaki M, Shimizu N, Shibasaki M, Narita M, Suzuki T (2013) Involvement of dopamine D2 receptor signal transduction in the discriminative stimulus effects of the kappa-opioid receptor agonist U-50,488H in rats. Behav Pharmacol 24:275–281
14. Comer SD, McNutt RW, Chang KJ, De Costa BR, Mosberg HI, Woods JH (1993) Discriminative stimulus effects of BW373U86: a nonpeptide ligand with selectivity for delta opioid receptors. J Pharmacol Exp Ther 267:866–874
15. Dykstra LA, Gmerek DE, Winger G, Woods JH (1987) Kappa opioids in rhesus monkeys. II. Analysis of the antagonistic actions of quadazocine and beta-funaltrexamine. J Pharmacol Exp Ther 242:421–427
16. Walker EA, Young AM (2002) Clocinnamox distinguishes opioid agonists according to relative efficacy in normal and morphine-treated rats trained to discriminate morphine. J Pharmacol Exp Ther 302:101–110
17. Harun N, Hassan Z, Navaratnam V, Mansor SM, Shoaib M (2015) Discriminative stimulus properties of mitragynine (kratom) in rats. Psychopharmacology (Berl) 232:2227–2238
18. Strickland JC, Rush CR, Stoops WW (2015) Mu opioid mediated discriminative-stimulus effects of tramadol: an individual subjects analysis. J Exp Anal Behav 103:361–374
19. Comer SD, France CP, Woods JH (1991) Training dose: influences in opioid drug discrimination. NIDA Res Monogr 116:145–161
20. Stolerman IP, Childs E, Ford MM, Grant KA (2011) Role of training dose in drug discrimination: a review. Behav Pharmacol 22:415–429
21. Zhang L, Walker EA, Sutherland J 2nd, Young AM (2000) Discriminative stimulus effects of two doses of fentanyl in rats: pharmacological selectivity and effect of training dose on agonist and antagonist effects of mu opioids. Psychopharmacology (Berl) 148:136–145
22. Bi J (2006) Sensory discrimination tests and measurements: statistical principles, procedures and tables. Blackwell, Ames
23. Emmerson PJ, Clark MJ, Mansour A, Akil H, Woods JH, Medzihradsky F (1996) Characterization of opioid agonist efficacy in a C6 glioma cell line expressing the mu opioid receptor. J Pharmacol Exp Ther 278:1121–1127

24. Peckham EM, Traynor JR (2006) Comparison of the antinociceptive response to morphine and morphine-like compounds in male and female Sprague–Dawley rats. J Pharmacol Exp Ther 316:1195–1201

25. Traynor JR, Nahorski SR (1995) Modulation by mu-opioid agonists of guanosine-5′-O-(3-[35S]thio)triphosphate binding to membranes from human neuroblastoma SH-SY5Y cells. Mol Pharmacol 47:848–854

26. Bickel WK, Bigelow GE, Preston KL, Liebson IA (1989) Opioid drug discrimination in humans: stability, specificity and relation to self-reported drug effect. J Pharmacol Exp Ther 251:1053–1063

27. Kamien JB, Bickel WK, Hughes JR, Higgins ST, Smith BJ (1993) Drug discrimination by humans compared to nonhumans: current status and future directions. Psychopharmacology (Berl) 111:259–270

28. Jones HE, Bigelow GE, Preston KL (1999) Assessment of opioid partial agonist activity with a three-choice hydromorphone dose-discrimination procedure. J Pharmacol Exp Ther 289:1350–1361

29. Remmers AE, Clark MJ, Mansour A, Akil H, Woods JH, Medzihradsky F (1999) Opioid efficacy in a C6 glioma cell line stably expressing the human kappa opioid receptor. J Pharmacol Exp Ther 288:827–833

30. Zhu J, Luo LY, Li JG, Chen C, Liu-Chen LY (1997) Activation of the cloned human kappa opioid receptor by agonists enhances [35S]GTPgammaS binding to membranes: determination of potencies and efficacies of ligands. J Pharmacol Exp Ther 282:676–684

31. Preston KL, Bigelow GE (2000) Effects of agonist–antagonist opioids in humans trained in a hydromorphone/not hydromorphone discrimination. J Pharmacol Exp Ther 295:114–124

32. Kreek MJ, Levran O, Reed B, Schlussman SD, Zhou Y, Butelman ER (2012) Opiate addiction and cocaine addiction: underlying molecular neurobiology and genetics. J Clin Invest 122:3387–3393

33. Yaksh TL, Wallace MS (2011) Opioids, analgesia and pain management. In: Brunton L, Chabner B, Knollman B (eds) Goodman and Gilman's the pharmacological basis of therapeutics, 12th edn. McGraw-Hill, New York

34. C.D.C. (2015) Injury prevention & control: prescription drug overdose. From http://www.cdc.gov/DrugOverdose/

35. Dart RC, Surratt HL, Cicero TJ, Parrino MW, Severtson SG, Bucher-Bartelson B, Green JL (2015) Trends in opioid analgesic abuse and mortality in the United States. N Engl J Med 372:241–248

36. Di Chiara G, Imperato A (1988) Opposite effects of mu and kappa opiate agonists on dopamine release in the nucleus accumbens and in the dorsal caudate of freely moving rats. J Pharmacol Exp Ther 244:1067–1080

37. Spanagel R, Herz A, Shippenberg TS (1992) Opposing tonically active endogenous opioid systems modulate the mesolimbic dopaminergic pathway. Proc Natl Acad Sci U S A 89:2046–2050

38. Colpaert FC, Niemegeers CJ (1975) On the narcotic cuing action of fentanyl and other narcotic analgesic drugs. Arch Int Pharmacodyn Ther 217:170–172

39. Jarbe TU (1978) Discriminative effects of morphine in the pigeon. Pharmacol Biochem Behav 9:411–416

40. Woods JH, Herling S, Valentino RJ, Hein DW, Coale EH Jr (1979) Narcotic drug discriminations by rhesus monkeys and pigeons. NIDA Res Monogr 27:128–134

41. Platt DM, Rowlett JK, Spealman RD (2001) Discriminative stimulus effects of intravenous heroin and its metabolites in rhesus monkeys: opioid and dopaminergic mechanisms. J Pharmacol Exp Ther 299:760–767

42. Colpaert FC, Niemegeers CJ, Janssen PA (1976) Fentanyl and apomorphine: asymmetrical generalization of discriminative stimulus properties. Neuropharmacology 15:541–545

43. Herling S, Woods JH (1981) Discriminative stimulus effects of etorphine in Rhesus monkeys. Psychopharmacology (Berl) 72:265–267

44. Preston KL, Bigelow GE, Bickel WK, Liebson IA (1989) Drug discrimination in human postaddicts: agonist–antagonist opioids. J Pharmacol Exp Ther 250:184–196

45. Gerak LR, France CP (1996) Discriminative stimulus effects of nalbuphine in rhesus monkeys. J Pharmacol Exp Ther 276:523–531

46. Negus SS, Picker MJ, Dykstra LA (1989) Kappa antagonist effects of buprenorphine in the rat drug-discrimination procedure. NIDA Res Monogr 95:518–519

47. Picker MJ, Craft RM, Negus SS, Powell KR, Mattox SR, Jones SR, Hargrove BK, Dykstra LA (1992) Intermediate efficacy mu opioids: examination of their morphine-like stimulus effects and response rate-decreasing effects in morphine-tolerant rats. J Pharmacol Exp Ther 263:668–681

48. Barrett AC, Smith ES, Picker MJ (2003) Use of irreversible antagonists to determine the relative efficacy of mu-opioids in a pigeon drug discrimination procedure: comparison of beta-funaltrexamine and clocinnamox. J Pharmacol Exp Ther 305:1061–1070

49. Burke TF, Woods JH, Lewis JW, Medzihradsky F (1994) Irreversible opioid antagonist effects of clocinnamox on opioid analgesia and mu receptor binding in mice. J Pharmacol Exp Ther 271:715–721

50. Comer SD, Burke TF, Lewis JW, Woods JH (1992) Clocinnamox: a novel, systemically-active, irreversible opioid antagonist. J Pharmacol Exp Ther 262:1051–1056

51. Zernig G, Burke T, Lewis JW, Woods JH (1996) Mechanism of clocinnamox blockade of opioid receptors: evidence from in vitro and ex vivo binding and behavioral assays. J Pharmacol Exp Ther 279:23–31

52. Carter LP, Griffiths RR (2009) Principles of laboratory assessment of drug abuse liability and implications for clinical development. Drug Alcohol Depend 105(Suppl 1):S14–25

53. Glennon RA, Young R, Negus SS, Banks ML (2011) Making the right choice: lessons from drug discrimination for research on drug reinforcement and drug self-administration. In: Glennon RA, Young R (eds) Drug discrimination: applications to medicinal chemistry and drug studies. Wiley, New York

54. Marusich JA, Lefever TW, Novak SP, Blough BE, Wiley JL (2013) Prediction and prevention of prescription drug abuse: role of preclinical assessment of substance abuse liability. Methods Rep RTI Press 1–14

55. Negus SS, Fantegrossi W (2008) Overview of conference on preclinical abuse liability testing: current methods and future challenges. Drug Alcohol Depend 92:301–306

56. O'Brien CP (2011) Drug addiction. In: Brunton L, Chabner B, Knollman B (eds) Goodman and Gilman's the pharmacological basis of therapeutics, 12th edn. McGraw-Hill, New York

57. Williams JT, Ingram SL, Henderson G, Chavkin C, von Zastrow M, Schulz S, Koch T, Evans CJ, Christie MJ (2013) Regulation of mu-opioid receptors: desensitization, phosphorylation, internalization, and tolerance. Pharmacol Rev 65:223–254

58. Colpaert FC, Kuyps JJ, Niemegeers CJ, Janssen PA (1976) Discriminative stimulus properties of fentanyl and morphine: tolerance and dependence. Pharmacol Biochem Behav 5:401–408

59. Galici R, McMahon LR, France CP (2005) Cross-tolerance and mu agonist efficacy in pigeons treated with LAAM or buprenorphine. Pharmacol Biochem Behav 81:626–634

60. Paronis CA, Holtzman SG (1994) Sensitization and tolerance to the discriminative stimulus effects of mu-opioid agonists. Psychopharmacology (Berl) 114:601–610

61. Young AM, Kapitsopoulos G, Makhay MM (1991) Tolerance to morphine-like stimulus effects of mu opioid agonists. J Pharmacol Exp Ther 257:795–805

62. Gold MS, Pottash AL, Sweeney DR, Kleber HD (1980) Efficacy of clonidine in opiate withdrawal: a study of thirty patients. Drug Alcohol Depend 6:201–208

63. Ignar DM, Kuhn CM (1990) Effects of specific mu and kappa opiate tolerance and abstinence on hypothalamo-pituitary-adrenal axis secretion in the rat. J Pharmacol Exp Ther 255:1287–1295

64. Rosen MI, McMahon TJ, Hameedi FA, Pearsall HR, Woods SW, Kreek MJ, Kosten TR (1996) Effect of clonidine pretreatment on naloxone-precipitated opiate withdrawal. J Pharmacol Exp Ther 276:1128–1135

65. McClung CA, Nestler EJ, Zachariou V (2005) Regulation of gene expression by chronic morphine and morphine withdrawal in the locus ceruleus and ventral tegmental area. J Neurosci 25:6005–6015
66. Seip-Cammack KM, Reed B, Zhang Y, Ho A, Kreek MJ (2012) Tolerance and sensitization to chronic escalating dose heroin following extended withdrawal in Fischer rats: possible role of mu-opioid receptors. Psychopharmacology (Berl) 225:127–140
67. Zhou Y, Bendor J, Hofmann L, Randesi M, Ho A, Kreek MJ (2006) Mu opioid receptor and orexin/hypocretin mRNA levels in the lateral hypothalamus and striatum are enhanced by morphine withdrawal. J Endocrinol 191:137–145
68. Hutcheson DM, Everitt BJ, Robbins TW, Dickinson A (2001) The role of withdrawal in heroin addiction: enhances reward or promotes avoidance? Nat Neurosci 4:943–947
69. Stinus L, Cador M, Zorrilla EP, Koob GF (2005) Buprenorphine and a CRF1 antagonist block the acquisition of opiate withdrawal-induced conditioned place aversion in rats. Neuropsychopharmacology 30:90–98
70. France CP, Woods JH (1989) Discriminative stimulus effects of naltrexone in morphine-treated rhesus monkeys. J Pharmacol Exp Ther 250:937–943
71. Becker GL, Gerak LR, Koek W, France CP (2008) Antagonist-precipitated and discontinuation-induced withdrawal in morphine-dependent rhesus monkeys. Psychopharmacology (Berl) 201:373–382
72. White DA, Holtzman SG (2003) Discriminative stimulus effects of acute morphine followed by naltrexone in the squirrel monkey. Psychopharmacology (Berl) 167:203–210
73. Schluger JH, Ho A, Borg L, Porter M, Maniar S, Gunduz M, Perret G, King A, Kreek MJ (1998) Nalmefene causes greater hypothalamic-pituitary-adrenal axis activation than naloxone in normal volunteers: implications for the treatment of alcoholism. Alcohol Clin Exp Res 22:1430–1436
74. Williams KL, Ko MC, Rice KC, Woods JH (2003) Effect of opioid receptor antagonists on hypothalamic-pituitary-adrenal activity in rhesus monkeys. Psychoneuroendocrinology 28:513–528
75. Katz JL (1986) Effects of clonidine and morphine on opioid withdrawal in rhesus monkeys. Psychopharmacology (Berl) 88:392–397
76. Oliveto A, Sevarino K, McCance-Katz E, Benios T, Poling J, Feingold A (2003) Clonidine and yohimbine in opioid-dependent humans responding under a naloxone novel-response discrimination procedure. Behav Pharmacol 14:97–109
77. Meert TF, Vermeirsch HA (2005) A preclinical comparison between different opioids: antinociceptive versus adverse effects. Pharmacol Biochem Behav 80:309–326
78. Walker EA, Makhay MM, House JD, Young AM (1994) In vivo apparent pA2 analysis for naltrexone antagonism of discriminative stimulus and analgesic effects of opiate agonists in rats. J Pharmacol Exp Ther 271:959–968
79. Gauvin DV, McComb M, Code R, Dalton JA, Baird TJ (2015) Abuse liability assessment of hydrocodone under current draft regulatory guidelines. J Pharmacol Toxicol Methods 75:118–129
80. Solinas M, Panlilio LV, Justinova Z, Yasar S, Goldberg SR (2006) Using drug-discrimination techniques to study the abuse-related effects of psychoactive drugs in rats. Nat Protoc 1:1194–1206
81. Mansour A, Fox CA, Meng F, Akil H, Watson SJ (1994) Kappa 1 receptor mRNA distribution in the rat CNS: comparison to kappa receptor binding and prodynorphin mRNA. Mol Cell Neurosci 5:124–144
82. Mathieu-Kia AM, Fan LQ, Kreek MJ, Simon EJ, Hiller JM (2001) Mu-, delta- and kappa-opioid receptor populations are differentially altered in distinct areas of postmortem brains of Alzheimer's disease patients. Brain Res 893:121–134
83. Simonin F, Gaveriaux-Ruff C, Befort K, Matthes H, Lannes B, Micheletti G, Mattei MG, Charron G, Bloch B, Kieffer B (1995) kappa-Opioid receptor in humans: cDNA and genomic cloning, chromosomal assignment, functional expression, pharmacology, and expression pattern in the central nervous system. Proc Natl Acad Sci U S A 92:7006–7010

84. Johnson MW, MacLean KA, Reissig CJ, Prisinzano TE, Griffiths RR (2011) Human psychopharmacology and dose-effects of salvinorin A, a kappa opioid agonist hallucinogen present in the plant Salvia divinorum. Drug Alcohol Depend 115:150–155

85. Pfeiffer A, Brantl V, Herz A, Emrich HM (1986) Psychotomimesis mediated by kappa opiate receptors. Science 233:774–776

86. Zhang Y, Butelman ER, Schlussman SD, Ho A, Kreek MJ (2005) Effects of the plant-derived hallucinogen salvinorin A on basal dopamine levels in the caudate putamen and in a conditioned place aversion assay in mice: agonist actions at kappa opioid receptors. Psychopharmacology (Berl) 179:551–558

87. Zhang Y, Butelman ER, Schlussman SD, Ho A, Kreek MJ (2004) Effect of the endogenous kappa opioid agonist dynorphin A(1–17) on cocaine-evoked increases in striatal dopamine levels and cocaine-induced place preference in C57BL/6J mice. Psychopharmacology (Berl) 172:422–429

88. Beardsley PM, Howard JL, Shelton KL, Carroll FI (2005) Differential effects of the novel kappa opioid receptor antagonist, JDTic, on reinstatement of cocaine-seeking induced by footshock stressors vs cocaine primes and its antidepressant-like effects in rats. Psychopharmacology (Berl) 183:118–126

89. Butelman ER, Yuferov V, Kreek MJ (2012) kappa-opioid receptor/dynorphin system: genetic and pharmacotherapeutic implications for addiction. Trends Neurosci 35:587–596

90. Schlosburg JE, Whitfield TW Jr, Park PE, Crawford EF, George O, Vendruscolo LF, Koob GF (2013) Long-term antagonism of kappa opioid receptors prevents escalation of and increased motivation for heroin intake. J Neurosci 33:19384–19392

91. Wang XM, Zhou Y, Spangler R, Ho A, Han JS, Kreek MJ (1999) Acute intermittent morphine increases preprodynorphin and kappa opioid receptor mRNA levels in the rat brain. Brain Res Mol Brain Res 66:184–187

92. Zhou Y, Leri F, Grella S, Aldrich J, Kreek MJ (2013) Involvement of dynorphin and kappa opioid receptor in yohimbine-induced reinstatement of heroin seeking in rats. Synapse 67:358–361

93. Chang C, Byon W, Lu Y, Jacobsen LK, Badura LL, Sawant-Basak A, Miller E, Liu J, Grimwood S, Wang EQ, Maurer TS (2011) Quantitative PK-PD model-based translational pharmacology of a novel kappa opioid receptor antagonist between rats and humans. AAPS J 13:565–575

94. Kreek MJ, Schluger J, Borg L, Gunduz M, Ho A (1999) Dynorphin A1-13 causes elevation of serum levels of prolactin through an opioid receptor mechanism in humans: gender differences and implications for modulation of dopaminergic tone in the treatment of addictions. J Pharmacol Exp Ther 288:260–269

95. Lowe SL, Wong CJ, Witcher J, Gonzales CR, Dickinson GL, Bell RL, Rorick-Kehn L, Weller M, Stoltz RR, Royalty J, Tauscher-Wisniewski S (2014) Safety, tolerability, and pharmacokinetic evaluation of single- and multiple-ascending doses of a novel kappa opioid receptor antagonist LY2456302 and drug interaction with ethanol in healthy subjects. J Clin Pharmacol 54:968–978

96. Roth BL, Baner K, Westkaemper R, Siebert D, Rice KC, Steinberg S, Ernsberger P, Rothman RB (2002) Salvinorin A: a potent naturally occurring nonnitrogenous kappa opioid selective agonist. Proc Natl Acad Sci U S A 99:11934–11939

97. Baker LE, Panos JJ, Killinger BA, Peet MM, Bell LM, Haliw LA, Walker SL (2009) Comparison of the discriminative stimulus effects of salvinorin A and its derivatives to U69,593 and U50,488 in rats. Psychopharmacology (Berl) 203:203–211

98. Butelman ER, Rus S, Prisinzano TE, Kreek MJ (2010) The discriminative effects of the kappa-opioid hallucinogen salvinorin A in nonhuman primates: dissociation from classic hallucinogen effects. Psychopharmacology (Berl) 210:253–262

99. Li JX, Rice KC, France CP (2008) Discriminative stimulus effects of 1-(2,5-dimethoxy-4-methylphenyl)-2-aminopropane in rhesus monkeys. J Pharmacol Exp Ther 324:827–833

100. Addy PH (2012) Acute and post-acute behavioral and psychological effects of salvinorin A in humans. Psychopharmacology (Berl) 220:195–204

101. Maqueda AE, Valle M, Addy PH, Antonijoan RM, Puntes M, Coimbra J, Ballester MR, Garrido M, Gonzalez M, Claramunt J, Barker S, Johnson MW, Griffiths RR, Riba J (2015) Salvinorin-A induces intense dissociative effects, blocking external sensory perception and modulating interoception and sense of body ownership in humans. Int J Neuropsychopharmacol 18:1–14

102. Mori T, Yoshizawa K, Shibasaki M, Suzuki T (2012) Discriminative stimulus effects of hallucinogenic drugs: a possible relation to reinforcing and aversive effects. J Pharmacol Sci 120:70–76

103. Krueger RF, Hopwood CJ, Wright AG, Markon KE (2014) Challenges and strategies in helping the DSM become more dimensional and empirically based. Curr Psychiatry Rep 16:515–521

104. NIMH-NIH (2011) NIMH research domain criteria (RDoC), 2013. From http://www.nimh.nih.gov/research-priorities/rdoc/nimh-research-domain-criteria-rdoc.shtml

105. Butelman ER, Ball JW, Kreek MJ (2002) Comparison of the discriminative and neuroendocrine effects of centrally penetrating kappa-opioid agonists in rhesus monkeys. Psychopharmacology (Berl) 164:115–120

106. Jewett DC, Mosberg HI, Woods JH (1996) Discriminative stimulus effects of a centrally administered, delta-opioid peptide (D-Pen2-D-Pen5-enkephalin) in pigeons. Psychopharmacology (Berl) 127:225–230

107. Peluso J, LaForge KS, Matthes HW, Kreek MJ, Kieffer BL, Gaveriaux-Ruff C (1998) Distribution of nociceptin/orphanin FQ receptor transcript in human central nervous system and immune cells. J Neuroimmunol 81:184–192

108. Cami-Kobeci G, Polgar W, Khroyan TV, Toll LR, Husbands SM (2011) Structural determinants of opioid and NOP receptor activity in derivatives of buprenorphine. J Med Chem

109. Chiou LC, Liao YY, Fan PC, Kuo PH, Wang CH, Riemer C, Prinssen EP (2007) Nociceptin/orphanin FQ peptide receptors: pharmacology and clinical implications. Curr Drug Targets 8:117–135

110. Ding H, Hayashida K, Suto T, Sukhtankar DD, Kimura M, Mendenhall V, Ko MC (2015) Supraspinal actions of nociceptin/orphanin FQ, morphine and substance P in regulating pain and itch in non-human primates. Br J Pharmacol 172:3302–3312

111. Ko MC, Woods JH, Fantegrossi WE, Galuska CM, Wichmann J, Prinssen EP (2009) Behavioral effects of a synthetic agonist selective for nociceptin/orphanin FQ peptide receptors in monkeys. Neuropsychopharmacology 34:2088–2096

112. Recker MD, Higgins GA (2004) The opioid receptor like-1 receptor agonist Ro 64-6198 (1S,3aS-8-2,3,3a,4,5,6-hexahydro-1H-phenalen-1-yl-1-phenyl-1,3,8-triaza-spiro[4.5]decan-4-one) produces a discriminative stimulus in rats distinct from that of a mu, kappa, and delta opioid receptor agonist cue. J Pharmacol Exp Ther 311:652–658

113. Saccone PA, Zelenock KA, Lindsey A, Zaks ME, Woods JH (2015) Characterization of the discriminative stimulus effects of the NOP agonist Ro 64-6198 in non-human primates. Drug Alcohol Depend 146, e86

114. Luttrell LM, Maudsley S, Bohn LM (2015) Fulfilling the promise of 'biased' GPCR agonism. Mol Pharmacol 88:579–588

115. White KL, Scopton AP, Rives ML, Bikbulatov RV, Polepally PR, Brown PJ, Kenakin T, Javitch JA, Zjawiony JK, Roth BL (2014) Identification of novel functionally selective kappa-opioid receptor scaffolds. Mol Pharmacol 85:83–90

116. Krivsky JA, Stoffel EC, Sumner JE, Inman BC, Craft RM (2006) Role of ventral tegmental area, periaqueductal gray and parabrachial nucleus in the discriminative stimulus effects of morphine in the rat. Behav Pharmacol 17:259–270

117. Shoaib M, Spanagel R (1994) Mesolimbic sites mediate the discriminative stimulus effects of morphine. Eur J Pharmacol 252:69–75

118. Yoshizawa K, Narita M, Saeki M, Isotani K, Horiuchi H, Imai S, Kuzumaki N, Suzuki T (2011) Activation of extracellular signal-regulated kinase is critical for the discriminative stimulus effects induced by U-50,488H. Synapse 65:1052–1061

119. Cuthbert BN (2014) Translating intermediate phenotypes to psychopathology: the NIMH Research Domain Criteria. Psychophysiology 51:1205–1206
120. Craft RM (2008) Sex differences in analgesic, reinforcing, discriminative, and motoric effects of opioids. Exp Clin Psychopharmacol 16:376–385
121. Neelakantan H, Ward SJ, Walker EA (2015) Discriminative stimulus effects of morphine and oxycodone in the absence and presence of acetic acid in male and female C57Bl/6 mice. Exp Clin Psychopharmacol 23:217–227
122. Craft RM, Kruzich PJ, Boyer JS, Harding JW, Hanesworth JM (1998) Sex differences in discriminative stimulus and diuretic effects of the kappa opioid agonist U69,593 in the rat. Pharmacol Biochem Behav 61:395–403
123. Clayton JA, Collins FS (2014) Policy: NIH to balance sex in cell and animal studies. Nature 509:282–283
124. Back SE, Payne RA, Waldrop AE, Smith A, Reeves S, Brady KT (2009) Prescription opioid aberrant behaviors: a pilot study of sex differences. Clin J Pain 25:477–484
125. Kuhn C (2015) Emergence of sex differences in the development of substance use and abuse during adolescence. Pharmacol Ther 153:55–78
126. Bond C, LaForge KS, Tian M, Melia D, Zhang S, Borg L, Gong J, Schluger J, Strong JA, Leal SM, Tischfield JA, Kreek MJ, Yu L (1998) Single-nucleotide polymorphism in the human mu opioid receptor gene alters beta-endorphin binding and activity: possible implications for opiate addiction. Proc Natl Acad Sci U S A 95:9608–9613
127. Morgan D, Cook CD, Picker MJ (1999) Sensitivity to the discriminative stimulus and antinociceptive effects of mu opioids: role of strain of rat, stimulus intensity, and intrinsic efficacy at the mu opioid receptor. J Pharmacol Exp Ther 289:965–975
128. Nielsen DA, Yuferov V, Hamon S, Jackson C, Ho A, Ott J, Kreek MJ (2009) Increased OPRM1 DNA methylation in lymphocytes of methadone-maintained former heroin addicts. Neuropsychopharmacology 34:867–873
129. Picetti R, Caccavo JA, Ho A, Kreek MJ (2012) Dose escalation and dose preference in extended-access heroin self-administration in Lewis and Fischer rats. Psychopharmacology (Berl) 220:163–172
130. Reed B, Butelman ER, Yuferov V, Randesi M, Kreek MJ (2014) Genetics of opiate addiction. Curr Psychiatry Rep 16:504
131. Bruchas MR, Land BB, Chavkin C (2010) The dynorphin/kappa opioid system as a modulator of stress-induced and pro-addictive behaviors. Brain Res 1314:44–55

Translational Value of Drug Discrimination with Typical and Atypical Antipsychotic Drugs

Joseph H. Porter, Kevin A. Webster, and Adam J. Prus

Abstract This chapter focuses on the *translational value* of drug discrimination as a preclinical assay for drug development. In particular, the importance of two factors, i.e., training dose and species, for drug discrimination studies with the atypical antipsychotic clozapine is examined. Serotonin receptors appear to be an important pharmacological mechanism mediating clozapine's discriminative cue in both rats and mice, although differences are clearly evident as antagonism of cholinergic muscarinic receptors is important in rats at a higher training dose (5.0 mg/kg) of clozapine, but not at a lower training dose (1.25 mg/kg). Antagonism of α_1 adrenoceptors is a sufficient mechanism in C57BL/6 and 129S2 mice to mimic clozapine's cue, but not in DBA/2 and B6129S mice, and only produces partial substitution in low-dose clozapine discrimination in rats. Dopamine antagonism produces partial substitution for clozapine in DBA/2, 129S2, and B6129S mice, but not in C57BL/6 mice, and partial substitution is seen with D_4 antagonism in low-dose clozapine drug discrimination in rats. Thus, it is evident that clozapine has a complex mixture of receptor contributions towards its discriminative cue based on the data from the four mouse strains that have been tested that is similar to the results from rat studies. A further examination of antipsychotic stimulus properties in humans, particularly in patients with schizophrenia, would go far in evaluating the translational value of the drug discrimination paradigm for antipsychotic drugs.

Keywords Antipsychotic drugs • Clozapine • Drug discrimination • Mouse strains • Training dose • Translational models

J.H. Porter (✉) and K.A. Webster
Department of Psychology, Virginia Commonwealth University, Richmond, VA 23284, USA
e-mail: jporter@vcu.edu

A.J. Prus
Department of Psychology, Northern Michigan University, Marquette, MI, USA

© Springer International Publishing AG 2017
Curr Topics Behav Neurosci (2018) 39: 193–212
DOI 10.1007/7854_2017_4
Published Online: 24 March 2017

Contents

This chapter, located within the volume *The Behavioural Neuroscience of Drug Discrimination* as part of the *Current Topics in Behavioral Neurosciences* series, reviews the major findings from the experimental literature to describe the current state of knowledge on the discriminative stimulus properties of antipsychotic drugs. As described below, much of that research has focused on clozapine (Clozaril®), which has remained as the "gold standard" among the atypical (second-generation) antipsychotics. One obvious goal of preclinical research is to develop assays that can be used for the development of new pharmacotherapeutic drugs that are more efficacious and have fewer side effects. While drug discrimination is not a model of human disorders like depression or schizophrenia, it does quantitatively measure the subjective effects of drugs (from different behavioral/therapeutic classifications) and their subjective (i.e., interoceptive) effects produced by antipsychotic drugs as well as by many others. Such subjective effects are primarily mediated by a drug's pharmacological action (typically, blockade or activation) at receptor sites. Thus, in addition to predicting the therapeutic efficacy of novel compounds, drug discrimination provides a universally straightforward and sensitive measure of how drug binding at the receptor level can influence conditioned behavioral events in vivo.

1   Treatments for Schizophrenia

Until 1952 and the discovery of the first antipsychotic drug chlorpromazine in France, there were no effective treatments for the symptoms of psychosis or other mental illnesses. Chlorpromazine was initially marketed as an antipsychotic in France in November 1952 as Largactil® ("large in action") and later in the USA in March 1954 as Thorazine® [1]. Chlorpromazine's ability to effectively treat psychotic symptoms marked the birth of psychopharmacology and provided a concrete link to organic roots of mental illnesses – e.g., the dopamine hypothesis of schizophrenia first proposed by Jacques M. van Rossum ([2]; for reviews see Baumeister and Francis [3] and Snyder et al. [4]). Perhaps the biggest impact of antipsychotic drug treatments was their ability to allow hospitalized schizophrenics to live relatively normal lives outside of mental hospitals and asylums. In fact, the therapeutic use of drugs to treat the symptoms of schizophrenia, depression, and anxiety initiated a significant reduction in institutionalized patients that began in the 1950s with a historic high of 559,000 patients in 1955 and continued over the next 30 years to 107,000 in 1988 [5]. One of the goals of psychopharmacology was to better understand the behavioral effects and neuropharmacological properties of these new therapeutic drugs that were rapidly being placed into clinical use throughout the 1950s and 1960s. It became apparent fairly early in clinical use that these first-generation ("typical") antipsychotic drugs (neuroleptics) produced extrapyramidal motor side effects (EPS) that resembled Parkinsonian symptoms, which appear to be due to antagonism of dopamine receptors. In fact, for many years clinicians believed that EPS and the therapeutic efficacy of the first-generation antipsychotics were explicatively linked. It was not until the atypical antipsychotic clozapine was found to be as (and perhaps more) effective for the treatment of schizophrenia that this belief was dispelled [6, 7]. One major difference in clozapine was that it differed from typical antipsychotics, displaying greater affinity for serotonin $5HT_2$ receptors relative to dopamine D_2 receptors [8].

The promise of atypical antipsychotics took a hit early in the 1970s when the early clinical use of clozapine was marked by a devastating clinical trial in Finland in which 17 patients (of about 3,000) developed agranulocytosis (a blood condition with reduced white blood cells) and eight of those patients died [9]. None-the-less, clozapine was eventually approved in the USA by the FDA on September 26, 1989 for use in treatment-resistant schizophrenia ([6]; www.accessdata.fda.gov). The introduction of clozapine for clinical use stimulated the development of a large number of these newer "atypical" (second-generation) antipsychotic drugs for the treatment of schizophrenia over the next 25 years; an initiative that greatly reduced EPS liability during treatment (although there are other significant side effects). Overall, there are advantages and disadvantages to the implementation of either atypical antipsychotic or typical antipsychotic, treatment strategies (e.g., [10]). Given the current state of medicinal chemistry, neuropsychopharmacology, and behavioral and molecular neuroscience, newer treatment strategies for schizophrenic and psychotic disorders are most certainly on the horizon, and drug discrimination will play a critical role in determining their pharmacotherapeutic efficacy and provide a better understanding of their in vivo behavioral effects.

2 Discriminative Stimulus Properties of Antipsychotic Drugs

Drug discrimination assays are useful in that they assess the subjective effect of a drug, usually referred to as a compound's "*discriminative cue*" or "*discriminative stimulus*" for that drug. Given that drugs belonging to similar therapeutic and/or behavioral classifications often share the same subjective effects in humans, as in other animals, drug discrimination studies offer a unique opportunity for translational approaches in nonhuman experimental preparations. Quantitatively, drug discrimination can be used to classify existing drugs and novel compounds and it can be further used to relate those subjective effects to specific receptor mechanisms in the brain. While there is clearly more research with drugs of abuse utilizing the drug discrimination paradigm, there is a substantial literature on the discriminative stimulus properties of psychotherapeutic drugs used in humans (e.g., see Chapter 11 – *The Discriminative Stimulus Properties of Drugs Used to Treat Depression and Anxiety* in this volume; [11]). The focus of the present chapter is on antipsychotic drugs and we provide a brief summary of the drug discrimination literature on antipsychotic drugs below, but we do not intend to review all of the literature in this field as comprehensive reviews are already available [12, 13]. Instead, this chapter will focus on the *translational value* of drug discrimination as a preclinical assay for drug development. The definition of translational research is somewhat vague and exactly how it differs from basic and applied research is debatable (see Fang and Casaderall [14]). In this chapter (and in Chapter 4 – Cross-species translational findings in the discriminative stimulus effects of ethanol) the term translation research is more focused on the ability to translate nonhuman animal studies findings to human studies and/or to the prevention or treatment of human disease. It should also be noted that the actual translation value of any basic research may not be known for years. As you will see in the sections to follow, the present chapter will focus on the role of training dose and cross-species comparisons.

The typical antipsychotic chlorpromazine was first tested in a discrimination task (three-compartment test chamber similar to a T-maze) by Stewart [15]. In this study, rats were trained to discriminate chlorpromazine (4.0 mg/kg, i.p.) from saline and it was found that other phenothiazines fully substituted for chlorpromazine. Barry et al. [16] were the first to successfully establish *two-lever* drug discrimination with 1.0 mg/kg chlorpromazine (i.p.) and found that non-brain penetrant quaternary chlorpromazine did not substitute for chlorpromazine, providing one of the first demonstrations that the discriminative stimulus properties of drugs are centrally mediated. Goas and Boston [17] later reported that 2.0 mg/kg chlorpromazine (p.o.) fully generalized to both the typical antipsychotic haloperidol and the atypical antipsychotic clozapine. More recently, Porter et al. [18] demonstrated that the chlorpromazine discriminative stimulus (1.0 mg/kg, i.p. in rats) generalized fully to the atypical antipsychotics clozapine and olanzapine and to the typical antipsychotic thioridazine, but only partially to the typical antipsychotic haloperidol and the dopamine D_2 antagonist raclopride. Chlorpromazine also has been used in a *three-lever* drug discrimination study by Porter et al. [19] in which rats were trained to discriminate between 5.0 mg/kg clozapine versus 1.0 mg/kg chlorpromazine versus vehicle. Interestingly, in that study the atypical antipsychotic clozapine substituted fully for the typical antipsychotic chlorpromazine at a lower ED_{50} (0.103 mg/kg) before the rats shifted to the clozapine lever to produce full generalization to clozapine ($ED_{50} = 1.69$ mg/kg). In contrast, chlorpromazine did not substitute for clozapine and engendered only chlorpromazine-appropriate responding ($ED_{50} = 0.196$ mg/kg) – thus, replicating the asymmetrical generalization between clozapine and chlorpromazine reported by Goas and Boston [17]. The atypical antipsychotic olanzapine had a pattern of substitution similar to clozapine. The typical antipsychotic haloperidol produced chlorpromazine-appropriate responding ($ED_{50} = 0.007$ mg/mg) only. These results confirmed that there is an overlap between the discriminative stimulus properties of clozapine and chlorpromazine, but not between clozapine and haloperidol; whereas, chlorpromazine and haloperidol share overlapping discriminative stimulus properties. These findings suggested that stimulus properties may be similar in some ways, but not identical, between typical and atypical antipsychotic drugs. Given that the subjective effects of drugs are mediated by specific receptor actions, it is not too surprising that clozapine and chlorpromazine share discriminative stimulus properties since their binding profiles are more similar than are clozapine's and haloperidol's binding profiles [20–23].

The typical antipsychotic haloperidol has proven to be much more difficult to establish as a discriminative stimulus. Colpaert et al. [24] trained four rats to discriminate 0.02 mg/kg haloperidol (s.c.); however, it required over 80 training sessions and no other drugs were tested for substitution. McElroy et al. [25] trained rats to discriminate 0.05 mg/kg haloperidol (i.p.) and reported that the phenothiazine chlorpromazine fully substituted for haloperidol and that amphetamine blocked haloperidol's discriminative stimulus indicating that dopamine antagonism was the underlying pharmacological mechanism for haloperidol's discriminative stimulus. One reason for the lack of studies on the discriminative stimulus properties of haloperidol is that it is a very difficult drug to establish as a discriminative cue versus saline. However, it has been used in drug–drug, two-lever discrimination studies (e.g., [26, 27]) and in drug–drug–vehicle, three-lever discrimination (e.g., [28, 29]) primarily to contrast dopamine antagonist with dopamine agonist receptor mechanisms. Thus, drug discrimination studies with typical antipsychotic drugs as the discriminative stimulus are fairly limited and have not proven to be very useful as a behavioral assay for drug discovery; instead, these studies focused on receptor mechanisms underlying their discriminative stimulus properties. In contrast, drug discrimination studies with atypical antipsychotic drugs are more prevalent and have more potential as useful assays for drug discovery.

As described above, clozapine was first established as a discriminative stimulus by Goas and Boston [17]. Later studies suggested that cholinergic receptor antagonism played an important role in the discriminative stimulus properties for clozapine in rats [30–33]; however, other studies suggested that serotonergic mechanisms in pigeons [34] and in C57BL/6 mice [35] mediated clozapine's discriminative cue. In addition, the training dose for clozapine in rats has been shown to be an important factor in drug discrimination studies [36–38]. Further, the panoply of receptors that clozapine (and most other atypical antipsychotics) binds to, coupled with a lack of clozapine substitution by ligands selective for these receptors, led to the notion that clozapine possesses a complex/compound discriminative cue involving multiple receptor actions [13, 31, 38, 39]. The importance of each of these factors (i.e., training dose and species) with clozapine drug discrimination will be more fully explored in sections that follow.

3   Importance of Training Dose for the Discriminative Stimulus Properties of Clozapine in Rats

A promising aspect to preclinical assays is their potential for aiding in drug discovery efforts (e.g., [40, 41]). A large number of drug discrimination studies have been conducted with the atypical antipsychotic drug clozapine. As reviewed by Porter and Prus [13], clozapine drug discrimination in rats has typically used a training dose of 5.0 mg/kg. In the many studies that have used this training dose, the ability of clozapine drug discrimination to distinguish typical from atypical antipsychotic drugs has been mixed and this assay does not appear to be reliably predictive of clozapine's "atypicality" with regard to its antipsychotic effects. Thus, use of a 5.0 mg/kg training dose with clozapine has proven to be of limited utility in terms of identifying the "atypical" characteristics of other antipsychotic drugs and for the development of novel atypical antipsychotic drugs. In contrast, clozapine drug discrimination with a lower training dose appears to be a more sensitive assay for distinguishing atypical from typical antipsychotic drugs (see also Lieberman et al. [42]).

It has been well established in the literature that the training dose of a drug influences its discriminative stimulus properties, and a lower training dose will typically increase sensitivity to the training drug as reflected by a lower ED_{50} value and a leftward shift of the generalization curve (see review by Stolerman et al. [43]). An early example of this was shown in a drug discrimination study with rats trained to discriminate either 1.75 or 5.6 mg/kg morphine from saline [44]. The lower morphine training dose produced approximately a 1/2 log unit leftward shift in the dose–response curve as compared to the higher, 5.6 mg/kg training dose. A dose of 0.3 mg/kg morphine produced over 90% morphine-appropriate responding in the 1.76 mg/kg training dose group; whereas, a dose of 1.0 mg/kg was required in the 5.6 mg/kg training dose group. A similar separation between the dose–response curves was evident when the narcotic antagonist naloxone (1.0 mg/kg) was coadministered with morphine and produced marked rightward shifts in both dose–response curves. In a study with the atypical antipsychotic drug clozapine, Goudie et al. [37] trained two groups of rats to discriminate either 2.0 mg/kg clozapine or 5.0 mg/kg clozapine from vehicle. As expected, the lower clozapine training dose engendered a significant leftward shift, as the ED_{50} was 0.58 mg/kg for the 2.0 mg/kg training dose and was 1.41 mg/kg for the 5.0 training dose (a 2.4-fold shift in the dose–response curve). They found that the atypical antipsychotic zotepine fully substituted to the 2.0 mg/kg training dose of clozapine but produced only a maximum of 50% CLZ-appropriate responding in the 5.0 mg/kg training dose group.

This increased sensitivity of lower training doses also is evident in a *three*-lever drug discrimination study in which rats were trained to discriminate 1.25 mg/kg clozapine (CLZ) from 5.0 mg/kg CLZ from vehicle [45, 46]. As can be seen in Fig. 1, the 1.25 mg/kg CLZ training dose produced a *significant* leftward shift in the dose–response curve (ED_{50} = 0.08 mg/kg; 95% CI = 0.04–0.16 mg/kg) as compared to the 5.0 mg/kg CLZ training dose–response curve (ED_{50} = 2.67 mg/kg; 95%

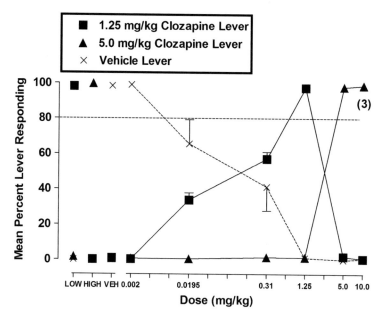

Fig. 1 Rats were trained in a three-lever drug discrimination procedure to discriminate 1.25 mg/kg clozapine (CLZ) from 5.0 mg/kg clozapine from vehicle (VEH). Mean percentage CLZ-lever responding (±SEM) for both training doses is shown on the y-axis. Full stimulus generalization was defined as equal to or greater than 80% CLZ-lever responding (*dashed line*). The number in *parenthesis* for the 10 mg/kg CLZ dose indicates the number of rats that met the response rate criteria (i.e., at least five responses per minute) and were therefore included in the mean percentage CLZ-lever responding calculation for that dose. For all other doses, $n = 12$ rats ([46]; reproduced with permission)

CI = 2.45–2.93 mg/kg). In the Prus et al. [45] study, it was found that the atypical antipsychotic olanzapine fully substituted for the 5.0 mg/kg CLZ training dose, but not for the 1.25 mg/kg CLZ training dose. In contrast, the atypical antipsychotics quetiapine and sertindole produced full substitution for the 1.25 mg/kg CLZ training dose, but not for the 5.0 mg/kg CLZ training dose. Similarly, the atypical risperidone produced strong, partial substitution (72% CLZ-appropriate responding) for the 1.25 mg/kg CLZ training dose and failed to substitute for the 5.0 mg/kg CLZ training dose.

The suggestion that training dose was an important factor for clozapine's discriminative stimulus was first demonstrated by Goudie et al. [37]. In an abstract, they reported the results of a study in which rats were trained to discriminate either 2.0 mg/kg CLZ or 5.0 mg/kg CLZ from vehicle. They found that generalization curves for clozapine, olanzapine, quetiapine, and JL13 (a clozapine congener) were all shifted to the left in the lower 2.0 mg/kg training dose group reflecting greater sensitivity for the lower clozapine training dose. Porter et al. [38] further examined the utility of using a lower clozapine training dose to determine if low-dose clozapine drug discrimination could distinguish typical and atypical antipsychotic drugs. In rats trained to discriminate 1.25 m/kg clozapine from vehicle, it was found that the atypic-al antipsychotics risperidone, sertindole, and olanzapine fully substituted (i.e., >80% CLZ-appropriate responding) for the 1.25 mg/kg training dose of clozapine, although partial substitution (i.e., >60% CLZ-appropriate responding) occurred with the atypical quetiapine (as noted above, Goudie et al. [37] reported full substitution with a 2.0 mg/kg CLZ training dose). In contrast, the typical antipsychotics haloperidol, chlorpromazine, and fluphenazine did not engender CLZ-appropriate responding (although it should be noted that thioridazine did produce partial substitution for clozapine). In another low-dose 1.25 mg/kg clozapine study, full substitution was shown for the atypical antipsy-chotic melperone [47]. Thus, a number of studies using lower training doses of clo-zapine in rats (1.25 or 2.0 mg/kg) have found that clozapine's discriminative stimulus generalizes to a greater number of atypical antipsychotic drugs than when a higher clozapine training dose (i.e., 5.0 mg/kg) is used. Thus, low-dose clozapine drug dis-crimination in rats has greater translational value than high-dose clozapine drug dis-crimination for development of new antipsychotic drugs. In the next section, we examine the underlying neuropharmacological mechanisms that mediate the discrim-inative stimulus properties of the 5.0 and 1.25 mg/kg training doses for clozapine in rat drug discrimination studies.

3.1 Neuropharmacological Mechanisms Mediating the Clozapine Discriminative Stimulus

3.1.1 5.0 mg/kg Clozapine Training Dose in Rats

Clozapine has a very diverse binding profile and differs from typical antipsychotics like haloperidol in that it displays a higher binding affinity for serotonin 5-HT_2 receptors than for dopamine D_2 receptors (see Meltzer et al. [8]). However, cloza-pine also has significant affinity for a number of other receptors including dopami-nergic D_4, serotonergic 5-HT2_C, 5-HT_6, 5-HT_7, cholinergic M_1, M_2, M_3, M_4, adrenergic α_1, α_2, and histamine H_1 receptors [20–23]. A number of studies have suggested that clozapine has a compound (complex) discriminative stimulus that is mediated by its activity at several of these receptors (see review by Porter and Prus [13]). However, in rats trained to discriminate 5.0 mg/kg clozapine the one receptor mechanism that has consistently emerged as important for mediating clozapine's discriminative cue is antagonism of muscarinic cholinergic receptors. Cholinergic antagonism was first proposed by Nielsen [33] who reported that scopolamine and atropine substituted for clozapine (5.76 mg/kg training dose). Kelley and Porter [32] found that there was a significant correlation ($r = 0.74, p < 0.01$) between the percen-tage of clozapine-appropriate lever responding and the percentage of scopolamine-appropriate lever responding in rats trained to discriminate either 5.0 mg/kg clozapine or 0.125 mg/kg scopolamine from vehicle (see Fig. 2).

The fact that clozapine and scopolamine displayed cross-generalization (i.e., both produced full substitution for the other drug) suggests that a common mechanism of

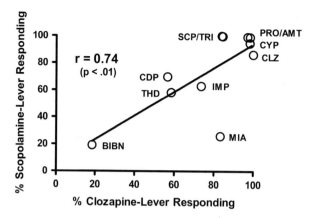

Fig. 2 Summary of results for drugs tested in rats trained to discriminate 5.0 mg/kg (i.p.) from vehicle and in rats trained to discriminate 1.25 mg/kg scopolamine from vehicle in two-lever drug discrimination. The highest percent of scopolamine-lever responding (*y*-axis) is shown as a function of the highest percent of clozapine-lever responding (*x*-axis) for each drug. The regression line for the data and the correlation coefficient also are shown. Abbreviations: *AMT* amitriptyline; *BIBN* BIBN-99; *CDP* chlordiazepoxide; *CLOZAPINE* clozapine; *CYP* cyproheptadine; *IMP* imipramine; *MIA* mianserin; *PMZ* promethazine; *SCP* scopolamine; *THD* thioridazine; *TRI* trihexyphenidyl (data adapted from Kelley and Porter [32] and published in Porter and Prus [13]; reproduced with permission)

action was shared by them (i.e., antagonism of cholinergic muscarinic receptors). Also, it was found that antagonism of cholinergic muscarinic M_1 receptors was more important than antagonism of M_2 receptors as the preferential M_1 muscarinic antagonist trihexyphenidyl fully substituted for clozapine and scopolamine; whereas, the M_2 muscarinic antagonist BIBN 99 did not substitute for either drug. These findings strongly suggest that antagonism of muscarinic cholinergic receptors (probably M_1) mediates the discriminative stimulus properties of 5.0 mg/kg clozapine. Antagonism of muscarinic receptors also has been shown in other studies to be important for clozapine's discriminative cue at this higher training dose [31, 47].

3.1.2 1.25 mg/kg Clozapine Training Dose in Rats

As discussed above, there are clear differences in the discriminative stimulus properties of 5.0 mg/kg clozapine and 1.25 mg/kg clozapine as reflected by the ability of these two different training doses to distinguish between typical and atypical antipsychotic drugs. While the discriminative cue of the 5.0 mg/kg training dose in rats appears to involve antagonism of muscarinic cholinergic receptors, different mechanisms appear to be responsible for the discriminative cue of the lower 1.25 mg/kg training dose. Recently, Prus et al. [48] examined the mechanisms underlying the discriminative stimulus properties of 1.25 mg/kg clozapine in a two-lever drug discrimination study in rats by testing a number of selective ligands. They found evidence for a compound/complex discriminative cue, as selective ligands at several receptors produced full or partial substitution for 1.25 mg/kg clozapine. Specifically, the 5-HT_{2A} inverse agonist M100907 and the two preferential $D_{4/5}\text{-HT}_2/\alpha_1$ receptor antagonists Lu37-114 and Lu37-254, and the α_1 adrenoceptor antagonist prazosin fully substituted for clozapine's discriminative stimulus. These findings suggest that serotonin 5-HT_{2A} inverse agonism and antagonism of dopamine D_4 and noradrenergic α_1 receptor antagonism mediate the discriminative stimulus properties of 1.25 mg/kg clozapine in rats and further confirm that clozapine produces a complex, compound discriminative stimulus.

These differences in the underlying neuropharmacological mechanisms that mediate the discriminative stimulus properties of clozapine at low and high training doses appear to be the primary factor in the ability of other antipsychotic drugs to substitute fully or partially for clozapine. While additional research will be necessary to determine the actual utility of using a low clozapine training dose for development of novel antipsychotic drugs, the research discussed above demonstrates the necessity of considering important controlling variables in the drug discrimination procedure (and of course all behavioral assays used for drug development). It may be that cholinergic mechanisms at the higher training dose are more related to the nontherapeutic properties of clozapine (and other atypical antipsychotics) than are the serotonergic, dopaminergic, and adrenergic mechanisms evident at the lower training dose. In the following sections, we will discuss another important variable for the translational value of clozapine drug discrimination – differences in the discriminative stimulus properties of clozapine across different species of animals.

4 Importance of Species Similarities/Differences for the Translational Value of Clozapine Drug Discrimination

As discussed in the preceding sections, training dose is an important variable for the translational value of the clozapine drug discrimination assay in rats. What is the translational value of clozapine drug discrimination in other species of animals? The vast majority of clozapine drug discrimination studies have been conducted with rats as the subjects. Using information from a review by Porter and Prus in 2009, a total of 34 articles were reviewed in which an antipsychotic was used as the training drug and 26 of those used clozapine. Only five of those studies used a species other than rats. One study used squirrel monkeys [49], one study used pigeons [34], two studies used C57BL/6 mice [35, 50], and one study used DBA/2 mice [51]. Since 2009, a PubMed/Medline search found three additional studies that did not use rats as the subjects in clozapine drug discrimination studies. Two of those studies used C57BL/6 mice [52, 53] and one used hybrid B6129 mice [54]. Below we will review and compare the similarities and differences of clozapine's discriminative stimulus across these different species and include some recent findings from our laboratory.

Pigeons were trained by Hoenicke et al. [34] to discriminate 1.0 mg/kg clozapine (i.m.) from saline in a two-key drug discrimination procedure. They reported that test compounds with antagonist activity at 5-HT$_2$ receptors fully substituted for clozapine's discriminative stimulus. This included the drugs cyproheptadine (anti-histamine), metergoline (psychoactive drug of the ergoline chemical class), mianserin (tetracyclic antidepressant), pizotifen (benzocycloheptene-based drug used for treating migraines), and fluperlapine (atypical antipsychotic). Hoenicke et al. also tested a large number of selective receptor ligands that were active at non-serotonergic systems and compounds from other therapeutic classifications and none of them substituted for clozapine. They also tested the typical antipsychotics chlorpromazine and thioridazine, neither of which substituted for clozapine. Unfortunately, they did not test other atypical antipsychotic drugs, as there was limited availability of such compounds at the time this study was conducted with pigeons. This finding obviously was in contrast to the earlier suggestion by Nielsen [33] that clozapine's discriminative stimulus in rats was mediated by antagonism of muscarinic cholinergic receptors for higher clozapine training doses (however, serotonergic mechanisms are more important at a low clozapine training dose in rats; see discussion above in Sect. 4.1.2). Also, Nielsen's study was with rats and Hoenicke et al.'s study was with pigeons. This was the first indication that species was an important variable to consider when studying the pharmacological mechanisms underlying the discriminative stimulus properties of clozapine and the possible utility of clozapine drug discrimination for development of novel antipsychotic drugs.

Carey and Bergman [49] trained squirrel monkeys to discriminate 1.0 mg/kg clozapine (i.m.) from saline in a two-lever drug discrimination procedure. The typical antipsychotics haloperidol, chlorpromazine, and thioridazine produced only minimal clozapine-appropriate responding (maximum of 44% clozapine-lever responding by 0.3 mg/kg chlorpromazine). Higher doses of these three drugs produced marked rate suppression. In contrast, the atypical antipsychotic drugs fluperlapine and quetiapine and the structurally related dibenzothiophene, perlapine all produced greater than 90% clozapine-appropriate responding. The atypical antipsychotics olanzapine and risperidone failed to substitute for clozapine, producing significant response rate suppression. However, when the dopamine D$_2$ agonist (+)-PHNO was coadministered with olanzapine, greater than 90% clozapine-appropriate responding was evident. Thus, clozapine drug discrimination in squirrel monkeys was able to distinguish atypical from typical antipsychotic drugs, albeit with some false negatives that may be attributed to the dopaminergic rate suppressant effects of higher doses for some of the antipsychotic drugs. Unfortunately, Carey and Bergman did not explore the underlying pharmacological mechanisms that mediate clozapine's discriminative stimulus in squirrel monkeys. The following section examines the discriminative stimulus properties of the atypical antipsychotic clozapine in several inbred mouse strains.

4.1   Discriminative Stimulus Properties of Clozapine in Mice

4.1.1   C57BL/6 Mice

The first study to examine the discriminative stimulus properties of clozapine in mice was by Philibin et al. [35]. In that study, male C57BL/6 mice were trained to discriminate 2.5 mg/kg (s.c.) from vehicle in a two-lever procedure (attempts to train 5.0 mg/kg clozapine were unsuccessful due to response rate suppression). Thirteen of 20 mice successfully acquired the clozapine discriminative cue and completed the clozapine dose-response curve with an $ED_{50} = 1.14$ mg/kg. It was shown that the serotonergic $5\text{-HT}_{2A/2B/2C}$ receptor antagonist ritanserin fully substituted for clozapine's discriminative stimulus and that the 5-HT_2 agonist quipazine produced a significant attenuation of clozapine's cue. Other selective ligands tested included the muscarinic receptor antagonist scopolamine, which produced partial substitution for clozapine (62% clozapine-appropriate responding). However, the indirect dopamine agonist amphetamine and the 5-HT_2 agonist quipazine did not produce clozapine-appropriate responding. These findings demonstrated that antagonism of serotonergic 5-HT_2 receptors is an important pharmacological mechanism underlying the clozapine discriminative cue in C57BL/6 mice. This finding was confirmed in a subsequent study in which male C57BL/6 mice also were trained to discriminate 2.5 mg/kg (s.c.) clozapine from vehicle [50]. The selective 5-HT_{2A} inverse agonist M100907 fully substituted for clozapine. In addition, the α_1-adrenoceptor antagonist prazosin also fully substituted for clozapine's discriminative stimulus. Thus, antagonism (or inverse agonism) of *either* serotonin 5-HT_2 receptors *or* α_1-adrenoceptors is sufficient to produce clozapine-appropriate responding in male C57BL/6 mice. Interestingly, it is also possible for drugs to share partial discriminative stimulus properties, likely through indirect mechanisms. Vunck et al. [52] demonstrated this in male C57BL/6 mice trained to discriminate either 2.5 mg/kg clozapine or 30 mg/kg N-methyl-D-aspartate (NMDA, glutamatergic agonist) from vehicle. Cross-generalization testing found partial substitution of clozapine for NMDA, but none for NMDA in clozapine-trained mice. However, when a non-generalizing dose of each training drug was tested in combination with the other drug, there was full and dose-dependent substitution in both groups (i.e., cross-generalization). Interesting, the α_1-adrenoceptor antagonist prazosin fully substituted for both clozapine and NMDA suggesting that any shared discriminative stimulus properties for clozapine and NMDA were likely mediated through α_1 adrenergic antagonism.

4.1.2   DBA/2 Mice

The discriminative stimulus properties of clozapine in male DBA/2 mice were compared to C57BL/6 mice by Porter et al. [51]. In contrast to C57BL/6 mice, antagonism of serotonergic 5-HT_2 receptors and α_1-adrenoceptors did *not* produce clozapine-appropriate responding. However, as shown in C57BL/6 mice, the muscarinic antagonist scopolamine produced partial substitution for clozapine (69% clozapine-appropriate responding). Interestingly, haloperidol, which is a potent dopamine antagonist, also produced partial substitution for clozapine in DBA/2 mice (68% clozapine-appropriate responding), but not in C57BL/6 mice (maximum of 48% clozapine-appropriate responding). Differences in brain dopamine systems between DBA/2 and C57BL/6 mice might account for the different substitution patterns of haloperidol. In support of this suggestion, it also was found that a higher dose of haloperidol (0.4 mg/kg) was required to fully suppress lever pressing in the DBA/2 mice as compared to C57BL/6 mice (0.2 mg/kg) (see Porter et al. [51] for fuller discussion).

4.1.3   129S Mice

In an unpublished study from our laboratory (Webster and Porter; data presented at meetings for the Society for Neuroscience [55, 56] and the European Society for Behavioural Pharmacology [57]), male 129S2 mice were trained to discriminate 1.25 mg/kg (s.c.) clozapine from saline. Initial attempts to train a 2.5 mg/kg dose of clozapine had to be abandoned because of response rate suppression. While the serotonergic 5-HT$_2$ receptor antagonist ritanserin did not substitute for clozapine (57% clozapine-appropriate responding), the more selective 5-HT$_{2A}$ inverse agonist M100907 produced partial substitution for clozapine (69% clozapine-appropriate responding). The α_1-adrenoceptor antagonist prazosin fully substituted for clozapine (83% clozapine-appropriate responding), which is similar to results observed in C57BL/6 mice, but not in DBA/2 mice. The muscarinic antagonist scopolamine occasioned partial substitution (68% clozapine-appropriate responding), similar to that seen in both C57BL/6 and DBA/2 mice. The dopamine antagonist haloperidol produced partial substitution (66% clozapine-appropriate responding), similar to that seen in DBA/2 mice. Thus, 129S2 mice displayed both similarities to and differences from C57BL/6 and DBA/2 mice. Also, 129S2 mice were clearly more sensitive to the rate-suppressing effects of clozapine as a lower training dose had to be used (DBA/2 mice were less sensitive – see previous section). Thus, some of the differences in substitution patterns for these selective ligands may reflect training dose effects, but this clearly was not the case for C57BL/6 and DBA/2 mice as both were trained at the 2.5 mg/kg dose of clozapine. Thus, the differences between these three inbred strains of mice can probably be attributed primarily to strain differences as opposed to training dose differences.

4.1.4   B6129S Hybrid Mice

A small number of male B6129S hybrid mice have recently been trained to discriminate 1.25 mg/kg clozapine (s.c.) from saline (Webster and Porter, unpublished data). Interestingly, the only selective ligand that produced full substitution for clozapine was the muscarinic antagonist scopolamine (86% clozapine-appropriate responding). The serotonergic 5-HT$_2$ receptor antagonist ritanserin (37% clozapine-appropriate responding) and the 5-HT2A inverse agonist M100907 did not substitute for clozapine. However, the α_1-adrenoceptor antagonist prazosin generated partial clozapine-appropriate responding (60% clozapine-appropriate responding), as did the dopamine antagonist haloperidol (73% clozapine-appropriate responding). Thus, the hybrid between C57BL/6 and 129S2 mice more closely resembled findings with rats trained to discriminate 5.0 mg/kg clozapine (see discussion in Sect. 3.1). Table 1 provides a summary of the results of substitution testing with selective receptor ligands in the four inbred mouse strains discussed in Sect. 4.1.

5   Subjective Properties of Antipsychotic Drugs in Humans

While research has been conducted on the discriminative stimulus properties of antipsychotic drugs in humans, unfortunately, no studies have used antipsychotics as the training drugs. Instead, antipsychotic drugs have been administered as a test compound in humans trained to discriminate a different substance. For example, Rush et al. [58] examined the effects of the atypical antipsychotic risperidone in humans trained to discriminate D-amphetamine. When coadministered with D-amphetamine,

Table 1 Receptor mechanisms mediating the discriminative stimulus properties of the atypical antipsychotic clozapine in four inbred mouse strains

Receptor mechanism	C57BL/6 mice[a]	DBA/2 mice[b]	129S2 mice[c]	B6129S mice[c]
5-HT$_2$				
Ritanserin (antagonist)	FULL	NO	NO	NO
5-HT$_{2A}$				
M100907 (inverse agonist)	FULL	NT	PARTIAL	NO
α_1 adrenoceptor				
Prazosin (antagonist)	FULL	NO	FULL	PARTIAL
Muscarinic				
Scopolamine (antagonist)	PARTIAL	PARTIAL	PARTIAL	FULL
Histamine H$_1$				
Pyrilamine (antagonist)	NO	NO	NO	NT
Dopamine				
Amphetamine (indirect agonist)	NO	NO	NO	NT
Dopamine D$_2$				
Haloperidol (antagonist)	NO	PARTIAL	PARTIAL	PARTIAL

The training dose for clozapine was 2.5 mg/kg in C57BL/6 and DBA/2 mice and 1.25 mg/kg in 129S2 and B6129S mice
FULL full substitution >80% clozapine-appropriate responding, *PARTIAL* partial substitution >60% to <80% clozapine-appropriate responding, *NO* no substitution <60% clozapine-appropriate responding
[a]Philibin et al. [35, 50]
[b]Porter et al. [51]
[c]Webster and Porter, unpublished data

risperidone (which serves as a D_2 receptor antagonist among other receptor actions) reduced D-amphetamine-appropriate responding from full generalization (>80%) to approximately 40%. The reduction in D-amphetamine-appropriate responding corresponded with related changes in reported subjective effects of D-amphetamine, including participants endorsing fewer words related to stimulant effects, a willingness to take the drug again, feelings of liking the drug, the level of good effects produced by the drug, etc.

In another study from this group, Lile et al. [59] tested the atypical antipsychotic drug aripiprazole in combination with cocaine in humans trained on a cocaine discrimination task. Coadministration of aripiprazole caused fewer doses of cocaine to be statistically greater placebo, suggesting an attenuation of cocaine's discriminative cue. When tested alone, aripiprazole did not engender cocaine-appropriate responding. Aripiprazole differs somewhat from other atypical antipsychotic drugs, in that it acts as a weak partial agonist at dopamine D_2 receptors, rather than as an antagonist. The study by Lile et al. [59] demonstrated that the degree of activation of D_2 receptors produced by aripiprazole alone was not sufficient to produce cocaine-like stimulus effects. Given that cocaine-like stimulus effects rely primarily on D_2 receptor activation, it is not surprising that aripiprazole did not substitute for cocaine as its effects are likely mediated by functional antagonism of D_2 receptors (i.e., the population of D_2 receptors occupied by aripiprazole but not activated by this compound). Changes in the subjective effects for cocaine when combined with aripiprazole were similar to the findings in the Rush et al. [58] study when risperidone was combined with amphetamine, with overall reduced subjective effects.

These studies in humans evaluating the stimulus effects of antipsychotic drugs can be more precisely described as a reference to psychostimulant studies and indicate that antipsychotic stimulus effects counter those produced by psychostimulants. Arguably, if an antipsychotic drug was used during discrimination training, then the discrimination would be based on the subjective (interoceptive) effects of the antipsychotic drug. Further, assessing the stimulus effects of antipsychotic drugs in patient populations for which the drugs were prescribed would provide additional benefits for the scientific value of data gathered on these compounds. Such stimulus effects might consist of or depend on: (1) individual differences in level of impairment, (2) the development of drug side effects, or (3) adaptations, such as tolerance or sensitization, to the interoceptive drug state after multiple administrations.

Moncrieff et al. [60] examined an internet site where users freely post comments on the effects they feel when taking antipsychotic medications. In their analysis of posted comments, the groups compared older antipsychotics (defined, in this case as chlorpromazine, trifluoperazine, or haloperidol) with risperidone and olanzapine. The more prevalent comments for all antipsychotic drugs regarded feelings of sedation, reduced motivation, and flattened emotions. A number of participants noted sexual dysfunction with risperidone, and olanzapine was particularly regarded as causing weight gain and increased appetite. The older antipsychotics were noted more often than risperidone or olanzapine for producing Parkinsonian effects.

As reviewed by Gerlach and Larsen [61], the vast majority of patients given long-term antipsychotic treatment displayed features of extrapyramidal motor symptoms.

These side effects were most prevalent among compounds with a strong affinity for dopamine D_2 receptors, including sulpiride, amisulpride, and risperidone [23]. Sedation was most common among patients treated with clozapine, but also relatively prevalent with olanzapine and quetiapine; all three of which are antagonists at histamine H_1 receptors, which have linked to sedation, although effects at other receptors may contribute to this effect.

Using an assessment of subjective well-being for antipsychotic treatment, Mizrahi et al. [62] correlated regional D_2 receptor occupancy produced by olanzapine or risperidone with total scores on the assessment. They found that patients indicated a diminished sense of overall well-being that negatively correlated with higher D_2 receptor occupancy in the striatum, temporal lobe, and insular cortex. An evaluation of the assessment's subscales indicated correlations between striatal D_2 receptor occupancy and mental functioning and between temporal lobe D_2 receptor occupancy and emotional regulation.

Thus, overall, only a limited number of studies have examined the discriminative stimulus properties of antipsychotic drugs in humans. Studies that have examined antipsychotic drugs tend to indicate that these drugs counteract the effects of psychostimulant drugs that cause the activation of dopamine receptors, which would be in keeping with the tendency of antipsychotic drugs to act as D_2 receptor antagonists [23]. Many of the self-reported subjective effects of antipsychotic drugs, which presumably would contribute to the discriminative stimulus properties of these drugs in humans, are associated with side effects. Reasonable links to receptor actions can be made to these subjective side effects in humans, with typical antipsychotic drug effects most often associated with D_2 receptor antagonism; whereas, atypical antipsychotic drug effects are linked to D_2 receptor antagonism along with other receptor mechanisms [23, 61]. For example, an adverse effect not described above but which offers an example of a non-dopaminergic mediated side effect is dry mouth. Dry mouth results from antagonism of cholinergic muscarinic receptors, which is found with some atypical antipsychotic drugs such as clozapine and olanzapine [63, 64].

6   Conclusions

Conclusions from the literature discussed in this chapter are somewhat limited given that only a small number of studies have actually examined clozapine's discriminative stimulus properties in mice, only one study in pigeons, one study in squirrel monkeys, with the rest of the studies using rats as the subject of choice. However, some basic conclusions can be reached about clozapine's discriminative stimulus across different species of animals. First, clozapine clearly has a robust discriminative stimulus as evidenced by the relative ease in establishing it as a training drug in both two- and three-lever drug discrimination paradigms across different species. Second, serotonin receptors appear to be an important pharmacological mechanism mediating clozapine's discriminative cue although differences are clearly evident as antagonism of cholinergic muscarinic receptors is important in rats at a higher training dose (5.0 mg/kg) of clozapine, but not at a lower training dose (1.25 mg/kg) species and even across inbred mouse strains. Third, in mice antagonism of muscarinic receptors appears to be less important in C57BL/6, DBA/2, and 129S2 mice (produces partial substitution for clozapine); however, it appears to mediate clozapine's cue in hybrid B6129S mice. Fourth, antagonism of α_1 adrenoceptors is a sufficient mechanism in C57BL/6 and 129S2 mice, but not in DBA/2 and B6129S mice, and only produces partial substitution in low-dose clozapine discrimination in rats. Fifth, dopamine antagonism produces partial substitution in DBA/2, 129S2, and B6129S mice, but not in C57BL/6 mice, and partial substitution with D_4 antagonism in low-dose clozapine drug discrimination in rats. Thus, it is evident that clozapine has a complex mixture of receptor contributions towards its discriminative cue based on the four mouse strains that have been tested, similar to the results from rat studies. Finally, we find evidence that antipsychotics produce subjective effects in humans that can be predicted from animal studies, although the limited number of studies examining discriminative stimuli of antipsychotics in humans makes this conclusion tentative.

While the majority of research has focused on rats, drug discrimination studies with mice offer an interesting opportunity for future research and hopefully will have greater translational value for future drug development, since the majority of genetic manipulations are in mice – not rats. Given the advances of using mice for behavioral pharmacogenetic approaches, an important first step is to establish behavioral, pharmacological, physiological, and biochemical comparisons across different inbred mouse strains. C57BL/6 and DBA/2 mice represent the most commonly used inbred mouse strains in behavioral pharmacology research (see Crawley et al. [65] for review) and they are important background strains for developing transgenic and knockout mouse models. The 129S2 and B6129S mouse strains also are frequently used in genetic mouse models. Thus, the research presented here for these four mouse strains represents a first step in establishing a necessary foundation of background information. More research is required to fully characterize the underlying mechanisms that characterize the similarities and differences across these inbred mouse strains and to determine how that relates to functional and anatomical characteristics (e.g., there are known differences in brain dopamine systems in C57BL/6 and DBA/2 mice; see Porter et al. [51] for fuller discussion). Hopefully, better characterization of these background mouse strains will aid in future drug development efforts for new and better antipsychotic medications. A further examination of antipsychotic stimulus properties in humans, particularly in patients with schizophrenia, would go far in evaluating the translational value of the drug discrimination paradigm for antipsychotic drugs.

References

1. Ban T (2007) Fifty years chlorpromazine: a historical perspective. Neuropsychiatr Dis Treat 3 (4):495–500
2. van Rossum JM (1966) The significance of dopamine receptor blockade for the action of neuroleptic drugs. In: Brill H (ed) Neuro-psycho-pharmacology. Excerpta Medica Foundation, Amsterdam, pp 321–329

3. Baumeister AA, Francis JL (2002) Historical development of the dopamine hypothesis of schizophrenia. J Hist Neurosci 11(3):265–277
4. Snyder SH, Banerjee SP, Yamamura HI et al (1974) Drugs, neurotransmitters and schizophrenia. Science 184:1243–1253
5. Shorter E (1997) A history of psychiatry. Wiley, New York, NY, p 208
6. Hippius (1999) A historical perspective of clozapine. J Clin Psychiatry 60(Suppl 12):22–23
7. Young CR, Longhurst JG, Bowers MB Jr, Mazure CM (1997) The expanding indications for clozapine. Exp Clin Psychopharmacol 5:216–234
8. Meltzer HY, Matsubara S, Lee JC (1989) Classification of typical and atypical antipsychotic drugs on the basis of dopamine D-1, D-2 and serotonin2 pKi values. J Pharmacol Exp Ther 251:238–246
9. Idänpään-Heikkilä J, Alhava E, Olkinuora M et al (1975) Clozapine and agranulocytosis. Lancet 2:611
10. Crossley NA (2010) Efficacy of atypical v. typical antipsychotics in the treatment of early psychosis: meta-analysis. Br J Psychiatry 196:434–439
11. Dekeyne A, Millan MJ (2003) Discriminative stimulus properties of antidepressant agents: a review. Behav Pharmacol 14:391–407
12. Goudie AJ, Smith JA (1999) Discriminative stimulus properties of antipsychotics. Pharmacol Biochem Behav 64(2):193–201
13. Porter JH, Prus AJ (2009) Discriminative stimulus properties of atypical and typical antipsychotic drugs: a review of preclinical studies. Psychopharmacology (Berl) 203:279–294
14. Fang FC, Casaderall A (2010) Editorial: lost in translation – basic science in the era of translational research. Infect Immun 78(2):563–566
15. Stewart J (1962) Differential responses based on the physiological consequences of pharmacological agents. Psychopharmacologia 3:132–138
16. Barry H III, Steenberg ML, Manian AA, Buckley JP (1974) Effects of chlorpromazine and three metabolites on behavioral responses in rats. Psychopharmacology (Berl) 34:351–360
17. Goas JA, Boston JE Jr (1978) Discriminative stimulus properties of clozapine and chlorpromazine. Pharmacol Biochem Behav 8:235–241
18. Porter JH, Covington HE III, Varvel SA, Vann RE, Warren TA (1998) Chlorpromazine as a discriminative stimulus in rats: generalization to typical and atypical antipsychotic. Drug Dev Res 48:38–44
19. Porter JH, Prus AJ, Vann RE, Varvel SA (2005) Discriminative stimulus properties of the atypical antipsychotic clozapine and the typical antipsychotic chlorpromazine in a three-choice drug discrimination procedure in rats. Psychopharmacology (Berl) 178:67–77
20. Arnt J, Skarsfeldt T (1998) Do novel antipsychotics have similar pharmacological characteristics? A review of the evidence. Neuropsychopharmacology 18:63–101
21. Bymaster FP, Calligaro DO, Falcone JF, Marsh RD, Moore NA, Tye NC, Seeman P, Wong DT (1996) Radioreceptor binding profile of the atypical antipsychotic olanzapine. Neuropsychopharmacology 14:87–96
22. Richelson E (1999) Receptor pharmacology of neuroleptics: relation to clinical effects. J Clin Psychiatry 60(Suppl 10):5–14
23. Schotte A, Janssen PF, Gommeren W, Luyten WH, Van Gompel P, Lesage AS, De Loore K, Leysen JE (1996) Risperidone compared with new and reference antipsychotic drugs: in vitro and in vivo receptor binding. Psychopharmacology (Berl) 124:57–73
24. Colpaert FC, Niemegeers CJE, Janssen PAJ (1976) Theoretical and methodological considerations on drug discrimination learning. Psychopharmacology (Berl) 46:169–177
25. McElroy JF, Stimmel JJ, O'Donnell JM (1989) Discriminative stimulus properties of haloperidol. Drug Dev Res 18:47–55
26. Haenlein M, Caul WF, Barrett RJ (1985) Amphetamine-haloperidol discrimination: effects of chronic drug treatment. Pharmacol Biochem Behav 23:949–952
27. Wiley JL, Porter JH (1993) Effects of serotonergic drugs in rats trained to discriminate clozapine from haloperidol. Bull Psychonomic Soc 31:94–96

28. Barrett RJ, Caul WF, Smith R (2005) Withdrawal, tolerance, and sensitization to dopamine mediated interoceptive cues in rats trained on a three-lever drug-discrimination task. Pharmacol Biochem Behav 81(1):1–8

29. Stadler JR, Caul WF, Barrett RJ (1999) Characterizing withdrawal in rats following repeated drug administration using an amphetamine-vehicle-haloperidol drug discrimination. Psychopharmacology (Berl) 143:219–226

30. Franklin SR, Tang AH (1994) Discriminative stimulus effects of clozapine in rats. Behav Pharmacol 5:113

31. Goudie AJ, Smith JA, Taylor A, Taylor MA, Tricklebank MD (1998) Discriminative stimulus properties of the atypical neuroleptic clozapine in rats: tests with subtype selective receptor ligands. Behav Pharmacol 9(8):699–710

32. Kelley BM, Porter JH (1997) The role of muscarinic cholinergic receptors in the discriminative stimulus properties of clozapine in rats. Pharmacol Biochem Behav 57(4):707–719

33. Nielsen EB (1988) Cholinergic mediation of the discriminative stimulus properties of clozapine. Psychopharmacology (Berl) 94:115–118

34. Hoenicke EM, Vanecek SA, Woods JH (1992) The discriminative stimulus effects of clozapine in pigeons: involvement of 5-hydroxytryptamine-1c and 5-hydroxytryptamine-2 receptors. J Pharmacol Exp Ther 263:276–284

35. Philibin SD, Prus AJ, Pehrson AL, Porter JH (2005) Serotonin receptor mechanisms mediate the discriminative stimulus properties of the atypical antipsychotic clozapine in C57BL/6 mice. Psychopharmacology (Berl) 180:49–56

36. Goudie AJ, Taylor A (1998) Olanzapine generalisation to the clozapine discriminative stimulus is determined by clozapine training dose. Br J Pharmacol 124S:54

37. Goudie AJ, Taylor MAI, Smith JA (1998) Stimulus properties of clozapine and related agents in rats at two clozapine training doses. J Psychopharmacol 12:A65

38. Porter JH, Varvel SA, Vann RE, Philibin SD, Wise LE (2000) Clozapine discrimination with a low training dose distinguishes atypical from typical antipsychotic drugs in rats. Psychopharmacology (Berl) 149(2):189–193

39. Goudie AJ, Taylor A (1998) Comparative characterisation of the discriminative stimulus properties of clozapine and other antipsychotics in rats. Psychopharmacology (Berl) 135:392–400

40. Geyer M, Ellenbroek B (2003) Animal behavior models of the mechanisms underlying antipsychotic atypicality. Prog Neuropsychopharmacol Biol Psychiatry 27:1071–1079

41. Moore NA, Tye NC, Axton MS, Risius FC (1992) The behavioral pharmacology of olanzapine, a novel "atypical" antipsychotic agent. J Pharmacol Exp Ther 262:545–551

42. Lieberman JA, Bymaster FP, Meltzer HY, Deutch AY, Duncan GE, Marx CE, Aprille JR, Dwyer DS, Li X-M, Mahadik SP, Duman RS, Porter JH, Modica-Napolitano JS, Newton SS, Csernansky JG (2008) Antipsychotic drugs: comparison in animal models of efficacy, neurotransmitter regulation, and neuroprotection. Pharmacol Rev 60:358–403

43. Stolerman IP, Childs E, Ford MM, Grant KA (2011) Role of training dose in drug discrimination: a review. Behav Pharmacol 22:414–429

44. Shannon HE, Holtzman SG (1979) Morphine training dose: a determinant of stimulus generalization to narcotic antagonists in the rat. Psychopharmacology (Berl) 61:239–244

45. Prus AJ, Philibin S, Pehrson AL, Porter JH (2005) Generalization to atypical antipsychotic drugs depends on training dose in rats trained to discriminate 1.25 mg/kg clozapine versus 5.0 mg/kg clozapine versus vehicle in a three-choice drug discrimination task. Behav Pharmacol 16(7):511–520

46. Prus AJ, Philibin SD, Pehrson AL, Porter JH (2006) Discriminative stimulus properties of the atypical antipsychotic drug clozapine in rats trained to discriminate 1.25 mg/kg clozapine vs. 5.0 mg/kg clozapine vs. vehicle. Behav Pharmacol 17(2):185–194

47. Prus AJ, Baker LE, Meltzer HY (2004) Discriminative stimulus properties of 1.25 and 5.0 mg/kg doses of clozapine in rats: examination of therole of dopamine, serotonin, and muscarinic receptor mechanisms. Pharmacol Biochem Behav 77(2):199–208

48. Prus AJ, Wise LE, Pehrson AL, Philibin SD, Bang-Andersen B, Arnt J, Porter JH (2016) Discriminative stimulus properties of 1.25 mg/kg clozapine in rats: mediation by serotonin 5-HT 2 and dopamine D 4 receptors. Brain Res 1648:298–305

49. Carey GJ, Bergman J (1997) Discriminative-stimulus effects of clozapine in squirrel monkeys: comparison with conventional and novel antipsychotic drugs. Psychopharmacology (Berl) 132:261–269

50. Philibin SD, Walentiny DM, Vunck SA, Prus AJ, Meltzer HY, Porter JH (2009) Further characterization of the discriminative stimulus properties of the atypical antipsychotic drug clozapine in C57BL/6 mice and a comparison to clozapine's major metabolite N-desmethylclozapine. Psychopharmacology (Berl) 203:303–315

51. Porter JH, Walentiny DM, Philibin SC, Vunck SA, Crabbe JC (2008) A comparison of the discriminative stimulus properties of the atypical antipsychotic drug clozapine in DBA/2 and C57BL/6 inbred mice. Behav Pharmacol 19:530–542

52. Vunck SA, Wiebelhaus JM, Arnt J, Porter JH (2011) Clozapine and N-methyl-D-aspartate have positive modulatory actions on their respective discriminative stimulus properties in C57BL/6 mice. Eur J Pharmacol 650:579–585

53. Wiebelhaus JM, Webster KA, Meltzer HY, Porter JH (2011) The metabolites N-desmethylclozapine and N-desmethylolanzapine produce cross-tolerance to the discriminative stimulus of the atypical antipsychotic clozapine in C57BL/6 mice. Behav Pharmacol 22:458–467

54. Farrell MS, McCorvy JD, Huang X-P, Urban DJ, White KL, Giguere PM, Doak AK, Bernstein AI, Stout KA, Park SM, Rodriguiz RM, Gray BW, Hyatt WS, Norwood AP, Webster KA, Gannon BM, Miller GW, Porter JH, Shoichet BK, Fantegrossi WE, Wetsel WC, Roth BL (2016) In vitro and in vivo characterization of the alkaloid nuciferine. PLoS One 11(3):1–27

55. Webster K, Brimmer G, Vunck SA, Wiebelhaus JM, Porter JH (2009) The discriminative stimulus properties of the atypical antipsychotic drug clozapine in 129S2 mice. Program no. 840.11. 2009 neuroscience meeting planner. Society for Neuroscience, Washington, DC

56. Webster KA, Almond LM, Porter JH (2010) Further characterization of the discriminative stimulus properties of the atypical antipsychotic drug clozapine in 129S2/SvHsd mice. Program no. 767.25. 2010 neuroscience meeting planner. Society for Neuroscience, Washington, DC

57. Porter JH, Webster KA, Philibin SD, Walentiny DM, Vunck SA (2011) Phenotype differences in clozapine's discriminative stimulus in C57BL/6, DBA/2, and 129S2 inbred mouse strains. Poster presented at the meeting of the European Behavioural Pharmacology Society in Amsterdam, Netherlands, August, 2011

58. Rush CR, Stoops WW, Hays LR, Glaser PEA, Hays LS (2003) Risperidone attenuates the discriminative-stimulus effects of d-amphetamine in humans. J Pharmacol Exp Ther 306:195–204

59. Lile JA, Stoops WW, Glaser PE, Hays LR, Rush CR (2011) Discriminative stimulus, subject-rated and cardiovascular effects of cocaine alone and in combination with aripiprazole in humans. J Psychopharmacol 25:1469–1479

60. Moncrieff J, Cohen D, Mason JP (2009) The subjective experience of taking antipsychotic medication: a content analysis of Internet data. Acta Psychiatr Scand 120(2):102–111

61. Gerlach J, Larsen EB (1999) Subjective experience and mental side-effects of antipsychotic treatment. Acta Psychiatr Scand 99:113–117

62. Mizrahi R, Rusjan P, Agid O, Graff A, Mamo DC, Zipursky RB et al (2007) Adverse subjective experience with antipsychotics and its relationship to striatal and extrastriatal D2 receptors: a PET study in schizophrenia. Am J Psychiatry 164:630–637

63. Bymaster FP, Felder CC, Tzavara E, Nomikos GG, Calligaro DO, McKinzie DL (2003) Muscarinic mechanisms of antipsychotic atypicality. Prog Neuropsychopharmacol Biol Psychiatry 27:1125–1143

64. Weston-Green K, Huang X-F, Lian J, Deng C (2012) Effects of olanzapine on muscarinic M3 receptor binding density in the brain relates to weight gain, plasma insulin and metabolic hormone levels. Eur Neuropsychopharmacol 22:364–373

65. Crawley JN, Belknap JK, Collins A, Crabbe JC, Frankel W, Henderson N et al (1997) Behavioral phenotypes of inbred mouse strains: implications and recommendations for molecular studies. Psychopharmacology (Berl) 132:107–124

The Discriminative Stimulus Properties of Drugs Used to Treat Depression and Anxiety

Adam J. Prus and Joseph H. Porter

Abstract Drug discrimination is a powerful tool for evaluating the stimulus effects of psychoactive drugs and for linking these effects to pharmacological mechanisms. This chapter reviews the primary findings from drug discrimination studies of antidepressant and anxiolytic drugs, including novel pharmacological mechanisms. The stimulus properties revealed from these animal studies largely correspond to the receptor affinities of antidepressant and anxiolytic drugs, indicating that subjective effects may correspond to either therapeutic or side effects of these medications. We discuss drug discrimination findings concerning adjunctive medications and novel pharmacologic strategies in antidepressant and anxiolytic research. Future directions for drug discrimination work include an urgent need to explore the subjective effects of medications in animal models, to better understand shifts in stimulus sensitivity during prolonged treatments, and to further characterize stimulus effects in female subjects. We conclude that drug discrimination is an informative preclinical procedure that reveals the interoceptive effects of pharmacological mechanisms as they relate to behaviors that are not captured in other preclinical models.

Keywords Animal models • Antidepressant • Anxiety • Anxiolytic • Depression • Discriminative stimulus • Drug discrimination • Operant • Preclinical model • Sedative • Stimulus properties

A.J. Prus (✉)
Department of Psychology, Northern Michigan University, Marquette, MI, USA
e-mail: aprus@nmu.edu

J.H. Porter
Department of Psychology, Virginia Commonwealth University, Richmond, VA, USA

© Springer International Publishing Switzerland 2016
Curr Topics Behav Neurosci (2018) 39: 213–242
DOI 10.1007/7854_2016_27
Published Online: 22 June 2016

Contents

1 Introduction

This chapter, located within the volume *The Behavioural Neuroscience of Drug Discrimination* as part of the *Current Topics in Behavioral Neurosciences* series, pertains to the utilization of drug discrimination as a way of examining the stimulus properties of antidepressants and anxiolytics. In this chapter, we review the major findings from the experimental literature to describe the current state of knowledge on the discriminative stimulus properties of these pharmacotherapeutics.

2 Treatments for Depression and Anxiety

The antitubercular drug iproniazid was the first serendipitously discovered compound to have antidepressant efficacy. Synthesized in the early 1950s [1], iproniazid's mechanism of actions included inhibition of monoamine oxidase (MAO) [2]. A series of studies evaluating iproniazid for treating tuberculosis reported improvements in mood, later leading to reports of reduced symptoms in depressed patients. While iproniazid was first marketed only as an antitubercular drug, it was swiftly incorporated as an off-label treatment for clinical depression. Because iproniazid was an irreversible inhibitor of both MAO_A and MAO_B [3], its clinical utility as an antidepressant was limited due to side effects, including hypertension and digestive complications. Modern MAO inhibitors include selective inhibitors of MAO_B, which is found in the central nervous system (CNS), or reversible inhibitors of MAO_A, which is found in brain and outside the CNS, including the liver and gastrointestinal tract (for review, see [4]).

The development of tricyclic antidepressants arose from efforts to replicate the success of chlorpromazine for the treatment of psychosis. Imipramine was synthesized from promethazine, a compound structurally similar to chlorpromazine. Imipramine proved ineffective for psychosis, but reductions in depressive symptoms in patients with depressive psychosis led to the notion of using imipramine as a novel antidepressant drug (see [4, 5]).

The pharmacological mechanisms of action for tricyclic antidepressant drugs are heterogeneous. Most tricyclic antidepressants inhibit reuptake of serotonin and norepinephrine, but have a weak affinity for dopamine transporters. Tricyclic antidepressant drugs also act as antagonists with a high affinity for the histamine H_1 receptor, and with a moderate affinity for serotonin $(5\text{-}HT)_{2A}$, $5\text{-}HT_{2C}$, and muscarinic receptors [6–10].

The success of tricyclic antidepressant drugs for treating depression led to development of zimelidine, a selective serotonin reuptake inhibitor (SSRI), by Astra AB Pharmaceuticals [11]. Marketed under the trade name Zelmid® in 1982, adverse side effects led Astra to remove zimelidine from the European market in 1983, paving the way for the development of fluoxetine by Eli Lilly. Fluoxetine and other SSRIs are more selective for the serotonin transporter versus the dopamine or norepinephrine transporter, but some SSRIs exhibit appreciable binding affinities for orthosteric monoamine neurotransmitter receptors. For example, fluoxetine exhibits a moderate affinity for $5\text{-}HT_{2A}$ receptors and a relatively high affinity for $5\text{-}HT_{2C}$ receptors [12]. Also, the SSRI paroxetine binds with high affinity for the serotonin transporter and has a moderate affinity for the norepinephrine transporter, and in addition, paroxetine binds with a moderate affinity for muscarinic cholinergic receptors [7]. On the other hand, the SSRI escitalopram exhibits selectivity for the serotonin transporter with a low affinity for monoamine receptors [13].

Many antidepressant drugs exhibit a high affinity for both serotonin and norepinephrine transporters, thus leading to the development of a combined serotonin–norepinephrine reuptake inhibitor (SNRI) as a promising line of antidepressant action. The first antidepressant to be marketed as an SNRI was venlafaxine by Wyeth in 1993. Receptor binding studies show a moderate to weak binding affinity of venlafaxine for the norepinephrine transporter and an approximately eightfold preference for the serotonin transporter [7, 12, 14]. Desvenlafaxine, the active metabolite of venlafaxine, has a similar binding profile and was later approved as an SNRI antidepressant drug [15]. Moreover, venlafaxine appears to lack an appreciable affinity for monoamine, and perhaps other types of receptors [12, 14, 16]. The SNRI duloxetine exhibits a high affinity for both serotonin and norepinephrine transporters, with approximately a tenfold greater affinity for the serotonin transporter over the norepinephrine transporter; and it exhibits a moderate affinity

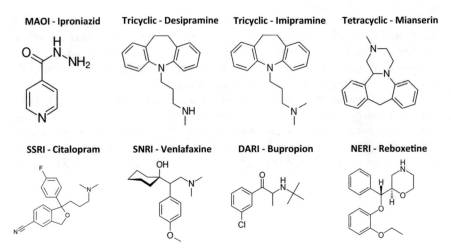

Fig. 1 Structures for antidepressant drugs

for the dopamine transporter and moderate to weak affinities for serotonin 5-HT$_{2A}$ and 5-HT$_6$ receptors [16].

Newer antidepressants with mechanisms of action that do not fit precisely into the categories described above are often considered as atypical antidepressants, multi-modal antidepressant compounds, or second generation antidepressants. The best known of these more recent and promising compounds is bupropion. Bupropion exhibits a moderate affinity for the dopamine transporter and a weak affinity for the norepinephrine and serotonin transporters as well as monoamine neurotransmitter receptors [12]. Reboxetine is also a selective norepinephrine reuptake inhibitor, although it possesses a modest affinity for serotonin transporters [14], and reboxetine is not usually considered an SNRI given its greater selectivity for the norepinephrine transporter. Vortioxetine exhibits a high affinity for the serotonin transporter, yet possesses a high affinity for a number of serotonin receptors, including 5-HT$_{1A}$, 5-HT$_{1B}$, 5-HT$_{3A}$, and 5-HT$_7$ receptors [17].

Thus, since the discovery of the first antidepressant drug iproniazid, all antide-pressants that have been marketed share mechanisms of action that enhance mono-amine neurotransmission (see Fig. 1 for examples of drug structures). The most common pharmacological antidepressant action is derived from directly elevating serotonin concentrations in brain, but drugs such as bupropion and reboxetine suggest that directly acting on serotonin neurotransmission may not be a sole requirement for producing antidepressant effects. Many, but not all, of the SSRIs bind to 5-HT$_{2C}$ receptors and various other serotonin receptors. The affinity of tricyclic antidepressant drugs for muscarinic and histamine receptors has tradition-ally been regarded as unbeneficial for clinical efficacy, and instead the primary cause of side effects, such as dry mouth, blurred vision, and constipation (linked to anticholinergic effects) or sedation (linked to their antihistaminergic effects), although recently more attention has been given to muscarinic antagonism as a

Barbiturate - Phenobarbital **Benzodiazepine - Diazepam** **5-HT$_{1A}$ Agonist - Buspirone**

Muscarinic Antagonist - Scopolamine   **NMDA Antagonist - Ketamine**

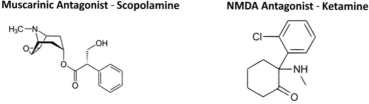

Fig. 2 Structures for antianxiety drugs and other drugs with antidepressant efficacy

possible therapeutic target for treating depression ([18]; see below). Also, glutamatergic targets for the treatment of depression have been the focus of much research since Berman et al. [19] demonstrated that a low, subanesthetic dose (i.v.) of the noncompetitive N-methyl-D-aspartate (NMDA) antagonist ketamine produced rapid and sustained antidepressant effects in depressed patients (for reviews see [5]; [20]).

Drugs that act on a number of different pharmacological mechanisms, including barbiturates, benzodiazepines, serotonin 5-HT$_{1A}$ rector agonists, and most recently antidepressant drugs that inhibit serotonin reuptake have been used to treat anxiety (see Fig. 2 for examples of drug structures). Section 3 will examine the stimulus properties of antidepressant drugs (which include drugs used as a first-line treatment for anxiety) and Sect. 4 will examine the discriminative stimulus properties of barbiturate, benzodiazepine, and serotonin 5-HT$_{1A}$ receptor agonist drug classes. Section 5 will present information on novel therapeutic targets for the treatment of depression, and finally Sect. 6 will provide an overview of adjunctive strategies for the treatment of depression.

3 Discriminative Stimulus Properties of Antidepressant Drugs

3.1 MAO Inhibitors

The first assessment of MAO inhibition in a drug discrimination study was by Huang and Ho [21] who found that the nonselective MAO$_{A/B}$ inhibitor iproniazid failed to substitute for d-amphetamine in rats. However, full substitution did occur after pretreatment with beta-phenylethylamine, which is a monoamine alkaloid and produces pharmacodynamic effects similar to d-amphetamine. The authors concluded that iproniazid and beta-phenylethylamine in combination had additive effects, together sufficient to mimic d-amphetamine's discriminative stimulus effect.

To our knowledge, only one study has established MAO inhibitors as a training drug in the drug discrimination paradigm. Overton [22] used a T-maze procedure to train the rats to discriminate the stimulus effects of a drug in order to escape a grid floor shock. Overton found that the nonselective MAO_A/MAO_B inhibitors iproniazid, nialamide, phenelzine, and tranylcypromine were readily established as discriminative stimuli (13–27 sessions), while nialamide required greater than 40 sessions to learn the discrimination. In other studies, MAO inhibitors have been evaluated as test compounds to determine if MAO inhibitors would mimic the discriminative stimulus properties of other training drugs like the psychostimulant cocaine. Cocaine, in particular, is useful for these studies, since cocaine enhances dopamine, norepinephrine, and serotonin levels [23–25].

Colpaert et al. [26] reported that the nonselective $MAO_{A/B}$ inhibitor tranylcypromine fully substituted for cocaine's discriminative stimulus in rats. In a subsequent study, these scientists also found that the selective MAO_B inhibitor deprenyl and the nonselective $MAO_{A/B}$ inhibitors pargyline and pheniprazine fully substituted for cocaine. However, the MAO_A inhibitor clorgyline (also known as clorgiline) and the nonselective MAO inhibitor nialamide did not [27]. Later studies questioned whether deprenyl's substitution for cocaine was actually due to increased levels of monoamines, as deprenyl was shown to produce amphetamine or methamphetamine as a metabolite [28].

Discriminative stimulus effects can be readily established by activating imidazoline I2 receptors, which are located on MAO_A and MAO_B enzymes. The activation of these receptors inhibits both enzymes [29, 30]. Jordan et al. [31] reported that the reversible MAO_A inhibitor moclobemide and the irreversible, nonselective $MAO_{A/B}$ inhibitor pargyline fully substituted for the high affinity imidazoline I2 ligand RX801077. In a later study, MacInnes and Handley [32] found that the selective reversible MAO_A inhibitor RO41-1049 also fully substituted for RX801077, while the reversible MAO_B inhibitors lazabemide and RO16-6491 did not. Thus, the interoceptive stimulus effects produced by I2 receptor-induced inhibition of MAO appear to be more associated with MAO_A inhibition, as compared to MAO_B inhibition.

Downstream receptor actions may mediate the discriminative stimulus effects of MAO inhibitors. Crissman and O'Donnell [33] trained rats to discriminate the β_1-adrenoceptor agonist isoproterenol. Worth noting here is that intracerebroventricular (ICV) administration was used during the initial discrimination training phase for this compound. ICV administration is not commonly used in drug discrimination studies due to the complications of keeping a cannula fixed in place for what are typically lengthy behavioral studies. Even so, once this discrimination was learned, rats were tested with a series of antidepressant drugs for their similarity with isoproterenol. They found that the nonselective $MAO_{A/B}$

Table 1 Receptor mechanisms producing full substitution for antidepressant drugs

Antidepressant type	Mechanisms producing full substitution
MAO inhibitors	Imidazoline I2 agonism
	β_1-adrenoceptor agonism
Tricyclic antidepressants	β_1-adrenoceptor agonism
	DA reuptake inhibitor
	NE reuptake inhibitor
	5-HT$_{1A}$ agonism
SSRI	5-HT$_{2C}$ agonism
	5-HT$_{1A}$ agonism
SNRI	β_1 receptor agonism
Bupropion (atypical antidepressant)	DA reuptake inhibitor
	Muscarinic receptor antagonism
	D$_1$ receptor agonism
	D$_2$ receptor agonism
Reboxetine (atypical antidepressant)	NK$_1$ receptor antagonism
	NE reuptake inhibitor

inhibitor phenelzine engendered full substitution for ICV isoproterenol. The authors interpreted these results as being due to enhanced norepinephrine concentrations that activated β_1 receptors.

Overall, we can reach the following general conclusions regarding the discriminative stimulus properties of MAO inhibitors. First, drugs that inhibit MAO$_B$ more selectively than MAO$_A$ produce stimulus effects similar to those produced by certain psychostimulant drugs. However, this conclusion does depend greatly on the particular MAO inhibitor that was tested and the finding that the substitution seen by deprenyl to methamphetamine and cocaine was likely due to deprenyl's active metabolites. This raises the question about the role of other MAO inhibitor active metabolites in the discriminative stimulus effects of these drugs. Second, stimulus effects elicited by directly inhibiting MAO via the imidazoline I2 receptor are similar to those produced by MAO$_A$ inhibitors. Third, enhanced monoamine neurotransmitter concentrations will lead to greater activation of receptors (e.g., β_1 noradrenergic receptors), which may engender properties that add to the overall discriminative stimulus properties of an MAO inhibitor. Table 1 provides a summary of general discriminative stimulus findings for MAO inhibitors and other antidepressant drugs.

3.2 Tricyclic Antidepressant Drugs

As noted earlier, tricyclic antidepressant drugs tend to inhibit norepinephrine and serotonin reuptake, along with exhibiting receptor binding to serotonin, histamine H$_1$, and muscarinic receptors. In addition to these receptor binding profiles, tricyclic antidepressant drugs also differ in their relative affinity for serotonin versus norepinephrine transporters. For example, imipramine displays a similar affinity for both serotonin and norepinephrine transporters, while desipramine exhibits over a 20-fold preferential affinity for norepinephrine transporters versus serotonin transporters [7]. In contrast, clomipramine has an approximately 20-fold greater affinity for serotonin transporters than for norepinephrine transporters [14].

Desipramine was the first tricyclic antidepressant drug to be established as a discriminative stimulus. Shearman et al. [34] trained rats to discriminate desipramine from vehicle for food reinforcement using two-lever operant chambers. Neither of the tricyclic antidepressant drugs imipramine or protriptyline substituted for desipramine's discriminative cue; however, amphetamine at a dose of 1.25 mg/kg produced partial substitution (67% drug lever selection) for desipramine. This finding is in agreement with later studies that found that cocaine can produce either full or partial substitution for desipramine [35, 36], although other studies have not [37, 38]. Thus, these drug discrimination studies generally confirm the inhibition of reuptake for serotonin and norepinephrine by desipramine. Other studies have shown that desipramine in combination with a sub-effective dose of cocaine produces full substitution for training doses of cocaine [37, 39]. In the study mentioned earlier by Crissman and O'Donnell [33], desipramine fully substituted for the β_1 receptor agonist isoproterenol, which was likely due to greater activation of these receptors from elevated concentrations of norepinephrine by desipramine.

One of the most studied tricyclic antidepressants in the drug discrimination paradigm is imipramine. As was the case with the MAO inhibitors, Overton [22] first demonstrated that rats could distinguish the stimulus effects of imipramine (40 mg/kg) from a non-drug state using a T-maze drug discrimination procedure in rats. In his study, the stimulus effects were readily discriminated as animals only required approximately 12 sessions to meet the training criterion. Subsequently, Schechter [40] used a two-lever operant chamber to train rats to discriminate imipramine from saline for food reinforcement. Schechter exposed the rats to different types of stressors (e.g., footshock, restraint) in order to produce something more analogous to a "depressive" state in humans. However, only the unstressed rats in this study were successfully trained to discriminate imipramine. Schechter found that the tricyclic antidepressants amitriptyline and desipramine fully substituted for imipramine in this group of rats.

Much of the research on imipramine's discriminative stimulus properties were conducted by Barrett's group in the early 1990s. Zhang and Barrett [41] trained pigeons to discriminate imipramine from vehicle using a two-key operant chamber for food reinforcement. As previously found in rats [40], amitriptyline fully substituted for imipramine. In addition, they reported that the psychostimulant cocaine and the antidepressant drug bupropion (dopamine and norepinephrine transporter inhibitor) also fully substituted for imipramine's discriminative stimulus. The mechanisms shared by imipramine, amitriptyline, and cocaine consist of reuptake inhibition of serotonin and norepinephrine, while imipramine and bupropion share an inhibition of norepinephrine reuptake. Direct receptor actions also might be shared by imipramine and amitriptyline, while enhanced concentrations of serotonin or norepinephrine likely led to downstream receptor actions.

In a follow-up to this study, Barrett and Zhang [42] reported that the 5-HT$_{1A}$ partial agonist 8-OH-DPAT mimicked the discriminative stimulus of imipramine, and that the discriminative stimulus effects of imipramine were blocked by administration of the 5-HT$_{1A}$ receptor antagonist NAN-190. They also found that the α_1 receptor antagonist prazosin blocked the discriminative stimulus effects of imipramine. Imipramine lacks any affinity for 5-HT$_{1A}$ receptors and neither of these drugs appears to exhibit appreciable binding affinities to the same receptors. Imipramine, however, does exhibit a high affinity for α_1 adrenoceptors in humans [6], but only a moderate affinity for α_1 receptors in rats [43] and chronic dosing with imipramine in rats does not produce upregulation of α_1 adrenoceptors [44]. This may account for the ability of prazosin to block the activation of α_1 receptors by imipramine.

Other well-studied tricyclic antidepressants in the drug discrimination literature are clomipramine, nortriptyline, and amitriptyline; however, none of these drugs have been used as the training drug in a discrimination study. As noted earlier in this chapter, Shearman et al. [34] did not find stimulus generalization from desipramine to protriptyline. Further, amitriptyline has been shown to fully substitute for imipramine [40, 41], for the atypical antipsychotic drug clozapine, and for the cholinergic muscarinic receptor antagonist scopolamine in rats [45], likely due to shared stimulus effects mediated by muscarinic receptor antagonism among these compounds. In pigeons, only partial substitution was found reported to clozapine by amitriptyline [46].

3.3 SSRI Antidepressant Drugs

As the primary mechanism of action for SSRIs is an elevation of extracellular serotonin concentrations due to inhibition of reuptake, it is worth considering the discriminative stimulus properties of "serotonin releasers" — drugs that elevate serotonin concentrations by a variety of different mechanisms but do not produce antidepressant effects. One of the primary serotonin releasers used in behavioral pharmacology is fenfluramine. Fenfluramine acts as an effective anorectic in humans (although it has serious side effects), which is likely due to activation of 5-HT$_{2C}$ receptors via elevated serotonin concentrations [47]. Fenfluramine increases extracellular serotonin release through an exocytosis-like mechanism at lower concentrations and by reversal of the serotonin transporter at higher concentrations [48], but does not bind to serotonin receptors. Goudie [49] established fenfluramine as a discriminative stimulus, using female rats in a two-lever operant task for food reinforcement, finding partial substitution of fenfluramine's metabolite norfenfluramine, which is now known to activate both 5-HT$_{2B}$ and 5-HT$_{2C}$ receptors [47] and produce greater increases in norepinephrine concentrations than fenfluramine [50]. Subsequent drug discrimination studies found full substitution by serotonin/norepinephrine releasers [51, 52], the serotonin receptor agonist quipazine (which exhibits a high affinity for 5-HT$_3$ and a moderate affinity for 5-HT$_{2B}$ and 5-HT$_{2C}$ receptors) [52], the 5-HT$_{2C/1B}$ receptor agonist *m*-

chlorophenylpiperazine (mCPP), and the selective 5-HT_{2C} receptor agonist MK-212 [53]. Partial stimulus generalization occurred to the 5-HT_{2A} preferring (over 5-HT_{2C}) receptor agonist 2,5-dimethoxy-4-iodoamphetamine (DOI), and full blockade of the fenfluramine discriminative stimulus occurred following pretreatment with the 5-HT_{2C} receptor antagonist SB206553, but not the 5-HT_{2C} receptor antagonist RS102221 [53]. Thus, it appears that activation of 5-HT_{2C} receptors may yield stimulus properties more similar to the discriminative stimulus properties of fenfluramine than does activation of other serotonin receptor subtypes. The prominence of 5-HT_{2C} mediated properties of the fenfluramine discriminative stimulus provides a compelling link to the primary receptor likely mediating fenfluramine's anorectic effects. Further, these findings offer some predictive value for the discriminative stimulus properties of SSRIs.

There are several difficulties to consider when establishing the discriminative stimulus properties of SSRIs. One of the first published attempts to establish fluoxetine as a training drug was reported by Marona-Lewicka and Nichols [54]. While this study successfully established the SSRIs citalopram and sertraline as discriminative stimuli in male rats, subjects could not reliably discriminate fluoxetine from vehicle after a year of training. The authors suggested that the long half-life of fluoxetine, which is close to a week in humans, precluded discriminating fluoxetine versus a true non-drug interoceptive state. Thus, fluoxetine is not an ideal candidate drug for discrimination training, but other SSRIs with half-lives of approximately 24 h or less in human – including citalopram, fluvoxamine, paroxetine, or sertraline – might be more easily established as discriminative stimuli [55].

Millan et al. [56] also established citalopram as a discriminative stimulus in rats after first determining doses of citalopram that elevate extracellular serotonin concentrations. Male rats successfully learned to discriminate citalopram (2.5 mg/kg) from vehicle in a two-choice procedure using food reinforcement. During generalization testing, both of the SSRIs, paroxetine and sertraline, fully substituted for citalopram. In a second study, Millan et al. [57] reported full stimulus generalization to the 5-HT_{2C} receptor agonist RO60-0175 and full stimulus blockade by the 5-HT_{2C} receptor antagonist SB242,084. Thus, like fenfluramine, 5-HT_{2C} receptor activation appears to mediate the discriminative stimulus properties of citalopram (also see review by [58]). Yet, unlike fenfluramine, citalopram, and indeed many other SSRIs with the exception of fluoxetine, causes weight gain in clinically depressed patients. In fact, weight gain is often attributed to SSRI treatment regimens [59].

Wolff and Leander [60] established the putative SSRI antidepressant, LY233708, as a discriminative stimulus in pigeons. LY233708 exhibited a relatively short half-life and serotonin concentrations returned to baseline levels a day after administration. Full stimulus generalization occurred to fluoxetine and citalopram, but not to the norepinephrine reuptake inhibitor nisoxetine. Full stimulus generalization also occurred from the 5-HT_{1A} receptor agonist 8-OH-DPAT, and pretreatment with the 5-HT_{1A} receptor antagonist WAY100,635 fully blocked the LY233708 discriminative cue.

Conditioned taste aversion (CTA) is a procedure with similar behavioral endpoints as classical drug discrimination assessments and has been useful for assessing the discriminative stimulus effects of SSRIs. As described by Riley and colleagues in Chap. 14 of this volume, a typical CTA drug discrimination paradigm utilizes the novelty of a drug effect paired with an unfamiliar (normally sweet-tasting) substance to reduce drinking of the unfamiliar substance during a subsequent test session. Yet, if the same or a similar drug were given during the preceding sessions that contained only tap water in the bottles (referred to as training sessions), then the drug given on the conditioning session would ineffectively pair with the novel tasting solution and lead to significant consumption of the solution on the test day.

Using the CTA drug discrimination procedure, Berendsen and Broekkamp [61] examined the discriminative stimulus effects of fluoxetine in male mice. Without "pre-exposure" drug administration (i.e., drugs given during training sessions), fluoxetine administration on the conditioning day led to significantly less consumption of the glucose solution during the test session. However, using fluoxetine as both a pre-exposure drug and a test drug led to increased consumption of the solution on the test day. A similar effect occurred when the $5-HT_{2C}$ receptor agonist MK212 was used as a pre-exposure drug, and fluoxetine as a test drug, demonstrating that fluoxetine and MK212 exhibited similar stimulus effects. This did not occur when the $5-HT_{2A/2C}$ receptor agonist DOI was used as the pre-exposure drug. A partial reduction of glucose solution intake was observed when the $5-HT_{1A}$ receptor partial agonist 8-OH-DPAT was used as the pre-exposure drug. Thus, these results tend to support the involvement of $5-HT_{2C}$ receptors, and to some extent $5-HT_{1A}$ receptors, mediating the discriminative stimulus effects of fluoxetine. Other SSRIs also have been studied using the CTA drug discrimination paradigm. For example, Gommans et al. [62] found that the SSRI fluvoxamine shared similar stimulus effects with $5-HT_{1A}$ receptor partial agonists, but not to $5-HT_{2A}$ or $5-HT_{2C}$ receptor agonists or to the SSRI fluoxetine in this study.

Overall, drug discrimination studies demonstrate that serotonin releasers (which do not appear to exhibit antidepressant effects) and SSRIs with clinical antidepressant efficacy can both be established as discriminative stimuli. However, operant drug discrimination procedures normally require daily training sessions – an obstacle when using compounds with long half-lives as training drugs. The CTA drug discrimination procedure appears to obviate this limitation, as this procedure allows for fluoxetine and other SSRIs to be established as discriminative stimuli. From serotonin releaser and SSRI drug discrimination studies, we learn that these compounds tend to exhibit generalization to each other, with occasional exceptions (e.g., [62]). In conclusion, the discriminative stimulus effects of SSRIs appear to be mediated primarily by activation of $5-HT_{2C}$ receptors and to a lesser extent by $5-HT_{1A}$ receptors, and probably not at all by $5-HT_{2A}$ receptors.

3.4 SNRI Antidepressant Drugs

The only study establishing an SNRI as a discriminative stimulus was conducted by Kayir et al. [63]. Using the CTA drug discrimination procedure, this group found that drugs that increased both serotonin (e.g., fluoxetine) and norepinephrine reuptake inhibition (e.g., reboxetine) fully substituted for the SNRI venlafaxine. Several studies have examined the stimulus effects of SNRIs in drug discrimination studies using these compounds as test agents rather than as training drugs. These studies generally support the idea that the stimulus properties of venlafaxine are elicited by enhanced serotonin and norepinephrine efflux. For example, venlafaxine has been shown to potentiate the discriminative stimulus effects of the hallucinogen LSD [64] and to fully substitute for the β_1 receptor agonist isoproterenol [33]. Similarly, the SNRI sibutramine has been shown to substitute for the psychostimulant cocaine [65].

3.5 Atypical Antidepressant Drugs

As noted earlier, atypical antidepressant drugs include those that do not fit the previously mentioned categories. While the best known drugs in this class are reboxetine (a selective norepinephrine reuptake inhibitor) and bupropion (a selective dopamine reuptake inhibitor), the first selective norepinephrine reuptake inhibitor to be studied using drug discrimination was the + enantiomer of oxaprotiline (tetracyclic family, related to the antidepressant maprotiline). Filip et al. [66] trained male rats to discriminate (+)-oxaprotiline from vehicle, finding full stimulus generalization to the tricyclic antidepressant drug desipramine.

Using a two-lever discrimination for food reinforcement in rats, Dekeyne et al. [67] first established reboxetine as a discriminative stimulus and used in vivo microdialysis to measure extracellular monoamine neurotransmitter concentrations to verify significant elevations of norepinephrine, but not serotonin, in the frontal cortex and hippocampus. Full stimulus generalization occurred from reboxetine to the SNRI venlafaxine and to antidepressants that preferentially block norepinephrine reuptake, desipramine, and maprotiline. Stimulus generalization did not occur to drugs lacking this mechanism. Millan and Dekeyne [68] later reported on an extensive evaluation of the discriminative stimulus properties of reboxetine, finding full stimulus generalization to all compounds that produced increases in norepinephrine concentrations via inhibition of norepinephrine reuptake. No appreciable degree of stimulus generalization occurred to SSRIs lacking an affinity for norepinephrine transporters. Stimulus generalization also did not occur to adrenoceptor agonists, although α_1 adrenoceptor antagonists fully, and an α_2 receptor antagonist partially, blocked the discriminative cue. Also, the NK1 receptor antagonist GR205,171 fully substituted for reboxetine. It is thought that NK1 receptor antagonists may elevate norepinephrine concentrations given that increased norepinephrine concentrations are observed in $NK1^{-/-}$ mice [68].

Howard and colleagues ([69]; portions of this study were reported previously in a book chapter, [70]) were the first to determine that the selective dopamine reuptake inhibitor bupropion could serve as a discriminative stimulus. In this study, bupropion's discriminative stimulus generalized to psychostimulants (d-amphetamine, cocaine, benzylpiperazine, methylphenidate, and caffeine) and to some antidepressant drugs (nortriptyline, viloxazine, and nomifensine), but not others (imipramine, amitriptyline, desipramine, and mianserin). Full stimulus generalization also occurred to the muscarinic receptor antagonist scopolamine, despite bupropion having no affinity for muscarinic receptors [71]. Interestingly, scopolamine also has been found to substitute for the tetracyclic antidepressant mianserin, which has minimal affinity for cholinergic receptors, although mianserin did not substitute for scopolamine [72]. Later, Blitzer and Becker [73] also reported full stimulus generalization to the psychostimulants amphetamine, cocaine, and caffeine, but were unable to antagonize bupropion's discriminative cue by antipsychotic drugs like haloperidol and thioridazine, which are potent D_2 receptor antagonists. Based on a lack of blockage by D_2 receptor antagonists, the authors concluded that bupropion's discriminative cue was not mediated by dopaminergic mechanisms, nor was it mimicked or blocked by adrenergic or serotonergic drugs.

However, some clarification of the discriminative stimulus properties of bupropion was made as more selective dopaminergic compounds became available [74]. In addition to verifying that the psychostimulant cocaine produced full stimulus generalization from bupropion, this study found that dopamine reuptake blockers fully substituted for bupropion, that full or partial substitution was observed for a series of dopamine D_1 and D_2 receptor agonists and that D_1 and D_2 receptor antagonists partially blocked the bupropion cue. Based on these results the authors concluded that the discriminative stimulus effects of bupropion are mediated by activation of D_1 and D_2 receptors. The finding that scopolamine also mimics bupropion's discriminative cue [71] warrants further investigation as it is possible that muscarinic mechanisms may play an indirect role in bupropion's discriminative stimulus properties and perhaps even in its antidepressant efficacy (see section below on use of scopolamine for the treatment of depression).

4   Discriminative Stimulus Properties of Anxiolytic drugs

Although various remedies for nervousness and anxiety were used for many years, including opiates, bromide salts, and alcohol, we recognize the discovery of barbiturates, the so-called "minor tranquilizers," as the first anxiolytic medications (antipsychotic drugs were called "major tranquilizers"). The first barbiturate effective in humans, barbital (Veronal®), was brought to market in 1904 and many of the other 49 barbiturates came soon after. While effective for anxiety, their propensity

Table 2 Receptor
mechanisms producing full
substitution for anxiolytic
drugs

Anxiolytic class	Mechanisms producing full substitution
Barbiturates	$GABA_A$ receptor positive modulators
Benzodiazepines	$GABA_A$ receptor positive modulators BZ I agonism BZ II agonism
Buspirone	$5\text{-}HT_{1A}$ agonism

for causing dependence and risk of lethal overdose led to serious public health concerns [75, 76].

Both barbiturates and benzodiazepines bind to allosteric sites near the Cl^- channel for $GABA_A$ receptors. Activation of these sites modulates the receptor to be more responsive to activation of the $GABA_A$ receptor site (i.e., positive modulation). Benzodiazepines have two sites: BZ I and BZ II. The BZ I type is found on the α_1 subunit containing the $GABA_A$ receptor binding site [77], while the BZ II site is found on α_2, α_3, and α_5 subunits [78, 79]. BZ I $GABA_A$ receptors are found in the thalamus, substantia nigra, and cerebellum, while BZ II $GABA_A$ receptors are found in cerebral cortex, hypothalamus, and amygdala [80]. The sleep aids zolpidem and eszopiclone show preferential binding to the BZ I site versus the BZ II site [81], which accounts for their sedative effects while lacking anxiolytic effects [82, 83].

Aside from drugs acting at $GABA_A$ receptors, the $5\text{-}HT_{1A}$ receptor agonist buspirone (BuSpar®) also is prescribed as an anxiolytic drug. Buspirone appears to be as effective as benzodiazepines for treating anxiety disorders [84], but treatment effects for anxiety do not appear until after 4–6 weeks of treatment. The delayed treatment response appears to be why many practitioners are skeptical about its effectiveness as an anxiolytic [85].

Today, first-line treatments for anxiety consist primarily of antidepressant drugs that inhibit serotonin reuptake. These agents avoid abuse concerns over benzodiazepine use and offer prophylactic effects toward avoiding anxiety episodes. Yet, we still refer to these drugs as antidepressant drugs, rather than anxiolytic drugs, to address their primary clinical use. Patients also may need to take benzodiazepines during antidepressant treatment on an "as needed" basis for addressing acute episodes of anxiety, such as a panic attack [86]. Drug discrimination studies involving antidepressant drugs were reviewed in Sect. 3 and this section will address the discriminative stimulus properties of barbiturate, benzodiazepine, and $5\text{-}HT_{1A}$ receptor agonist drug classes. Table 2 shows a general summary of the discriminative stimulus properties for anxiolytic drugs.

4.1 Barbiturates

Most of what we know about the discriminative stimulus properties of barbiturates were sorted out in the 1970s and early 1980s. Hirschhorn and Winter [87] first established a barbiturate as a discriminative stimulus by training rats to discriminate barbital from saline. This study found that its stimulus properties were not similar to those produced by 5-HT$_2$ receptor agonists including the hallucinogens mescaline or LSD. York [88] later reported that male rats trained to discriminate either barbital or phenobarbital from saline did not evoke either partial or full generalization to ethanol. In male pigeons, Herling et al. [89] found full stimulus generalization occurring from pentobarbital to other barbiturates as well as to benzodiazepines. The CNS stimulant and pro-convulsant bemegride blocked discriminative control by pentobarbital, and stimulus generalization did not occur from pentobarbital to opioid agonists or anticonvulsants. Stimulus generalization also did not occur to the GABA$_A$ agonist and GABA$_A$-rho partial agonist muscimol or to the GABA$_B$ receptor agonist baclofen. Similar findings were shown in male rhesus monkeys trained to discriminate pentobarbital from vehicle [90]. Non-rate suppressant doses of the benzodiazepine antagonist Ro 15-4513 did not block the discriminative stimulus effects of phenobarbital in male mice [91].

Barbiturates elicit discriminative stimulus effects that are most similar to those generated by other barbiturates, but benzodiazepines elicit surprisingly similar interoceptive effects. These stimulus effects likely occur at doses capable of engendering anxiolytic effects, as pentobarbital has been shown to increase rates of punished responding in a drug discrimination procedure in pigeons [92]. Licata et al. [93] provided a refinement on our understanding of the stimulus effects produced by barbiturates by showing that the GABA$_A$ receptor positive modulator L-838,417, which is selective for alpha 2, 3, and 5 subunits (those corresponding with the BZ II site), fully substituted for the barbiturates amobarbital and pentobarbital in squirrel monkeys.

4.2 Benzodiazepines

Not surprisingly, the discriminative stimulus properties of benzodiazepines appear to be mediated primarily by benzodiazepine receptors, and their discriminative stimulus properties also are similar to stimulus effects produced by other GABA$_A$ positive modulators. Colpaert et al. [94] first evaluated the discriminative stimulus effects of a benzodiazepine by training male rats to discriminate chlordiazepoxide from vehicle in a two-lever operant task. Full stimulus generalization occurred to all of the benzodiazepines tested and to all of the barbiturates tested. Neither partial nor full generalization occurred to the GABA$_B$ agonist baclofen. Haug [95] found that the discriminative stimulus effects of diazepam were completely blocked by the convulsant pentylenetetrazol (PTZ), which binds to the picrotoxin site within the GABA$_A$ channel, and partially blocked by the GABA$_A$ orthosteric antagonist bicuculline. Young and Glennon [96] found full stimulus generalization from diazepam to at least 17 other benzodiazepines. This later study found a nearly perfect positive correlation between each benzodiazepine's ED$_{50}$ dose for diazepam-lever responding and the lowest effective therapeutic dose in humans.

Woudenberg and Slangen [97] found that midazolam fully generalized to other benzodiazepines tested, but also generalization did not occur to buspirone. Benzodiazepines have been evaluated as discriminative stimuli in humans too, as reviewed in Chap. 13 of this volume. For example, Johanson found that humans could discriminate diazepam from placebo and that the discriminative stimulus effects of diazepam were similar to those of lorazepam [98] and triazolam [99], but not to those produced by buspirone [99].

The imidazopyridine hypnotic zolpidem, which is a nonbenzodiazepine, binds preferentially to BZ I sites and does not substitute for chlordiazepoxide in male rats [100]. In baboons, however, full stimulus generalization occurs from lorazepam or from the barbiturate pentobarbital to zolpidem [101]. In male rats, full generalization occurred from zolpidem's discriminative stimulus to the BZ I preferring agonist zopiclone [81], to benzodiazepines, and to a barbiturate.

Mintzer et al. [102] demonstrated that the discriminative stimulus effects between zolpidem and the benzodiazepine triazolam can be differentiated by using a three-choice drug discrimination procedure in human subjects. Such a discrimination might be expected to focus on BZ II-elicited stimulus properties by the benzodiazepine and the BZ I-elicited stimulus properties by zolpidem. Yet, when asked to fill out subjective effect questionnaires for comparing zolpidem to triazolam, participants endorsed terms such as "blurred vision," "dry mouth," and "nervous" to the zolpidem condition and the majority of the remaining terms endorsed were equivalent between these two drugs. However, a number of subjective effect terms on the questionnaires differentiated these drugs from placebo. Rush et al. [103] confirmed that humans can discriminate zolpidem from placebo using a two-choice drug discrimination procedure and that full stimulus generalization occurs to benzodiazepines and barbiturates.

One of the aims of current benzodiazepine drug discrimination research is to assess stimulus effects for particular alpha subtypes of the $GABA_A$ receptor. Licata et al. [93] used a highly selective BZ II agonist, L-838,417, as a training drug in male squirrel monkeys, finding full stimulus generalization to benzodiazepines and barbiturates, and also to the BZ I preferring agonists zolpidem or zopiclone. Further research in this area will require more selective BZ I receptor ligands.

4.3 Buspirone

Buspirone functions as an agonist at serotonin $5-HT_{1A}$ receptors, with a binding affinity of approximately 20 nM [104]. Buspirone does bind with a moderate to weak affinity for $5-HT_6$ (~400 nM) [105] and $5-HT_7$ (376 nM) [106] receptors and has a somewhat stronger affinity (78.0 nM) for dopamine D_4 receptors [107]. Hendry et al. [108] first discovered that buspirone, as the training drug, elicits stimulus effects that differ from those of benzodiazepines and barbiturates. A later study confirmed that buspirone did not share discriminative stimulus properties with benzodiazepines in baboons and rats [109]. Moreover, humans can differentiate the discriminative stimulus effects of acute buspirone versus diazepam and placebo in a three-choice procedure [110].

Stimulus effects elicited by 5-HT$_{1A}$ receptor agonism are highly relevant to the buspirone cue, as buspirone fully substitutes in rats [111] and in pigeons [112] that were trained to discriminate the 5-HT$_{1A}$ agonist 8-OH-DPAT. This substitution is symmetrical as 8-OH-DPAT fully substitutes for buspirone in pigeons [113]. When drugs cross-generalize to each other, this is usually a strong indication of similar underlying mechanisms mediating their discriminative stimulus properties [114]. Sanger [115] did report full stimulus generalization occurring from the α_2 adrenoceptor antagonist idazoxan to buspirone in male rats, but as other 5-HT$_{1A}$ receptor agonists fully substitute for idazoxan [115, 116], the shared discriminative stimulus properties between buspirone and idazoxan are likely mediated by activity at 5-HT$_{1A}$ receptors.

There is some evidence for a dopaminergic component partially mediating the buspirone discriminative cue. Buspirone has been shown to fully block an apomorphine discriminative stimulus in squirrel monkeys (two males, one female) [117] and d-amphetamine in rhesus monkeys (two males, two females) [118]. Further, buspirone partially attenuates a cocaine discriminative cue in male rats [119] and partially generalizes to the D$_2$ preferring antagonist haloperidol [120]. Rijnders and Slangen [121] reported similar findings, showing partial generalization from buspirone to haloperidol, sulpiride, and the D$_2$ receptor antagonist R 79598.

5 Discriminative Stimulus Properties of Novel Therapeutic Targets for the Treatment of Depression

While the monoamine hypothesis has been firmly anchored for over 50 years as the predominate theory explaining the underlying neuropharmacology of depression (see reviews by [4, 5]), there has been a great deal of interest in recent years in targeting other, non-aminergic mechanisms as novel therapeutic targets for the treatment of depression [122]. In fact, cholinergic [123, 124] and glutamatergic [4, 5] processes are being considered as novel antidepressant treatment strategies. While the drug discrimination literature has historically focused on the abuse liability of cholinergic- and glutamatergic-acting drugs, there are emerging experimental programs devoted to relating cholinergic and glutamatergic compounds with antidepressant mechanisms of action. We will address each of these areas below. Table 3 provides a summary of these findings.

Table 3 Novel antidepressant strategies

Pharmacological mechanism	Rationale	Stimulus substitution findings
Anticholinergic (muscarinic)	Clinical data showing improvements in unipolar and bipolar depression Rapid antidepressant effects May specifically involve muscarinic M_2 receptors	Muscarinic receptor antagonism in CNS
Glutamatergic NMDA receptor antagonism	Clinical data showing improvements in unipolar and bipolar depression Rapid antidepressant effects Long-lasting antidepressant effects	No substitution by a glycine site partial agonist

5.1   Anticholinergic Drugs as Antidepressants

A number of studies have touted the possible antidepressant effects of anticholinergic drugs. As early as 1981, the anticholinergic drug biperiden (Akineton®) was administered to patients with major depressive disorder [125]. That study reported a significant improvement in symptoms, but had to be discontinued after 3 weeks in two patients due to side effects. However, a later study using a double-blind procedure failed to find a reliable antidepressant effect for biperiden [126]. More promising results have been reported for the anticholinergic scopolamine. A recent systematic review of the literature by Jaffe et al. [127] found seven studies that evaluated mood and depression after a low dose of intravenous scopolamine administration (ranging from 3 to 5 days of administration, either on consecutive or intermittent days). Based on their review of the available literature, they concluded that "Scopolamine is an effective and rapid antidepressant in both unipolar and bipolar depression. . .." with patients exhibiting significant reductions in depressive symptoms with 3 days after the first administration of scopolamine. They noted that no patients dropped out of the studies due to secondary effects, although subjective confusion was typically reported by patients 2 h after infusion of scopolamine. In a placebo-controlled clinical trial scopolamine induced drowsiness, blurred vision, dry mouth, light-headedness, and reduced blood pressure, but these side effects were well tolerated and no subjects dropped out of the clinical trial. Scopolamine infusions produced a rapid and robust antidepressant response in the patients. Je Jeon et al. [124] examined the role of muscarinic receptors in the pathophysiology of mood disorders and concluded that a body of recent evidence supports the role of muscarinic cholinergic receptors in the pathophysiology of both major depressive disorders and bipolar disorders. Specifically, they argue that targeting the cholinergic M_2 receptors might produce more rapid and robust clinical effects, especially if new drugs could be developed that have minimal or no peripheral cholinergic effects (see also [18, 128, 148]). Clearly, drug discrimination studies could play an important role in delineating the receptor pharmacology of novel therapeutic compounds. While the drug discrimination research on anticholinergic drugs has not focused on this particular topic, there have been a number of studies that have examined the discriminative stimulus properties of anticholinergic drugs and a few that lend some support to the idea of therapeutic properties for anticholinergic drugs.

The cholinergic antagonist atropine was one of the first drugs studied in a two-lever drug discrimination assay and, as would be expected, scopolamine fully substituted for atropine's discriminative stimulus [129]. Interestingly, they also demonstrated that atropine's discriminative cue was centrally mediated as atropine methyl bromide, which is a quaternary compound that does not readily cross the blood–brain barrier, and did not produce any atropine-appropriate responding. Later, Jung et al. [130] examined the discriminative stimulus properties of scopolamine as the training drug. He reported similar results about scopolamine's discriminate cue being centrally mediated (scopolamine methylbromide, which does not readily cross the blood–brain barrier, did not substitute for scopolamine) and reported that scopolamine's cue was mediated by antagonism of muscarinic receptors. This finding was not surprising given scopolamine's very potent and selective binding at all five (M_1–M_5) muscarinic cholinergic receptors (<2.1 nM, obtained from PDSD database on 20 March 2016). Unfortunately, there has been no systematic examination of the ability of various antidepressant drugs to mimic scopolamine's discriminative cue. However, scopolamine has been a test drug in other studies in which antidepressants or antipsychotic drugs have been established as the training drug. For example, Jones et al. [69] established the atypical antidepressant bupropion (see earlier discussion in this chapter about bupropion) as a discriminative stimulus in a two-lever discrimination procedure in rats. They found that a dose of 0.5 mg/kg scopolamine produced full substitution for bupropion in 2 of 6 rats (responding was significantly diminished in the other 4), albeit no reliable substitution was evident at doses above and below 0.5 mg/kg. This variability in results for scopolamine was attributed to disruptions in response rates. These findings suggest that scopolamine might possibly share some discriminative stimulus properties with bupropion, but response disruption obviously prevents any definite conclusions. One approach would be to use a low, non-generalizing dose of bupropion in combination with low, non-generalizing doses of scopolamine; if these compounds share discriminative stimulus properties, then a greater level of stimulus generalization would be expected. More research in this area is needed to test this hypothesis.

Kelley et al. [72] established a two-lever drug discrimination with the tetracyclic antidepressant mianserin (4.0 mg/kg) in one group of rats and scopolamine (0.25 mg/kg) in another. An asymmetrical generalization between mianserin and scopolamine was observed. Scopolamine produced mianserin-appropriate responding in the mianserin-trained rats, but mianserin did not produce scopolamine-appropriate responding in the scopolamine-trained rats. While the underlying mechanisms responsible for mianserin's discriminative stimulus properties have not been delineated, it is based at least in part by antagonism of certain serotonergic receptors [131–133]. In particular, mianserin exhibits a high affinity for serotonin 5-HT_{2A}, 5-HT_{2B}, and 5-HT_{2C} receptors [134]. However, given that the very selective muscarinic cholinergic antagonist scopolamine fully substituted for mianserin, there does appear to be a cholinergic component to mianserin's discriminative cue.

5.2   Glutamatergic Drugs as Antidepressants

Since Berman and colleagues conducted a proof of concept study in 2000 demonstrating that low dose intravenous ketamine could produce both rapid and sustained antidepressant effects, there has been a surge of recent interest into examining glutamatergic drugs as potentially novel therapeutic targets for the treatment of depression. While a lot of this research has focused on the therapeutic use of ketamine in the clinic, a large number of studies have been examining the underlying glutamatergic mechanisms that are responsible for ketamine's antidepressant effects [19]. There are a number of more recent studies that have examined glutamatergic mechanisms as targets for antidepressant effects, and several are currently in clinical trials (see reviews by [5], [135]). As was the case with scopolamine and other anticholinergic drugs, the drug discrimination literature on ketamine and other noncompetitive NMDA antagonists is considerably lacking, having focused more on the abuse liability of these drugs rather than their potential for having therapeutic effects. For example, GLYX-13 (Rapastinel®) is a "functional" partial agonist at the NMDA receptor glycine site (it does not bind directly to the glycine site) currently in phase III clinical trials for use as an adjunctive therapy in treatment-resistant major depressive disorder. In a recent study rats were trained to discriminate 10 mg/kg ketamine from saline vehicle [136]. Then varying doses of GLYX-13 were tested for generalization to ketamine. In doses up to 156 mg/kg, GLYX-13 did not generate ketamine-associated responding (i.e., it did not substitute for ketamine) and did not produce a suppression of response rates. Based on these results, the authors concluded that Glyx-13 did not share discriminative stimulus properties with ketamine and exhibited no sedative or abuse-related side effects. These results profile a preliminary, yet highly promising, role for glutamatergic-acting agents with highly selective antidepressant effects.

To the best of our knowledge, there have been no other published studies in which ketamine was the discriminative stimulus and current (or potential) antidepressant or anxiolytic drugs have been tested for generalization. This clearly represents a future area of research that is needed. Drug discrimination studies could also play a valuable role in helping to delineate the possible role of mechanisms (either direct or indirect) that mediate ketamine's antidepressant effects at the NMDA receptor.

6 Discriminative Stimulus Properties of Adjunctive Strategies for Antidepressant Drugs

The atypical antipsychotics aripiprazole (Abilify®) and quetiapine (Seroquel®) are used as adjunctive treatments for depressive disorders [137], and quetiapine is often used as a first-line treatment for bipolar disorder [86, 138, 139] and its antidepressant effects may be due in part to its active metabolite N-desalkylquetiapine (norquetiapine) (see [140, 141]). Jensen et al. [141] have shown that N-desalkylquetiapine has its highest binding affinity at histamine H_1 receptors and displays a moderate affinity at the norepinephrine reuptake transporter, serotonin receptors, the α_{1B} adrenoceptor, and muscarinic receptors (M_1, M_3, and M_5). While it is not known which of these receptor mechanisms may play a role in quetiapine's antidepressant effects, the discriminative stimulus properties of quetiapine have been examined by Goudie and colleagues. The atypical antipsychotics clozapine, olanzapine, and risperidone fully substituted for quetiapine, but the typical antipsychotics haloperidol, chlorpromazine, and loxapine, and the atypical antipsychotic amisulpride did not [142]. In a subsequent study the underlying receptor mechanisms for quetiapine's discriminative cue were examined [143]. The only selective ligand that fully substituted for quetiapine was the muscarinic antagonist, scopolamine (87% drug lever responding), but partial substitution was seen with the α_1-adrenoceptor antagonist prazosin, the presumed preferential dopamine D_3 receptor antagonist PNU 91194A, and the 5-$HT_{2A/2B/2C}/H_1/M_{1-5}$ antagonist cyproheptadine. Based on these results, Goudie et al. concluded that quetiapine's discriminative stimulus properties reflect a "compound" cue involving several receptors, but clearly, muscarinic antagonism was sufficient, although not necessary, to mimic quetiapine's discriminative cue. Given that the parent drug quetiapine has very low affinity for muscarinic receptors, this result can probably be attributed to the muscarinic antagonism exhibited by quetiapine's metabolite N-desalkylquetiapine [141]. In addition to N-desalkylquetiapine, active metabolites for a number of other drugs also have potential as antidepressant drugs [144]. Thus, the drug discrimination assay can be used to help elucidate the receptor mechanisms of these (and other) drugs and to help in the development of future, novel medications for the treatment of depression.

7 Conclusion

The drug discrimination procedure has been extensively used to evaluate the stimulus properties of antidepressant and anxiolytic drugs. We find drug discriminative data on all classes of antidepressants and anxiolytics, including non-anxiolytic hypnotics that bind to benzodiazepine receptors. Moreover, drugs from most of these different classes have been evaluated in humans, lending an assessment of the translational value of these procedures. In general, the stimulus properties of drugs shown in animals tend to be displayed in humans. While attempts have been made in humans to qualitatively identify the characteristics of the discriminative stimulus effects of drugs, the most reliable predictor of a drug's discriminability remains its action at the receptor level.

A potential shortcoming of these drug discrimination studies, and a shortcoming that is shared by the vast majority of behavioral pharmacology studies, is the use of only male subjects. From the few studies included in this chapter that included both male and female subjects, there does not appear to be differences in the stimulus of the drugs tested. Yet, given that depression and anxiety are more prevalent in women than men [145], much may be gained by learning more about potential sex differences in the stimulus properties of anxiolytic and antidepressant drugs.

Finally, it is worth noting that all of the current prescribed antidepressant drugs and the anxiolytic drug buspirone require weeks of chronic administration in order for therapeutic efficacy to occur. While drug discrimination training does indeed require weeks to meet a high accuracy criterion, the intermittent nature of treatment and placebo days: (1) may not be sufficient for a drug to produce drug effects unique to *chronic* administration, and (2) may engender altered sensitivity to some of the behavioral effects and interoceptive effects of compounds commonly used to treat affective disorders [146]. In fact, it is necessary for a drug's effects to be absent during vehicle test sessions, as demonstrated by a failed to attempt to establish fluoxetine as a discriminative stimulus. For example, drug discrimination studies with antidepressants and buspirone likely only represent acute activity at CNS receptors. Future drug discrimination studies with antidepressants utilizing a procedure that involves a chronic dosing regimen would provide much needed predictive and face validity to the preclinical literature. Many behavioral probes of affective responses are advantageous with regard to their swift and high throughput nature [147]. Clearly, the overwhelming strength of using conditioning procedures for assessing affective measures of drug action, as with drug discrimination, stems from their durability and reproducibility.

References

1. Fox HH, Gibas JT (1953) Synthetic tuberculostats. V. Alkylidene derivatives of isonicotinyhydrazine. J Org Chem 18:983–989
2. Zeller EA, Barsky J, Fouts JR, Kirchheimer WF, Van Orden LS (1952) Influence of isonicotinic acid hydrazide (INH) and 1-isonicotinyl-2-isopropyl hydrazide (IIH) on bacterial and mammalian enzymes. Experientia 8:349–350
3. Youdim MBH (1972) Multiple forms of monoamine oxidase and their properties advances in biochemical psychopharmacology. Raven, New York, pp 67–77
4. Lopez-Munoz F, Alamo C (2009) Monoaminergic neurotransmission: the history of the discovery of antidepressants from 1950s until today. Curr Pharm Des 15:1563–1586
5. Hillhouse TM, Porter JH (2015) A brief history of the development of antidepressant drugs: from monoamines to glutamate. Exp Clin Psychopharmacol 23:1–21
6. Cusack B, Nelson A, Richelson E (1994) Binding of antidepressants to human brain receptors: focus on newer generation compounds. Psychopharmacology (Berl) 114:559–565

7. Owens MJ, Morgan WN, Plott SJ, Nemeroff CB (1997) Neurotransmitter receptor and transporter binding profile of antidepressants and their metabolites. J Pharmacol Exp Ther 283:1305–1322

8. Richelson E, Nelson A (1984) Antagonism by antidepressants of neurotransmitter receptors of normal human brain in vitro. J Pharmacol Exp Ther 230:94–102

9. Tatsumi M, Groshan K, Blakely RD, Richelson E (1997) Pharmacological profile of antidepressants and related compounds at human monoamine transporters. Eur J Pharmacol 340:249–258

10. Wander TJ, Nelson A, Okazaki H, Richelson E (1986) Antagonism by antidepressants of serotonin S1 and S2 receptors of normal human brain in vitro. Eur J Pharmacol 132:115–121

11. Benkert O, Laakmann G, Ott L, Strauss A, Zimmer R (1977) Effect of zimelidine (H 102/09) in depressive patients. Arzneimittelforschung 27:2421–3

12. Sánchez C, Hyttel J (1999) Comparison of the effects of antidepressants and their metabolites on reuptake of biogenic amines and on receptor binding. Cell Mol Neurobiol 19:467–489

13. Owens JM, Knight DL, Nemeroff CB (2002) Second generation SSRIS: human monoamine transporter binding profile of escitalopram and R-fluoxetine. Encéphale 28:350–355

14. Millan MJ, Gobert A, Lejeune F, Newman-Tancredi A, Rivet J-M, Auclair A, Peglion J-L (2001) S33005, a novel ligand at both serotonin and norepinephrine transporters: I. Receptor binding, electrophysiological, and neurochemical profile in comparison with venlafaxine, reboxetine, citalopram, and clomipramine. J Pharmacol Exp Ther 298:565–580

15. Deecher DC, Beyer CE, Johnston G, Bray J, Shah S, Abou-Gharbia M, Andree TH (2006) Desvenlafaxine succinate: a New serotonin and norepinephrine reuptake inhibitor. J Pharmacol Exp Ther 318:657–665

16. Bymaster FP, Dreshfield-Ahmad LJ, Threlkeld PG, Shaw JL, Thompson L, Nelson DL, Hemrick-Luecke SK, Wong DT (2001) Comparative affinity of duloxetine and venlafaxine for serotonin and norepinephrine transporters in vitro and in vivo, human serotonin receptor subtypes, and other neuronal receptors. Neuropsychopharmacology 25:871–880

17. Bang-Andersen B, Ruhland T, Jorgensen M, Smith G, Frederiksen K, Jensen KG, Zhong H, Nielsen SM, Hogg S, Mork A, Stensbol TB (2011) Discovery of 1-[2-(2,4-dimethylphenylsulfanyl)phenyl]piperazine (Lu AA21004): a novel multimodal compound for the treatment of major depressive disorder. J Med Chem 54:3206–3221

18. Hasselmann H (2014) Scopolamine and depression: a role for muscarinic antagonism? CNS Neurol Disord Drug Targets 13:673–683

19. Berman RM, Cappiello A, Anand A, Oren DA, Heninger GR, Charney DS, Krystal JH (2000) Antidepressant effects of ketamine in depressed patients. Biol Psychiatry 47:351–354. doi: 10.1016/s0006-3223(99)00230-9

20. Naughton M, Clarke G, O'Leary OF, Cryan JF, Dinan TG (2014) A review of ketamine in affective disorders: current evidence of clinical efficacy, limitations of use and pre-clinical evidence on proposed mechanisms of action. J Affect Disord 156:24–35

21. Huang J, Ho BT (1974) The effect of pretreatment with iproniazid on the behavioral activities of beta-phenylethylamine in rats. Psychopharmacology (Berl) 35:77–81

22. Overton DA (1982) Comparison of the degree of discriminability of various drugs using the T-maze drug discrimination paradigm. Psychopharmacology (Berl) 76:385–395

23. Chen JG, Sachpatzidis A, Rudnick G (1997) The third transmembrane domain of the serotonin transporter contains residues associated with substrate and cocaine binding. J Biol Chem 272:28321–28327

24. Jones SR, Garris PA, Wightman RM (1995) Different effects of cocaine and nomifensine on dopamine uptake in the caudate-putamen and nucleus accumbens. J Pharmacol Exp Ther 274:396–403

25. Ritz MC, Cone EJ, Kuhar MJ (1990) Cocaine inhibition of ligand binding at dopamine, norepinephrine and serotonin transporters: a structure-activity study. Life Sci 46:635–645

26. Colpaert FC, Niemegeers CJ, Janssen PA (1979) Discriminative stimulus properties of cocaine: neuropharmacological characteristics as derived from stimulus generalization experiments. Pharmacol Biochem Behav 10:535–546

27. Colpaert FC, Niemegeers CJ, Janssen PA (1980) Evidence that a preferred substrate for type B monoamine oxidase mediates stimulus properties of MAO inhibitors: a possible role for beta-phenylethylamine in the cocaine cue. Pharmacol Biochem Behav 13:513–517

28. Yasar S, Justinova Z, Lee SH, Stefanski R, Goldberg SR, Tanda G (2006) Metabolic transformation plays a primary role in the psychostimulant-like discriminative-stimulus effects of selegiline [(R)-(-)-deprenyl]. J Pharmacol Exp Ther 317:387–94

29. Carpene C, Collon P, Remaury A, Cordi A, Hudson A, Nutt D, Lafontan M (1995) Inhibition of amine oxidase activity by derivatives that recognize imidazoline I2 sites. J Pharmacol Exp Ther 272:681–688

30. Ozaita A, Olmos G, Boronat MA, Lizcano JM, Unzeta M, Garcia-Sevilla JA (1997) Inhibition of monoamine oxidase A and B activities by imidazol(ine)/guanidine drugs, nature of the interaction and distinction from I2-imidazoline receptors in rat liver. Br J Pharmacol 121:901–912

31. Jordan S, Jackson HC, Nutt DJ, Handley SL (1996) Discriminative stimulus produced by the imidazoline I2 site ligand, 2 -BFI. J Psychopharmacol 10:273–278

32. MacInnes N, Handley SL (2002) Characterization of the discriminable stimulus produced by 2-BFI: effects of imidazoline I(2)-site ligands, MAOIs, beta-carbolines, agmatine and ibogaine. Br J Pharmacol 135:1227–1234

33. Crissman AM, O'Donnell JM (2002) Effects of antidepressants in rats trained to discriminate centrally administered isoproterenol. J Pharmacol Exp Ther 302:606–611

34. Shearman G, Miksic S, Lal H (1978) Discriminative stimulus properties of desipramine. Neuropharmacology 17:1045–1048

35. Baker LE, Riddle EE, Saunders RB, Appel JB (1993) The role of monoamine uptake in the discriminative stimulus effects of cocaine and related compounds. Behav Pharmacol 4:69–79

36. Spealman RD (1995) Noradrenergic involvement in the discriminative stimulus effects of cocaine in squirrel monkeys. J Pharmacol Exp Ther 275:53–62

37. Cunningham KA, Callahan PM (1991) Monoamine reuptake inhibitors enhance the discriminative state induced by cocaine in the rat. Psychopharmacology (Berl) 104:177–180

38. Tella SR, Goldberg SR (2001) Subtle differences in the discriminative stimulus effects of cocaine and GBR-12909. Prog Neuropsychopharmacol Biol Psychiatry 25:639–656

39. Kleven MS, Koek W (1998) Discriminative stimulus properties of cocaine: enhancement by monoamine reuptake blockers. J Pharmacol Exp Ther 284:1015–1025

40. Schechter MD (1983) Discriminative stimulus control with imipramine: transfer to other antidepressants. Pharmacol Biochem Behav 19:751–754

41. Zhang L, Barrett JE (1991) Imipramine as a discriminative stimulus. J Pharmacol Exp Ther 259:1088–1093

42. Barrett JE, Zhang L (1991) Involvement of 5-HT$_{1A}$ activity in the discriminative stimulus effects of imipramine. Pharmacol Biochem Behav 38:407–410

43. U'Prichard DC, Greenberg DA, Snyder SH (1977) Binding characteristics of a radiolabeled agonist and antagonist at central nervous system alpha noradrenergic receptors. Mol Pharmacol 13:454–73

44. Li PP, Warsh JJ, Sibony D, Chiu A (1988) Assessment of rat Brain Alpha$_1$-adrenoceptor binding and activation of inositol phospholipid turnover following chronic imipramine treatment. Neurochem Res 13:1111–1118

45. Kelley BM, Porter JH (1997) The role of muscarinic cholinergic receptors in the discriminative stimulus properties of clozapine in rats. Pharmacol Biochem Behav 57:707–719

46. Hoenicke EM, Vanecek SA, Woods JH (1992) The discriminative stimulus effects of clozapine in pigeons: involvement of 5-hydroxytryptamine1C and 5-hydroxytryptamine2 receptors. J Pharmacol Exp Ther 263:276–284

47. Setola V, Dukat M, Glennon RA, Roth BL (2005) Molecular determinants for the interaction of the valvulopathic anorexigen norfenfluramine with the 5-HT2B receptor. Mol Pharmacol 68:20–33
48. Newman ME, Shapira B, Lerer B (1998) Evaluation of central serotonergic function in affective and related disorders by the fenfluramine challenge test: a critical review. Int J Neuropsychopharmacol 1:49–69
49. Goudie AJ (1977) Discriminative stimulus properties of fenfluramine in an operant task: an analysis of its cue function. Psychopharmacology (Berl) 53:97–102
50. Rothman RB, Clark RD, Partilla JS, Baumann MH (2003) (+)-Fenfluramine and its major metabolite, (+)-norfenfluramine, are potent substrates for norepinephrine transporters. J Pharmacol Exp Ther 305:1191–1199
51. McElroy JF, Feldman RS (1984) Discriminative stimulus properties of fenfluramine: evidence for serotonergic involvement. Psychopharmacology (Berl) 83:172–178
52. White FJ, Appel JB (1981) A neuropharmacological analysis of the discriminative stimulus properties of fenfluramine. Psychopharmacology (Berl) 73:110–115
53. McCreary AC, Filip M, Cunningham KA (2003) Discriminative stimulus properties of (+/-)-fenfluramine: the role of 5-HT2 receptor subtypes. Behav Neurosci 117:212–221
54. Marona-Lewicka D, Nichols DE (1998) Drug discrimination studies of the interoceptive cues produced by selective serotonin uptake inhibitors and selective serotonin releasing agents. Psychopharmacology (Berl) 138:67–75
55. Marken PA, Munro JS (2000) Selecting a selective serotonin reuptake inhibitor: clinically important distinguishing features. J Clin Psychiatry 2:205–210
56. Millan MJ, Gobert A, Girardon S, Dekeyne A (1999) Citalopram elicits a discriminative stimulus in rats at a dose selectively increasing extracellular levels of serotonin vs. dopamine and noradrenaline. Eur J Pharmacol 364:147–150
57. Millan MJ, Girardon S, Dekeyne A (1999) 5-HT$_{2C}$ receptors are involved in the discriminative stimulus effects of citalopram in rats. Psychopharmacology (Berl) 142:432–444
58. Dekeyne A, Millan MJ (2003) Discriminative stimulus properties of antidepressant agents: a review. Behav Pharmacol 14:391–407
59. Uguz F, Sahingoz M, Gungor B, Aksoy F, Askin R (2015) Weight gain and associated factors in patients using newer antidepressant drugs. Gen Hosp Psychiatry 37:46–48
60. Wolff MC, Leander JD (1999) The discriminative stimulus properties of LY233708, a selective serotonin reuptake inhibitor, in the pigeon. Psychopharmacology (Berl) 146:275–279
61. Berendsen HH, Broekkamp CL (1994) Comparison of stimulus properties of fluoxetine and 5-HT receptor agonists in a conditioned taste aversion procedure. Eur J Pharmacol 253:83–89
62. Gommans J, Bouwknecht JA, Hijzen TH, Berendsen HH, Broekkamp CL, Maes RA, Olivier B (1998) Stimulus properties of fluvoxamine in a conditioned taste aversion procedure. Psychopharmacology (Berl) 140:496–502
63. Kayir H, Alici T, Goktalay G, Yildirim M, Ulusoy GK, Ceyhan M, Celik T, Uzbay TI (2008) Stimulus properties of venlafaxine in a conditioned taste aversion procedure. Eur J Pharmacol 596:102–106
64. Winter JC, Helsley S, Fiorella D, Rabin RA (1999) The acute effects of monoamine reuptake inhibitors on the stimulus effects of hallucinogens. Pharmacol Biochem Behav 63:507–513
65. Awasaki Y, Nojima H, Nishida N (2011) Application of the conditioned taste aversion paradigm to assess discriminative stimulus properties of psychostimulants in rats. Drug Alcohol Depend 118:288–294
66. Filip M, Chojnacka-Wojcik E, Przegalinski E (1993) Discriminative stimulus properties of (+)-oxaprotiline in rats. Pol J Pharmacol 45:151–156
67. Dekeyne A, Gobert A, Iob L, Cistarelli L, Melon C, Millan MJ (2001) Discriminative stimulus properties of the selective norepinephrine reuptake inhibitor, reboxetine, in rats. Psychopharmacology (Berl) 158:213–218

68. Millan MJ, Dekeyne A (2007) Discriminative stimulus properties of the selective norepi-nephrine reuptake inhibitor, reboxetine, in rats: a characterization with alpha/beta-adrenoceptor subtype selective ligands, antidepressants, and antagonists at neuropeptide receptors. Int J Neuropsychopharmacol 10:579–593

69. Jones CN, Howard JL, McBennett ST (1980) Stimulus properties of antidepressants in the rat. Psychopharmacology (Berl) 67:111–118

70. Howard JL, Jones CN, McBennett ST (1978) Discriminative stimulus properties of antide-pressants. In: Colpaert FC, Rosecrans JA (eds) Stimulus properties of drugs: ten years of progress. Elsevier/North-Holland Biomedical Press, Amsterdam, pp 157–166

71. Andersen PH (1989) The dopamine inhibitor GBR 12909: selectivity and molecular mech-anism of action. Eur J Pharmacol 166:493–504

72. Kelley BM, Porter JH, Varvel SA (1995) Mianserin as a discriminative stimulus in rats: asymmetrical cross-generalization with scopolamine. Psychopharmacology (Berl) 120:491–493

73. Blitzer RD, Becker RE (1985) Characterization of the bupropion cue in the rat: lack of evidence for a dopaminergic mechanism. Psychopharmacology (Berl) 85:173–177

74. Terry P, Katz JL (1997) Dopaminergic mediation of the discriminative stimulus effects of bupropion in rats. Psychopharmacology (Berl) 134:201–212

75. Lopez-Munoz F, Ucha-Udabe R, Alamo C (2005) The history of barbiturates a century after their clinical introduction. Neuropsychiatr Dis Treat 1:329–343

76. Tone A (2005) Listening to the past: history, psychiatry, and anxiety. Can J Psychiatry 50:373–380

77. Sigel E, Buhr A (1997) The benzodiazepine binding site of GABAA receptors. Trends Pharmacol Sci 18:425–429

78. Klepner CA, Lippa AS, Benson DI, Sano MC, Beer B (1979) Resolution of two biochemi-cally and pharmacologically distinct benzodiazepine receptors. Pharmacol Biochem Behav 11:457–462

79. Pritchett DB, Luddens H, Seeburg PH (1989) Type I and type II GABA$_A$-benzodiazepine receptors produced in transfected cells. Science 245:1389–1392

80. Wisden W, Laurie DJ, Monyer H, Seeburg PH (1992) The distribution of 13 GABAA receptor subunit mRNAs in the rat brain. I. Telencephalon, diencephalon, mesencephalon. J Neurosci 12:1040–1062

81. Dämgen K, Lüddens H (1999) Zaleplon displays a selectivity to recombinant GABAA receptors different from zolipdem, zopiclone and benzodiazepines. Neurosci Res Commun 25:139–148

82. Sanger DJ (2004) The pharmacology and mechanisms of action of new generation, non-benzodiazepine hypnotic agents. CNS Drugs 18:9–15

83. Wieland HA, Luddens H (1994) Four amino acid exchanges convert a diazepam-insensitive, inverse agonist-preferring GABAA receptor into a diazepam-preferring GABA$_A$ receptor. J Med Chem 37:4576–4580

84. Goldberg HL, Finnerty RJ (1979) The comparative efficacy of buspirone and diazepam in the treatment of anxiety. Am J Psychiatry 136:1184–1187

85. Khouzam HR, Emes R (2002) The use of buspirone in primary care. J Psychosoc Nurs Ment Health Serv 40:34–41

86. Coplan JD, Aaronson CJ, Panthangi V, Kim Y (2015) Treating comorbid anxiety and depression: psychosocial and pharmacological approaches. World J Psychiatry 5:366–378

87. Hirschhorn ID, Winter JC (1975) Differences in the stimulus properties of barbital and hallucinogens. Pharmacol Biochem Behav 3:343–347

88. York JL (1978) A comparison of the discriminative stimulus effects of ethanol, barbital, and phenobarbital in rats. Psychopharmacology (Berl) 60:19–23

89. Herling S, Valentino RJ, Winger GD (1980) Discriminative stimulus effects of pentobarbital in pigeons. Psychopharmacology (Berl) 71:21–28

90. Winger G, Herling S (1982) Discriminative stimulus effects of pentobarbital in rhesus monkeys: tests of stimulus generalization and duration of action. Psychopharmacology (Berl) 76:172–176

91. Rees DC, Balster RL (1988) Attenuation of the discriminative stimulus properties of ethanol and oxazepam, but not of pentobarbital, by Ro 15-4513 in mice. J Pharmacol Exp Ther 244:592–598

92. McMillan DE, Li M, Hardwick WC (1997) Discriminative stimulus effects and antipunishment effects of drugs measured during the same session. Pharmacol Biochem Behav 56:161–166

93. Licata SC, Platt DM, Ruedi-Bettschen D, Atack JR, Dawson GR, Van Linn ML, Cook JM, Rowlett JK (2010) Discriminative stimulus effects of L-838,417 (7-tert-butyl-3-(2,5-difluoro-phenyl)-6-(2-methyl-2H-[1,2,4]triazol-3-ylmethoxy)- [1,2,4]triazolo[4,3-b]pyridazine): role of GABA(A) receptor subtypes. Neuropharmacology 58:357–364

94. Colpaert FC, Desmedt LK, Janssen PA (1976) Discriminative stimulus properties of benzo-diazepines, barbiturates and pharmacologically related drugs; relation to some intrinsic and anticonvulsant effects. Eur J Pharmacol 37:113–123

95. Haug T (1983) Neuropharmacological specificity of the diazepam stimulus complex: effects of agonists and antagonists. Eur J Pharmacol 93:221–227

96. Young R, Glennon RA (1987) Stimulus properties of benzodiazepines: correlations with binding affinities, therapeutic potency, and structure activity relationships (SAR). Psycho-pharmacology (Berl) 93:529–533

97. Woudenberg F, Slangen JL (1989) Discriminative stimulus properties of midazolam: com-parison with other benzodiazepines. Psychopharmacology (Berl) 97:466–470

98. Johanson CE (1991) Discriminative stimulus effects of psychomotor stimulants and benzo-diazepines in humans. NIDA Res Monogr 116:181–196

99. Johanson CE (1991) Discriminative stimulus effects of diazepam in humans. J Pharmacol Exp Ther 257:634–643

100. Depoortere H, Zivkovic B, Lloyd KG, Sanger DJ, Perrault G, Langer SZ, Bartholini G (1986) Zolpidem, a novel nonbenzodiazepine hypnotic. I. Neuropharmacological and behavioral effects. J Pharmacol Exp Ther 237:649–658

101. Griffiths RR, Sannerud CA, Ator NA, Brady JV (1992) Zolpidem behavioral pharmacology in baboons: self-injection, discrimination, tolerance and withdrawal. J Pharmacol Exp Ther 260:1199–1208

102. Mintzer MZ, Frey JM, Griffiths RR (1998) Zolpidem is differentiated from triazolam in humans using a three-response drug discrimination procedure. Behav Pharmacol 9:545–559

103. Rush CR, Baker RW, Rowlett JK (2000) Discriminative-stimulus effects of zolpidem, triazolam, pentobarbital, and caffeine in zolpidem-trained humans. Exp Clin Psychopharmacol 8:22–36

104. Millan MJ, Lejeune F, Gobert A (2000) Reciprocal autoreceptor and heteroreceptor control of serotonergic, dopaminergic and noradrenergic transmission in the frontal cortex: relevance to the actions of antidepressant agents. J Psychopharmacol 14:114–38

105. Plassat JL, Amlaiky N, Hen R (1993) Molecular cloning of a mammalian serotonin receptor that activates adenylate cyclase. Mol Pharmacol 44:229–236

106. Lovenberg TW, Baron BM, de Lecea L, Miller JD, Prosser RA, Rea MA, Foye PE, Racke M, Slone AL, Siegel BW et al (1993) A novel adenylyl cyclase-activating serotonin receptor (5-HT7) implicated in the regulation of mammalian circadian rhythms. Neuron 11:449–458

107. Tallman JF, Primus RJ, Brodbeck R, Cornfield L, Meade R, Woodruff K, Ross P, Thurkauf A, Gallager DW (1997) I. NGD 94-1: identification of a novel, high-affinity antagonist at the human dopamine D4 receptor. J Pharmacol Exp Ther 282:1011–1019

108. Hendry JS, Balster RL, Rosecrans JA (1983) Discriminative stimulus properties of buspirone compared to central nervous system depressants in rats. Pharmacol Biochem Behav 19:97–101

109. Ator NA, Griffiths RR (1986) Discriminative stimulus effects of atypical anxiolytics in baboons and rats. J Pharmacol Exp Ther 237:393–403
110. Frey JM, Mintzer MZ, Rush CR, Griffiths RR (1998) Buspirone is differentiated from diazepam in humans using a three-response drug discrimination procedure. Psychopharmacology (Berl) 138:16–26
111. Cunningham KA, Callahan PM, Appel JB (1987) Discriminative stimulus properties of 8-hydroxy-2-(di-n-propylamino)tetralin (8-OHDPAT): implications for understanding the actions of novel anxiolytics. Eur J Pharmacol 138:29–36
112. Nader MA, Hoffmann S, Gleeson S, Barrett JE (1989) Further characterization of the discriminative stimulus effects of buspirone using monoamine agonists and antagonists in the pigeon. Behav Pharmacol 1:57–67
113. Mansbach RS, Barrett JE (1987) Discriminative stimulus properties of buspirone in the pigeon. J Pharmacol Exp Ther 240:364–9
114. Schuster CR, Balster RL (1977) The discriminative stimulus properties of drugs. In: Thompson T, Dews PB (eds) Advances in behavioral pharmacology, vol 1. Academic Press, New York, pp 85–138
115. Sanger DJ (1989) Discriminative stimulus effects of the alpha 2-adrenoceptor antagonist idazoxan. Psychopharmacology (Berl) 99:117–121
116. Prus AJ, Zornio PA, Schuck CJ, Heerts T, Jacobson SM, Winiarski DA (2010) Discriminative stimulus properties of idazoxan: mediation by both $\alpha 2$ adrenoceptor antagonism and 5-HT1A receptor agonism. Drug Dev Res 71:261–267
117. Kamien JB, Woolverton WL (1990) Buspirone blocks the discriminative stimulus effects of apomorphine in monkeys. Pharmacol Biochem Behav 35:117–120
118. Nader MA, Woolverton WL (1994) Blockade of the discriminative stimulus effects of d-amphetamine in rhesus monkeys with serotonin 5-HT(1A) agonists. Behav Pharmacol 5:591–598
119. Callahan PM, Cunningham KA (1997) Modulation of the discriminative stimulus properties of cocaine: comparison of the effects of fluoxetine with 5-HT$_{1A}$ and 5-HT$_{1B}$ receptor agonists. Neuropharmacology 36:373–381
120. Ator NA (1991) Discriminative stimulus effects of the novel anxiolytic buspirone. Behav Pharmacol 2:3–14
121. Rijnders HJ, Slangen JL (1993) The discriminative stimulus properties of buspirone involve dopamine-2 receptor antagonist activity. Psychopharmacology (Berl) 111:55–61
122. Berton O, Nestler EJ (2006) New approaches to antidepressant drug discovery: beyond monoamines. Nat Rev Neurosci 7:137–51
123. Howland RH (2009) The antidepressant effects of anticholinergic drugs. Psychopharmacology (Berl) 47(6):17–20
124. Je Jeon W, Dean B, Scarr E, Gibbons A (2015) The role of muscarinic receptors in the pathophysiology of mood disorders: a potential novel Treatment? Curr Neuropharmacol 13:739–749
125. Kasper S, Moises HW, Beckmann H (1981) The anticholinergic biperiden in depressive disorders. Pharmacopsychiatria 14:195–198
126. Gillin JC, Lauriello J, Kelsoe JR, Rapaport M, Golshan S, Kenny WM et al (1995) No antidepressant effect of biperiden compared with placebo in depression: a double-blind 6-week clinical trial. Psychiatry Res 58:99–105
127. Jaffe RJ, Novakovic V, Peselow ED (2013) Scopolamine as an antidepressant: a systematic review. Clin Neuropharmacol 36:24–26
128. Machado-Vieira R, Henter ID, Zarate Jr CA (in press) New targets for rapid antidepressant action. Prog Neurobiol
129. Kubena RK, Barry H (1969) Generalization by rats of alcohol and atropine stimulus characteristics in other drugs. Psychopharmacology (Berl) 15:196–206
130. Jung M, Pèrio A, Worms P, Biziere K (1988) Characterization of the scopolamine stimulus in rats. Psychopharmacology (Berl) 95:195–199

131. Friedman RL, Barrett RJ, Sanders-Bush E (1984) Discriminative stimulus properties of quipazine: mediation by serotonin 2 binding sites. J Pharmacol Exp Ther 228:628–635
132. Yamamoto T, Clark R, Woods JH (1984) Mianserin: discriminative stimulus effects in pigeons. Fed Proc 43:572
133. Yamamoto T, Walker EA, Woods JH (1991) Agonist and antagonist properties of serotonergic compounds in pigeons trained to discriminate either quipazine or 5-hydroxytrytophan. J Pharmacol Exp Ther 258:999–1007
134. Knight AR, Misra A, Quirk K, Benwell K, Revell D, Kennett G, Bickerdike M (2004) Pharmacological characterization of the agonist radioligand binding site of 5-HT$_{2A}$, 5-HT$_{2B}$ and 5-HT$_{2C}$ receptors. Naunyn Schmiedebergs Arch Pharmacol 370:114–123
135. Machado-Vieira R, Henter ID, Zarate CA Jr (2015) New targets for rapid antidepressant action. Prog Neurobiol S0301-0082(15)30038-1. doi: 10.1016/j.pneurobio.2015.12.001. [Epub ahead of print]
136. Burgdorf J, Zhang X-I, Nicholson KL, Balster RL, Leander JD, Stanton PK, Gross AL, Kroes RA, Moskalm JR (2013) GLYX-13, a NMDA receptor glycine-site functional partial agonist, induces antidepressant-like effects without ketamine-like side effects. Neuropsychopharmacology 38:729–742
137. Wang P, Si T (2013) Use of antipsychotics in the treatment of depressive disorders. Shanghai Arch Psychiatry 25:134–140
138. Ketter TA, Miller S, Dell'Osso B, Wang PW (2016) Treatment of bipolar disorder: review of evidence regarding quetiapine and lithium. J Affect Disord 191:256–273
139. Sanford M, Keating GM (2012) Quetiapine: a review of its use in the management of bipolar depression. CNS Drugs 26(5):435–460
140. Hillhouse TM, Shankland Z, Matazel KS, Keiser AA, Prus AJ (2014) The quetiapine active metabolite N-desalkylquetiapine and the neurotensin NTS$_1$ receptor agonist PD149163 exhibit antidepressant-like behavioral effects in male Sprague-Dawley rats. Exp Clin Psychopharmacol 22:548–556
141. Jensen NH, Rodriguiz RM, Caron MG, Wetsel WC, Rothman RB, Roth BL (2008) N-Desalkylquetiapine, a potent norepinephrine reuptake inhibitor and partial 5-HT1a agonist, as a putative mediator of quetiapine's antidepressant activity. Neuropsychopharmacology 33:2303–2312
142. Smith JA, Goudie AJ (2002) Discriminative stimulus properties in rats of the novel antipsychotic quetiapine. Exp Clin Psychopharmacol 10:376–384
143. Goudie AJ, Smith JA, Millan MJ (2004) Characterization of the effects of receptor-selective ligands in rats discriminating the novel antipsychotic quetiapine. Psychopharmacology (Berl) 171:212–222
144. López-Muñoz F, Alamo C (2013) Active metabolites as antidepressant drugs: the role of norquetiapine in the mechanism of action of quetiapine in the treatment of mood disorders. Front Psychiatry 4:102
145. Kessler RC, Petukhova M, Sampson NA, Zaslavsky AM, Wittchen H-U (2012) Twelve-month and lifetime prevalence and lifetime morbid risk of anxiety and mood disorders in the United States. Int J Methods Psychiatr Res 21:169–184
146. Post RM (1980) Intermittent versus continuous stimulation: effect of time interval on the development of sensitization or tolerance. Life Sci 26:1275–1282
147. Crabbe JC, Wahlsten D, Dudek BC (1999) Genetics of mouse behavior: interactions with laboratory environment. Science 284:1670–1672
148. Drets WC, Furey ML (2010) Replication of scopolamine's antidepressant efficacy in major depressive disorder: a randomized, placebo-controlled clinical trial. Biol Psychiatry 67:432–438

Part III
Approaches to Drug Discrimination

Pharmacokinetic–Pharmacodynamic (PKPD) Analysis with Drug Discrimination

S. Stevens Negus and Matthew L. Banks

Abstract Discriminative stimulus and other drug effects are determined by the concentration of drug at its target receptor and by the pharmacodynamic consequences of drug-receptor interaction. For in vivo procedures such as drug discrimination, drug concentration at receptors in a given anatomical location (e.g., the brain) is determined both by the dose of drug administered and by pharmacokinetic processes of absorption, distribution, metabolism, and excretion that deliver drug to and from that anatomical location. Drug discrimination data are often analyzed by strategies of dose-effect analysis to determine parameters such as potency and efficacy. Pharmacokinetic–Pharmacodynamic (PKPD) analysis is an alternative to conventional dose-effect analysis, and it relates drug effects to a measure of drug concentration in a body compartment (e.g., venous blood) rather than to drug dose. PKPD analysis can yield insights on pharmacokinetic and pharmacodynamic determinants of drug action. PKPD analysis can also facilitate translational research by identifying species differences in pharmacokinetics and providing a basis for integrating these differences into interpretation of drug effects. Examples are discussed here to illustrate the application of PKPD analysis to the evaluation of drug effects in rhesus monkeys trained to discriminate cocaine from saline.

Keywords Acute tolerance • Cocaine • Drug discrimination • Hysteresis • Pharmacodynamics • Pharmacokinetics • Prodrug

S.S. Negus (✉) and M.L. Banks
Department of Pharmacology and Toxicology, Virginia Commonwealth University, Richmond, VA 23298, USA
e-mail: sidney.negus@vcuhealth.org

© Springer International Publishing Switzerland 2016
Curr Topics Behav Neurosci (2018) 39: 245–260
DOI 10.1007/7854_2016_36
Published Online: 27 August 2016

Contents

1   Introduction

Drugs produce their effects by interacting with receptor targets, and drug discrimination is one behavioral procedure that is useful for investigating determinants of this interaction. In conceptualizing drug-receptor interactions in whole organisms, it is convenient to think of the receptors as relatively fixed in anatomical space, whereas each dug molecule embarks on a journey from its site of administration, through the body to the receptor upon which it acts, and then back out of the body. Pharmacokinetics and pharmacodynamics are subdisciplines within the field of pharmacology that address two facets of this journey. Pharmacokinetics (PK) is concerned with the processes that govern a drug's path through the body and its resulting concentration in different body compartments. Pharmacodynamics (PD), in contrast, is concerned with the physiological and behavioral consequences produced by that subset of drug molecules that find and occupy receptors during their journey through the body.

The relationship between PK and PD is described by PKPD analysis that relates drug concentration to drug effect. This type of analysis provides an alternative to conventional "dose-effect" analysis of drug effects, and they have value for at least three reasons [1]. First, drug effects are ultimately determined by drug concentration at the receptors upon which the drug acts, and that concentration is determined not only by the drug dose administered, but also by the PK processes that deliver that dose to and from the receptors. "Dose" is a measure of the amount of drug determined prior to its delivery, often in units of drug mass relative to the mass of the organism (e.g., mg/kg). Dose is precisely controlled by the experimenter, and it often serves as the principal independent variable in analysis of data from in vivo studies. For example, the "dose-effect curve" is a common mode of data presentation used to estimate critical drug features such as potency and efficacy. However, after a dose is administered, the drug must be absorbed into the body from the site of its administration (e.g., absorbed from gastrointestinal tract into the blood stream after oral delivery) and distributed from that site to the sites where receptors are located (e.g., distributed by the circulatory system from the gastrointestinal tract to brain). Moreover, drug molecules are subject to degradation via metabolism by enzymes and to removal from the body via excretion by routes such as urine, feces, or exhaled air. Together, these PK processes of absorption, distribution, metabolism, and excretion convert a drug dose administered at a single anatomical site and a single point in time into a dynamic tide of drug concentrations that rises and then falls throughout the body over time. These changing drug concentrations through time can then be related to changing drug effects through time to yield a richer data set than can be achieved by a reference to only a single drug dose administered at the beginning of an experiment. The most precise assessment of this relationship between drug concentration and drug effect would ideally measure drug concentrations at the site of receptors that mediate the measured effect. In practice, measurement of drug concentration at the receptor is often difficult, and the site of receptors might be unknown or broadly distributed. Accordingly, a common compromise is to measure drug concentrations in more accessible compartments (e.g., venous blood or cerebrospinal fluid) that usefully approximate drug concentrations across broad areas within the organism.

A second advantage of PKPD analysis is that it permits evaluation of the relationship between drug effect and concentrations not only of the administered drug, but also of drug metabolites. All drugs are subject to at least some degree of metabolism in the body, and in many cases, these metabolites are active and may contribute to the overall effect produced by an administered drug dose. An extreme example of this phenomenon is prodrugs, which are compounds designed to be metabolized in the body to active metabolites that then produce the drug's intended effect [2]. When samples of blood or cerebrospinal fluid are collected and analyzed for concentrations of the administered drug, they can also be analyzed for concentrations of known or suspected metabolites, and changing drug effects over time can be related to changing concentrations of the metabolites as well as of the parent drug.

A third advantage of PKPD analysis is that it provides a basis not only for evaluating changing drug effects over time within an organism, but also for evaluating variable drug effects between organisms [3]. Thus, the administration of a given drug dose in mg/kg units often produces different effects across subjects within a species or across subjects of different species in translational studies. One factor that may contribute to such between-subject or between-species variability in drug effect is variability in PK processes. For example, metabolism may proceed at different rates or yield different metabolites in different subjects, and these differences in metabolism will result in different temporal profiles of drug and metabolite concentrations and associated behavioral and physiological effects despite use of the same administered dose. Use of drug and metabolite concentration, rather than drug dose, as the primary independent variable can reveal PK differences across subjects or species and provide a basis for integrating these differences into interpretation of drug effects.

The remainder of this chapter will illustrate strategies for using PKPD analysis in drug discrimination research using results from studies in rhesus monkeys trained to discriminate cocaine from saline.

2   PKPD Analysis of the Discriminative Stimulus Effects of Cocaine

2.1   PKPD Analysis in Rhesus Monkeys

Cocaine produces reliable discriminative stimulus effects in rhesus monkeys and other species, and these effects are both dose- and time-dependent. As one example, Fig. 1a shows the time course of the cocaine training dose in rhesus monkeys trained to discriminate 0.4 mg/kg intramuscular cocaine from saline in a two-key, food-reinforced drug discrimination procedure [4]. During training sessions, either cocaine or saline was administered 10 min before a 5-min response period, and only responding on the injection-appropriate lever produced food. During time-course test sessions (separate test sessions for each pretreatment time), the cocaine training dose was administered 1, 3, 5, 10, 20, 30, 60, or 100 min before 5-min response periods, during which responding on either key produced food. Under these conditions, the discriminative stimulus effects of cocaine displayed a rapid onset of action, peaking within 3 min, and had a relatively short duration of action, with effects declining after 20 min and no longer apparent after 100 min. Figure 1b shows venous plasma levels of cocaine from these same monkeys. Samples were collected separately from behavioral studies, and for plasma collection, subjects were anesthetized with ketamine, equipped with a temporary catheter in the saphenous vein, and placed into a primate restraint chair. The training dose of 0.4 mg/kg cocaine was administered intramuscularly as in behavioral sessions, and samples were collected at the same times as the onset of response periods in behavioral sessions. Venous cocaine levels peaked after 10 min and then declined. Figure 1c directly compares the time course of discriminative stimulus effects and venous cocaine levels after administration of 0.4 mg/kg cocaine, and for this figure, "Time" on the X-axis is represented on a log scale to facilitate comparison of effects that occurred early as well as later after cocaine administration. This comparison shows that both the onset and offset of cocaine-induced discriminative stimulus effects occurred earlier than the rise and fall in venous cocaine levels. Lastly, Fig. 1d shows a plot of discriminative stimulus effect as a function of venous cocaine levels over time, and arrows show the sequence in which data points were collected from first to last. This plot shows a variable relationship over time between venous cocaine levels and levels of cocaine-appropriate responding. For example, similar venous cocaine levels of 35–40 ng/ml were associated with nearly 100% cocaine-appropriate responding after 3 min but with less than 25% cocaine-appropriate responding after 60 min. This type of data display is known as a "hysteresis loop," with the term "hysteresis" denoting a changing relationship over time between drug concentration and drug effect, and the term "loop" denoting the circular shape of the graph. Moreover, the direction of the loop can also be specified, and in this case, the loop is clockwise (i.e., the trajectory of data points over time flows in a clockwise direction).

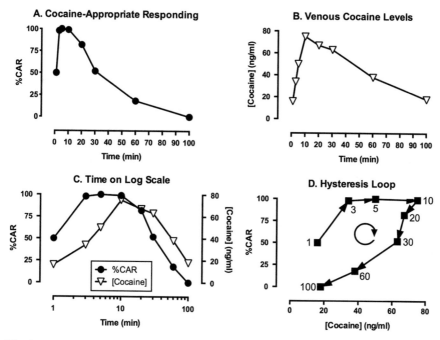

Fig. 1 PKPD analysis of discriminative stimulus effects produced by 0.4 mg/kg intramuscular cocaine in rhesus monkeys. (**a**) Time course of discriminative stimulus effects expressed as % Cocaine-Appropriate Responding (%CAR). (**b**) Time course of venous cocaine concentrations expressed in units of ng/ml of plasma. (**c**) Comparison of the time course of discriminative stimulus effects and venous cocaine levels with time expressed on a log scale. (**d**) Concentration-effect relationship between venous cocaine levels and %CAR and the resulting clockwise hysteresis loop. *Numbers* indicate the time in minutes after cocaine injection, and the clockwise circular *arrow* indicates the clockwise flow of data in the hysteresis loop. Adapted from Lamas et al. [4]

2.2   Relationship to PKPD Analysis in Humans

Hysteresis loops are common in PKPD analyses, and the presence and direction of the loop (clockwise or counterclockwise) can be used to draw inferences about PK and PD processes that contribute to drug effects in whole organisms [5]. At the most superficial level of analysis, the clockwise hysteresis loop observed in Fig. 1d for the relationship between venous cocaine levels and cocaine-induced discriminative stimulus effects in monkeys indicates that discriminative stimulus effects declined faster than venous drug concentrations. Before addressing the implications of this finding in more depth, it is first useful to note that these results in monkeys agree with the observation of clockwise hysteresis loops for cocaine-induced subjective effects in humans [6, 7]. For example, Evans et al. evaluated the time course of venous cocaine levels and a range of subjective effects after either smoked cocaine (25 or 50 mg) or intravenous cocaine (16 or 32 mg) in human subjects experienced with both routes of cocaine administration [7]. Of relevance to this review, both smoked and intravenous cocaine yielded clockwise hysteresis loops relating venous cocaine levels to subjective effects such as "Stimulated," "High," and "Drug Liking." In addition to this qualitative similarity, results of these studies in monkeys and humans could also be compared quantitatively. Specifically, intramuscular administration of 0.4 mg/kg cocaine in monkeys (equivalent to 28 mg in a 70 kg human) produced venous cocaine concentrations that were similar in magnitude, though with a slightly delayed time course, to those produced by 25–50 mg of smoked cocaine in humans, and both produced about half of the peak venous cocaine concentrations produced in humans by intravenous 16–32 mg cocaine. The delayed time course is consistent with the slower rate of drug absorption by the intramuscular route of cocaine administration used in monkeys than by the inhalation and intravenous routes used in humans. However, the finding that similar venous cocaine levels produced discriminative stimulus effects in monkeys and subjective effects in humans provides additional evidence for similarities between discriminative effects of drugs in animals and subjective drug effects in humans.

2.3   PK Factors in the Clockwise Hysteresis Loop for Cocaine

In addition to providing a nuanced basis for evaluating translation of drug effects across species, concentration-effect curves and hysteresis loops can also provide additional insights into the pharmacological determinants of drug effects in general and discriminative stimulus effects in particular. Table 1 lists some of the processes that may contribute to clockwise or counterclockwise hysteresis loops. In the case

Table 1 Factors that may contribute to clockwise and counterclockwise hysteresis loop

---

I. Clockwise hysteresis
  A. Pharmacokinetic factors
   1. Slower distribution to site of drug concentration measurement than to site of drug action
   2. Generation of an active antagonistic metabolite
  B. Pharmacodynamic factors (acute tolerance/tachyphylaxis)[a]
   1. Rates of receptor binding or signal transduction much faster than rates of drug distribution
   2. Desensitization/downregulation of receptors or downstream signaling pathways over time
   3. Recruitment of negative feedback processes

II. Counterclockwise hysteresis
  A. Pharmacokinetic factors
   1. Faster distribution to site of drug concentration measurement than to site of drug action
   2. Generation of an active agonist metabolite (e.g., by a prodrug)
  B. Pharmacodynamic factors[a]
   1. Rates of receptor binding or signal transduction much slower than rates of drug distribution
   2. Sensitization/upregulation of receptors or downstream signaling pathways over time
   3. Recruitment of positive feedback processes

---

[a]Listed PD factors apply for drugs that are agonists at their target receptor. For drugs that function as antagonists or inhibitors, different mechanisms would apply. See text for a discussion of mechanisms that might apply for cocaine

of cocaine discrimination shown in Fig. 1, at least two factors appear to contribute to the clockwise hysteresis loop that relates venous cocaine levels to discriminative stimulus effects.

First, regarding PK, recall that a drug is absorbed from its site of administration and distributed through a circuit of compartments before it is ultimately metabolized and/or excreted. For example, Fig. 2 shows that intramuscular cocaine is absorbed into the blood stream in muscle and transferred by veins to the heart and cardiopulmonary circulatory system where blood is oxygenated. After oxygenated blood containing cocaine returns to the heart, it is pumped via the aorta and systemic arteries to sites throughout the body, including sites of drug action such as brain, before being collected in veins and returned to the heart and cardiopulmonary system. Metabolism and excretion can occur at multiple points along this circuit. For cocaine, metabolism occurs largely via esterases in blood and liver, and excretion of cocaine and metabolites occurs largely via the kidneys into urine [9]. For the study shown in Fig. 1, cocaine-induced discriminative stimulus effects are thought to be mediated largely by binding of cocaine to dopamine transporters at one location (i.e., brain; [10, 11]), and cocaine concentrations were determined in plasma isolated from a different location (i.e., venous blood samples collected from the saphenous vein). As depicted in Fig. 2, intramuscular cocaine would be distributed to its site of action in brain before it would reach the saphenous vein and other systemic veins, and this lag in drug distribution from the site of cocaine action to the site of blood collection could contribute to the lag between expression of discriminative stimulus effects and the later increases in venous cocaine levels. Results from the study by Evans et al. in humans support this possibility [7]. In addition to measuring subjective effects and venous cocaine levels after cocaine administration in humans, these authors also measured arterial cocaine levels, and thereby sampled cocaine concentrations from compartments that would be reached before as well as after access of cocaine to its sites of action in brain. Arterial cocaine levels peaked at approximately tenfold higher concentrations than venous levels, these peaks were reached more quickly in arteries than in veins (15 s vs. 4 min for both smoked and intravenous cocaine), and the arterial concentration-effect curve resulted in a counterclockwise rather than a clockwise hysteresis loop. A similar finding for

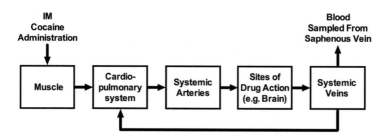

Fig. 2 Schematic of drug distribution after intramuscular drug injection. For the studies shown in Figs. 1 and 3, drug and metabolite concentrations were determined from plasma of blood samples collected from the saphenous vein

counterclockwise vs. clockwise hysteresis loops has also been found for arterial vs. venous concentration-effect curves for the short-acting opioid remifentanil, which is also rapidly metabolized by esterases in blood [12]. These results are consistent with drug distribution first to arteries, then to sites of action in brain, and lastly to veins. Additionally, rapid metabolism to inactive metabolites contributes to markedly lower concentrations reaching venous blood, thereby accentuating the expression of clockwise hysteresis loops that relate centrally mediated drug effects to venous drug levels.

2.4 PD Factors in the Clockwise Hysteresis Loop for Cocaine

Although PK factors likely bear primary responsibility for the clockwise hysteresis loop relating venous cocaine levels to discriminative stimulus subjective effects for cocaine, PD factors may also contribute. In particular, clockwise hysteresis loops are suggestive of acute tolerance to drug effects. "Tolerance" is a descriptive rather than an explanatory term, and in the context of PKPD analysis, it indicates a decrease in effect produced by a given drug concentration over time without implicating a particular mechanism. Possible mechanisms that may contribute to the phenomenon of acute tolerance (also known as "tachyphylaxis") are listed in Table 1. Insofar as cocaine produces its discriminative stimulus effects primarily by blocking dopamine transporters and increasing extracellular dopamine levels in brain regions such as nucleus accumbens [10, 11, 13], possible mechanisms of acute tolerance to the discriminative stimulus effects of cocaine could result from upregulation of dopamine transporters to facilitate dopamine clearance, decreases in dopamine release due to feedback inhibition of dopamine neuronal activity, and/or desensitization or downregulation of postsynaptic dopamine receptors responding to elevated dopamine levels. The precise mechanisms that might confer acute tolerance during the time course of effects pursuant to a single cocaine administration remain to be fully elucidated. However, acute tolerance has been observed for many cocaine effects in experimental designs that involve two sequential cocaine treatments. For example, pretreatment with an active cocaine dose in humans decreased the cardiovascular and subjective effects of a second cocaine dose administered 60 min later [14], and in rhesus monkeys, two similarly spaced cocaine injections resulted in a smaller increase in extracellular dopamine levels in nucleus accumbens after the second injection [15].

3 PKPD Analysis of the Cocaine-Like Discriminative Stimulus Effects of Lisdexamfetamine and Phendimetrazine

3.1 Lisdexamfetamine

Lisdexamfetamine is a prodrug for D-amphetamine in which the amino acid L-lysine is coupled to the nitrogen of amphetamine [16, 17]. It is approved for treatment of attention-deficit hyperactivity disorder and binge-eating disorder, and it is also under consideration as a maintenance medication for treatment of cocaine abuse [8, 18]. Lisdexamfetamine is thought to be inactive as a parent drug, but it is metabolized in blood to lysine and the active metabolite amphetamine by peptidase enzymes associated with red blood cells [19]. Administration of amphetamine itself substitutes for the discriminative stimulus effects of cocaine across a wide range of conditions (e.g., [20]), and Fig. 3 shows results from a study that examined effects of lisdexamfetamine in rhesus monkeys trained to discriminate 0.32 mg/kg intramuscular cocaine from saline in a procedure otherwise identical to the one described above for studies with cocaine [8]. Lisdexamfetamine produced a dose- and time-dependent substitution for cocaine, and Fig. 3a shows the time course of cocaine-like discriminative stimulus effects produced by a dose of 3.2 mg/kg lisdexamfetamine, together with venous plasma levels of lisdexamfetamine and D-amphetamine. The discriminative stimulus effects of lisdexamfetamine had a slow onset and long duration of action. Venous levels of lisdexamfetamine were highest at the initial measurement at 10 min and declined rapidly to low levels, whereas venous amphetamine levels peaked more slowly and declined more gradually over a period of 2 days. The delayed appearance of amphetamine is consistent with the conclusion that amphetamine is a metabolite of lisdexamfetamine.

Figure 3b shows the hysteresis loop relating discriminative stimulus effects to venous levels of the parent compound lisdexamfetamine, and this hysteresis loop differs from that for cocaine in Fig. 1d in two ways. First, the initial rise in %CAR was not associated with a parallel rise in venous lisdexamfetamine levels. Rather, the highest lisdexamfetamine levels measured at 10 min were associated with low levels of cocaine-appropriate responding, and the onset of discriminative stimulus effects was associated with a drop in venous lisdexamfetamine levels. Second, the hysteresis loop flowed in a counterclockwise rather than in a clockwise direction. These two phenomena together are consistent with the conclusion that lisdexamfetamine is an inactive prodrug being converted to an active metabolite [5].

Figure 3c shows the hysteresis loop relating discriminative stimulus effects to venous plasma levels of amphetamine. In contrast to the plot for the parent drug, the plot for amphetamine did show rising plasma levels during the onset of discriminative stimulus effects during the first 30 min after drug administration, suggesting that amphetamine is indeed functioning as an active metabolite of lisdexamfetamine. However, as with the parent drug, the overall hysteresis loop

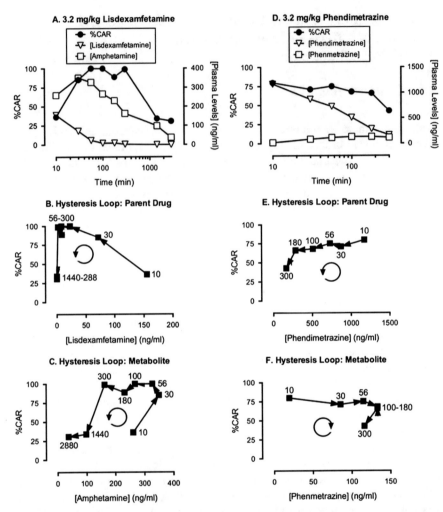

Fig. 3 PKPD analysis of cocaine-like discriminative stimulus effects produced by intramuscular lisdexamfetamine and phendimetrazine in rhesus monkeys. (**a, d**) Time course of discriminative stimulus effects (expressed as % Cocaine-Appropriate Responding; %CAR) and venous plasma levels of the parent drug and metabolite (ng/ml) for lisdexamfetamine (**a**) or phendimetrazine (**d**). Note that time in minutes is shown on a log scale. (**b, e**) Hysteresis loops for venous levels of the parent drug and %CAR for lisdexamfetamine (**b**) and phendimetrazine (**e**). (**c, f**) Hysteresis loops for venous levels of the metabolite and % CAR for lisdexamfetamine (**c**, metabolite = amphetamine) and phendimetrazine (**f**, metabolite = phenmetrazine). Numbers in (**b, c, e,** and **f**) indicate the time in minutes after parent drug injection, and the circular *arrows* indicate the clockwise or counterclockwise flow of data in the hysteresis loop. Adapted from Banks et al. [8]

for amphetamine also flowed in a counterclockwise direction. A counterclockwise hysteresis loop was also reported for the relationship for venous amphetamine levels to locomotor activity and mesolimbic dopamine release in rats after lisdexamfetamine administration [21]. This observation has been interpreted to suggest that amphetamine levels accumulate in systemic vasculature in general, and systemic veins in particular, more quickly than in brain to produce centrally mediated effects [8, 21]. More specifically, in reference to Fig. 2, these findings suggest that most of the conversion of lisdexamfetamine to amphetamine occurs in systemic veins. This conclusion would be consistent with (1) the requirement for peptidases in red blood cells to accomplish this metabolism, (2) the higher percentage of total blood volume in veins vs. arteries, and (3) the consequent longer residence time for any one circulating blood constituent (e.g., a red blood cell or drug molecule) in veins vs. arteries. Any amphetamine generated from lisdexamfetamine in veins would then require recirculation for delivery to brain. Moreover, rates of amphetamine delivery from vasculature across the blood–brain barrier and into neural tissue may also be limited [21], and this would produce a further delay between the time course of venous amphetamine levels and the time course of discriminative stimulus effects.

Two other points warrant mention. First, the venous amphetamine levels associated with cocaine-like discriminative stimulus effects in monkeys are much higher after lisdexamfetamine administration than after administration of amphetamine itself. For example, Fig. 3a shows that the dose of 3.2 mg/kg lisdexamfetamine sufficient to produce full substitution yielded a peak venous amphetamine levels of more than 300 ng/ml, whereas a dose of 0.32 mg/kg amphetamine sufficient to produce full substitution produced peak venous amphetamine levels of less than 100 ng/ml (M.L. Banks and S.S. Negus; unpublished results), and an oral dose of 20 mg amphetamine sufficient to produce significant subjective effects in humans yielded peak venous plasma levels of approximately 40 ng/ml [22]. One likely explanation for this difference is that venous levels after amphetamine administration likely underestimate the arterial drug levels initially delivered to the site of action (e.g., see above for cocaine), whereas venous levels after lisdexamfetamine are likely very similar to arterial levels delivered to the site of action (because amphetamine is generated largely in the systemic venous compartment). Direct evaluation of this hypothesis would be useful by comparing venous and arterial levels of amphetamine after lisdexamfetamine administration. In a second and related point, counterclockwise hysteresis loops relating venous amphetamine levels and centrally mediated behavioral effects after lisdexamfetamine administration differ from the finding of clockwise hysteresis loops after administration of amphetamine itself. For example, oral amphetamine in humans results in clockwise hysteresis loops that relate venous amphetamine levels to subjective effects [22], and we have similarly found that intramuscular amphetamine in rhesus monkeys produces clockwise hysteresis loops that relate venous amphetamine levels to cocaine-like discriminative stimulus effects (M.L. Banks and S.S. Negus, unpublished results). This distinction in rotational direction for hysteresis loops for amphetamine administered either directly or generated via metabolism of lisdexamfetamine illustrates one manifestation of PK differences that can be produced by different formulations of the same drug. In this case, the implication is that administration of amphetamine itself results in distribution of drug to sites of drug action before delivery to systemic veins, whereas administration of lisdexamfetamine results in generation of amphetamine in systemic veins prior to its delivery to sites of drug action.

3.2 Phendimetrazine

Phendimetrazine is approved for clinical use as an appetite suppressant for the treatment of obesity [23], and like lisdexamfetamine, it is also under consideration as a maintenance medication for the treatment of cocaine use disorder [24]. Phendimetrazine is metabolized to the compound phenmetrazine, and although both drugs interact with dopamine and norepinephrine transporters, the metabolite has high potency and functions as an amphetamine-like transporter substrate that promotes release of dopamine and norepinephrine, whereas the parent compound is more than 100-fold less potent and functions as a cocaine-like transporter inhibitor that prevents dopamine and norepinephrine reuptake [25]. The low potency of phendimetrazine at monoamine transporters suggested that it might function as a relatively inactive prodrug for the active metabolite phenmetrazine, similar to the function of lisdexamfetamine as a prodrug for amphetamine. This hypothesis was tested in PKPD studies in cocaine-discriminating rhesus monkeys [26]. For the purposes of the discussion below, phendimetrazine will be referred to as PDM, and phenmetrazine will be referred to as PM, because the spellings of the full drug names are similar and easily confused.

Initial studies indicated that administration of PM directly produced dose- and time-dependent substitution for cocaine and increases in venous PM levels, and the hysteresis plot relating venous PM concentration to cocaine-appropriate responding rotated in a clockwise direction similar to that described above for cocaine and amphetamine. PDM also produced dose- and time-dependent substitution for cocaine, and Fig. 3d shows results with a dose of 3.2 mg/kg PDM. Figure 3d also shows that this PDM dose produced time-dependent increases in venous levels of both PDM and PM. PDM levels peaked at the earliest time point at levels greater than 1,000 ng/ml, whereas PM levels rose more slowly and peaked at tenfold lower levels of approximately 100 ng/ml. The delayed emergence of PM after PDM administration is consistent with the status of PM as a metabolite of PDM. Moreover, venous PM levels were similar after administration of behaviorally active doses either of PM itself or of PDM, consistent with the conclusion that PM was functioning as an active metabolite sufficient to mediate behavioral effects of PDM. However, the PKPD profile of PDM and its metabolite PM differed from the profile for lisdexamfetamine and its metabolite amphetamine in two ways as illustrated by the hysteresis plots.

First, Fig. 3e shows the hysteresis loop that relates venous PDM levels to cocaine-appropriate responding. As with lisdexamfetamine, the direction of rotation for this hysteresis loop was counterclockwise; however, in contrast to results with lisdexamfetamine, the highest venous levels of PDM were associated with the highest levels of cocaine-appropriate responding. Although earlier time points were not assessed, these results indicate that the onset of cocaine-appropriate responding was associated with the period of rising PDM levels.

Second, Fig. 3f shows the hysteresis loop that relates PM levels to cocaine-appropriate responding. In contrast to the findings for amphetamine after lisdexamfetamine administration, the hysteresis loop for PM after PDM administration rotated in a clockwise direction. Of particular importance, high levels of cocaine-appropriate responding were observed at the earliest time point when PM levels were low, and the period of rising PM levels was associated not with onset of cocaine-appropriate responding, but rather with a period of sustained cocaine-appropriate responding. At later time points, there was a decrease in both venous PM levels and in cocaine-appropriate responding.

Taken together, these results were not consistent with the conclusion that PDM was an inactive parent drug for the active metabolite PM. Rather, these findings suggest that both PDM and PM were active, and the time course of cocaine-like discriminative stimulus effects after PDM administration reflected an initial phase of cocaine-like effects mediated by the parent drug PDM followed by a later phase of cocaine-like effects mediated by the metabolite PM.

4 Conclusions

PKPD analysis is an alternative to conventional dose-effect analysis of in vivo drug effects, and it focuses on the relationship of drug-induced behavioral or physiological effects to drug and metabolite concentrations in the body rather than to drug dose. Hysteresis loops are one manifestation of PKPD analysis, and these loops describe the time course of the potentially variable relationship between drug/metabolite concentration and drug effect over time. PKPD analysis, including analysis of hysteresis loops, can play a valuable role in interpretation of drug effects and PKPD relationships for the purposes of drug assessment and translational research in pharmacology. This chapter provided examples of the application of PKPD analysis to studies of the discriminative stimulus effects of drugs.

Acknowledgements Supported by NIH Grant R01DA026946.

References

1. van der Graaf PH, Benson N (2011) Systems pharmacology: bridging systems biology and pharmacokinetics-pharmacodynamics (PKPD) in drug discovery and development. Pharm Res 28:1460–1464
2. Huttunen KM, Raunio H, Rautio J (2011) Prodrugs--from serendipity to rational design. Pharmacol Rev 63:750–771

3. Bueters T, Ploeger BA, Visser SA (2013) The virtue of translational PKPD modeling in drug discovery: selecting the right clinical candidate while sparing animal lives. Drug Discov Today 18:853–862
4. Lamas X, Negus SS, Hall E, Mello NK (1995) Relationship between the discriminative stimulus effects and plasma concentrations of intramuscular cocaine in rhesus monkeys. Psychopharmacology (Berl) 121:331–338
5. Louizos C, Yanez JA, Forrest ML, Davies NM (2014) Understanding the hysteresis loop conundrum in pharmacokinetic/pharmacodynamic relationships. J Pharm Pharm Sci 17:34–91
6. Ellefsen KN, Concheiro M, Pirard S, Gorelick DA, Huestis MA (2016) Pharmacodynamic effects and relationships to plasma and oral fluid pharmacokinetics after intravenous cocaine administration., Drug Alcohol Depend
7. Evans SM, Cone EJ, Henningfield JE (1996) Arterial and venous cocaine plasma concentrations in humans: relationship to route of administration, cardiovascular effects and subjective effects. J Pharmacol Exp Ther 279:1345–1356
8. Banks ML, Hutsell BA, Blough BE, Poklis JL, Negus SS (2015) Preclinical assessment of lisdexamfetamine as an agonist medication candidate for cocaine addiction: effects in rhesus monkeys trained to discriminate cocaine or to self-administer cocaine in a cocaine versus food choice procedure. Int J Neuropsychopharmacol 18
9. Jones RT (1997) Pharmacokinetics of cocaine: considerations when assessing cocaine use by urinalysis. NIDA Res Monogr 175:221–234
10. Kleven MS, Anthony EW, Woolverton WL (1990) Pharmacological characterization of the discriminative stimulus effects of cocaine in rhesus monkeys. J Pharmacol Exp Ther 254:312–317
11. Witkin JM, Nichols DE, Terry P, Katz JL (1991) Behavioral effects of selective dopaminergic compounds in rats discriminating cocaine injections. J Pharmacol Exp Ther 257:706–713
12. Hermann DJ, Egan TD, Muir KT (1999) Influence of arteriovenous sampling on remifentanil pharmacokinetics and pharmacodynamics. Clin Pharmacol Ther 65:511–518
13. Desai RI, Paronis CA, Martin J, Desai R, Bergman J (2010) Monoaminergic psychomotor stimulants: discriminative stimulus effects and dopamine efflux. J Pharmacol Exp Ther 333:834–843
14. Fischman MW, Schuster CR, Javaid J, Hatano Y, Davis J (1985) Acute tolerance development to the cardiovascular and subjective effects of cocaine. J Pharmacol Exp Ther 235:677–682
15. Bradberry CW (2000) Acute and chronic dopamine dynamics in a nonhuman primate model of recreational cocaine use. J Neurosci 20:7109–7115
16. Blick SK, Keating GM (2007) Lisdexamfetamine. Paediatr Drugs 9:129–135, discussion 136–128
17. Heal DJ, Buckley NW, Gosden J, Slater N, France CP, Hackett D (2013) A preclinical evaluation of the discriminative and reinforcing properties of lisdexamfetamine in comparison to d-amfetamine, methylphenidate and modafinil. Neuropharmacology 73C:348–358
18. Mooney ME, Herin DV, Specker S, Babb D, Levin FR, Grabowski J (2015) Pilot study of the effects of lisdexamfetamine on cocaine use: A randomized, double-blind, placebo-controlled trial. Drug Alcohol Depend 153:94–103
19. Hutson PH, Pennick M, Secker R (2014) Preclinical pharmacokinetics, pharmacology and toxicology of lisdexamfetamine: a novel D-amphetamine pro-drug. Neuropharmacology 87:41–50
20. de la Garza RD, Johanson CE (1983) The discriminative stimulus properties of cocaine in the rhesus monkey. Pharmacol Biochem Behav 19:145–148
21. Rowley HL, Kulkarni R, Gosden J, Brammer R, Hackett D, Heal DJ (2012) Lisdexamfetamine and immediate release d-amphetamine - differences in pharmacokinetic/pharmacodynamic relationships revealed by striatal microdialysis in freely-moving rats with simultaneous determination of plasma drug concentrations and locomotor activity. Neuropharmacology 63:1064–1074

22. Brauer LH, Ambre J, De Wit H (1996) Acute tolerance to subjective but not cardiovascular effects of D-amphetamine in normal, healthy men. J Clin Psychopharmacol 16:72–76
23. Bray GA, Ryan DH (2011) Drug treatment of obesity. Psychiatr Clin North Am 34:871–880
24. Howell LL, Negus SS (2014) Monoamine transporter inhibitors and substrates as treatments for stimulant abuse. Adv Pharmacol 69:129–176
25. Rothman RB, Katsnelson M, Vu N, Partilla JS, Dersch CM, Blough BE, Baumann MH (2002) Interaction of the anorectic medication, phendimetrazine, and its metabolites with monoamine transporters in rat brain. Eur J Pharmacol 447:51–57
26. Banks ML, Blough BE, Fennell TR, Snyder RW, Negus SS (2013) Role of phenmetrazine as an active metabolite of phendimetrazine: evidence from studies of drug discrimination and pharmacokinetics in rhesus monkeys. Drug Alcohol Depend 130:158–166

Human Drug Discrimination: Elucidating the Neuropharmacology of Commonly Abused Illicit Drugs

B. Levi Bolin, Joseph L. Alcorn, Anna R. Reynolds, Joshua A. Lile, William W. Stoops, and Craig R. Rush

Abstract Drug-discrimination procedures empirically evaluate the control that internal drug states have over behavior. They provide a highly selective method to investigate the neuropharmacological underpinnings of the interoceptive effects of drugs in vivo. As a result, drug discrimination has been one of the most widely used assays in the field of behavioral pharmacology. Drug-discrimination procedures have been adapted for use with humans and are conceptually similar to preclinical drug-discrimination techniques in that a behavior is differentially reinforced contingent on the presence or absence of a specific interoceptive drug stimulus. This chapter provides a basic overview of human drug-discrimination procedures and reviews the extant literature concerning the use of these procedures to elucidate the underlying neuropharmacological mechanisms of commonly abused illicit drugs (i.e., stimulants, opioids, and cannabis) in humans. This chapter is not intended to review every available study that used drug-discrimination procedures in humans. Instead, when possible, exemplary studies that used a stimulant, opioid, or Δ^9-tetrahydrocannabinol (the primary psychoactive constituent of cannabis) to assess the discriminative-stimulus effects of drugs in humans are reviewed

B.L. Bolin, J.L. Alcorn, and A.R. Reynolds
Department of Behavioral Science, University of Kentucky College of Medicine, 140 Medical Behavioral Science Building, Lexington, KY 40536-0086, USA

J.A. Lile, W.W. Stoops, and C.R. Rush (✉)
Department of Behavioral Science, University of Kentucky College of Medicine, 140 Medical Behavioral Science Building, Lexington, KY 40536-0086, USA

Department of Psychology, University of Kentucky College of Arts and Sciences, 110 Kastle Hall, Lexington, KY 40506-0044, USA

Department of Psychiatry, University of Kentucky College of Medicine, 3470 Blazer Parkway, Lexington, KY 40509, USA
e-mail: crush2@email.uky.edu

© Springer International Publishing Switzerland 2016
Curr Topics Behav Neurosci (2018) 39: 261–296
DOI 10.1007/7854_2016_10
Published Online: 4 June 2016

for illustrative purposes. We conclude by commenting on the current state and future of human drug-discrimination research.

Keywords Abuse potential • Amphetamines • Cannabis • Cocaine • Drug discrimination • Humans • Medications development • Neuropharmacology • Opioids • Pharmacotherapy • Subject-rated effects • Substance abuse • THC

Contents

1   Introduction

Drug-discrimination procedures empirically evaluate the control internal drug states have over behavior. They provide a highly selective method to investigate the neuropharmacological underpinnings of the interoceptive effects of drugs in vivo. As a result, drug discrimination has been one of the most widely used assays in the field of behavioral pharmacology. Since the publication of one of the earliest studies to suggest the control of behavior by the presence or absence of the interoceptive-stimulus effects of alcohol in rats [1], there has been substantial work investigating the discriminative-stimulus effects of drugs spanning more than four decades (e.g., [2]). Drug-discrimination procedures have also been adapted for use with humans and remain conceptually similar to preclinical drug-discrimination procedures in that a behavior is differentially reinforced contingent on the presence or absence of a specific interoceptive drug stimulus (see Chap. 1; also see [3]). A PubMed search using the quoted search phrase "drug discrimination" yields 1,284 peer-reviewed publications dating back to the mid-1940s (i.e., [4]). Of the total number of published drug-discrimination studies, those concerning human drug discrimination comprise approximately 16% (i.e., 205 reports). Figure 1 shows the total number of drug-discrimination publications per year since 1973 and the relative proportion of those concerning human drug discrimination.

As noted above and described in previous chapters, the interoceptive-stimulus effects of drugs and the ensuing stimulus control of behavior have been widely studied in non-human laboratory animals using drug-discrimination procedures.

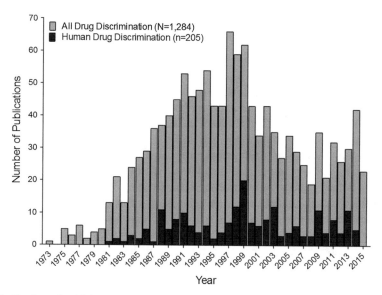

Fig. 1 Number of published drug-discrimination reports per year from 1973 to 2015. The total number of drug-discrimination publications is shown in *light gray* bars. The relative number of published drug-discrimination studies involving human participants is shown in *dark gray* bars. *X*-axis: Publication Year. *Y*-axis: Number of Publications

Below, the extant literature that assessed the discriminative-stimulus effects of stimulants, opioids, and Δ^9-tetrahydrocannabinol (Δ^9-THC; the primary pharmacological constituent in cannabis) in humans is reviewed. Since the adaptation of drug-discrimination procedures for use with humans, a number of reviews have been published. These reviews focused on: (a) the relationship between the discriminative-stimulus and subjective effects of drugs (e.g., [5–7]); (b) the concordance between preclinical and human drug-discrimination experiments [8]; and (c) the neuropharmacological selectivity of drug-discrimination procedures relative to subjective drug-effect questionnaires [9]. Although the present chapter provides some general discussion of these previously reviewed topics, it differs from earlier reviews in that it primarily focuses on the utility of human drug-discrimination procedures to elucidate the underlying neuropharmacological mechanisms of commonly abused illicit drugs (i.e., stimulants, opioids, and cannabis). This chapter is not intended to review every available study that used human drug-discrimination procedures. Instead, when possible, studies that used a stimulant, opioid, or Δ^9-THC to assess the discriminative-stimulus effects of drugs in humans are reviewed for illustrative purposes. Lastly, we conclude by commenting on the current state and future of human drug-discrimination research.

1.1 Subject Recruitment and Selection

Potential subjects are typically recruited through formal advertisements in local newspapers, online classified ads (e.g., Craigslist), flyers posted in public areas, and by word-of-mouth referral. Volunteers who may qualify upon initial screening complete a rigorous in-person screening that includes a complete medical history, physical health screen, and psychiatric assessment. Volunteers also provide basic demographic information (e.g., age, sex, and socioeconomic status) and complete a battery of questionnaires that assess drug-use history and severity as well as symptomology for other clinically relevant conditions such as depression and attention-deficit hyperactivity disorder. Responses on these instruments are used to determine whether volunteers satisfy the study inclusion criteria or meet criteria that would exclude them from participation (e.g., active disease process, psychiatric disorder, and prescribed medication(s) contraindicated with the study medication). Given the substantial time commitment required by human drug-discrimination studies, another important consideration is whether a potential subject is able to dedicate the time necessary to complete the study. A physician reviews all screening materials to determine whether the volunteer is physically and psychologically eligible for participation. Thorough physical and mental health screening is absolutely imperative to ensure subject safety in any study involving the administration of pharmacological agents to human subjects.

The discriminative-stimulus effects of various drugs have been assessed in normal healthy volunteers (e.g., [10, 11]), drug-dependent individuals (e.g., [12, 13]), and individuals with a history of drug dependence who are currently abstinent/detoxified (e.g., [14]). However, there are no published studies in which the discriminative-stimulus effects of particular drugs have been prospectively compared between these populations. Several factors should be considered when selecting the most appropriate population of subjects given the specific research question(s) and the primary aim(s) of the study. For example, participants with an extensive history of substance abuse may be most appropriate in the context of testing whether a novel compound has potential for abuse itself or may effectively attenuate the discriminative-stimulus effects of a drug with known abuse potential. An important caveat, however, is that their extensive drug-use history may complicate interpretation of the results because of differences in expectancies, conditioning history, and tolerance [15]. Although there are advantages and disadvantages to using various populations, research in individuals with and without histories of substance abuse is necessary to gain a more complete understanding of the neuropharmacological mechanisms that underlie the discriminative-stimulus effects of drugs [15].

Test Environment and Experimental Materials and Methods

The test environment and experimental materials required to conduct a human drug-discrimination experiment generally consists of a test room containing a desk, chair, a computer with a mouse, numeric keypad and programming to present the drug-discrimination task and record the data, and equipment that is used to monitor participants' vital signs. Although the use of a computer is more typical, pen and paper could also be used for task presentation and data collection. The room may also be equipped with a television and other recreational materials (e.g., magazines, books, games, and craft supplies) that volunteers may use when not engaged in experimental activities.

In one example of a two-choice drug-discrimination task, the volunteer is presented with two response options (e.g., Drug A and Not Drug A) on the computer screen and instructed to indicate which drug condition that they think they received by distributing 100 points between the two options using the numeric keypad. For example, if a volunteer is relatively confident that they received Drug A, they might allocate 80 points to the Drug A option and 20 points to the Not Drug A option. Volunteers complete the drug-discrimination task multiple times at regular intervals throughout the session: usually every 30 min to an hour depending on the pharmacokinetics of the drug(s) under study. The total number of points allocated to the correct response option out of all possible points is exchanged for money at a constant rate. For example, points have been exchanged for money at rate of $0.04–$0.08 per point in previous drug-discrimination studies conducted in our laboratory [16, 17]. Participants can earn $20–$40 per session but the specific rate with which points are exchanged for money (i.e., $0.04 vs. $0.08) does not appear to significantly alter performance on the task [16, 17].

The use of money as the reinforcer in human drug-discrimination studies is a primary difference from preclinical drug-discrimination studies. In preclinical studies, subjects are often food restricted so that food reinforcers effectively maintain behavior. Another notable difference between preclinical and human drug-discrimination studies is that some human studies do not utilize a formal schedule of reinforcement, at least as typically conceptualized, and reinforcement is withheld until the end of the session when subjects are paid. In contrast, responding by animals is typically maintained by a fixed-ratio schedule of reinforcement and reinforcers are delivered or withheld upon completion of each response requirement.

1.2   Human Drug Discrimination: Procedural Overview

This section of the chapter provides a general experimental overview and highlights the basic methodological elements of human drug-discrimination procedures. Notable procedural variations between drug-discrimination studies, more complex drug-discrimination procedures, and the advantages and limitations of these approaches are then discussed. As noted above, the methods used in human drug-discrimination studies are very similar to those used in preclinical drug-discrimination research. Although a standardized human drug-discrimination procedure has not been established, these experiments often consist of three phases that are completed in a fixed order: (1) Sampling Phase; (2) Acquisition Phase; and (3) Test Phase.

Sampling Phase During the sampling phase, participants complete several experimental sessions to acquaint them with the interoceptive-stimulus effects of the training dose. The training dose is usually identified to participants by a specific code (e.g., Drug A or Red Drug). Participants may also complete sampling sessions during which they receive placebo. In this case, placebo is identified with a unique code (e.g., Not Drug A; Drug B; or Blue Drug). During the sampling sessions, participants are verbally instructed to attend to the effects of the drug because correctly identifying the drug they received will determine the amount of monetary compensation that they earn in future sessions.

Acquisition Phase Following the sampling phase, an acquisition phase (sometimes referred to as the test-of-acquisition or control phase) is conducted in which the training dose and placebo are administered once per day across several sessions (e.g., 4–12 total sessions) in random order. During each session in this phase, volunteers ingest drug or placebo under blinded conditions and then complete the drug-discrimination task along with subjective drug-effect questionnaires periodically for several hours after drug administration. Although participants are asked to identify which treatment they received on the drug-discrimination task periodically throughout the session, the correct treatment code (i.e., Drug A vs. Not Drug A; Drug A vs. Drug B; and Red Drug vs. Blue Drug) is not revealed to the participant until the conclusion of the session. The percentage of correct responses (i.e., correct identification of the treatment) is then converted to money and the participant is told immediately how much bonus money they earned during the experimental session. The performance criterion for having acquired the discrimination is predetermined (e.g., 80% correct responding on four consecutive days), and only those participants that meet the criterion in a specified number of sessions (e.g., 12) advance beyond the acquisition phase. The extensive training associated with human drug-discrimination procedures provides participants with similar recent behavioral and pharmacological histories, which is thought to reduce variability both within and across participants.

Test Phase The final phase is the test phase, during which the discriminative-stimulus effects of different doses of the training drug, novel drugs, or drug combinations are determined. Sessions involving the administration of doses or drugs other than the training condition are deemed to be "test sessions." Participants are not told the purpose of test sessions, nor do they know when these sessions are scheduled until completing the session. As is the case in preclinical studies, there is no correct response per se during these test sessions, so participants usually receive all of the available money that is contingent on correctly identifying the drug condition that was administered. Test-of-acquisition sessions that are identical to those in the acquisition phase are interspersed among test sessions to ensure that participants continue to accurately discriminate the training dose versus placebo. Additional sessions are inserted to re-establish accurate discrimination if the participant fails to correctly identify the training condition they received during a test-of-acquisition session conducted during the test phase. The number of test-of-acquisition sessions included in the test phase varies but is usually fewer than the total number of test sessions (e.g., 25–50%).

In general, there are two strategies in the choice of drug conditions administered in the test phase with the goal of elucidating the neuropharmacological mechanisms that mediate the discriminative-stimulus effects of the training drug. The first is the use of substitution procedures, in which a range of doses of other drugs is tested to determine if they share discriminative-stimulus effects with the training drug. Based on the drugs that produce significant drug-appropriate responding, inferences can be made regarding the neuropharmacological mechanisms that mediate the effects of the training drug. The second approach is to determine a dose–response curve for the training drug alone and in combination with pharmacologically selective compounds. These compounds can be administered concurrently with the training drug or one given as a pretreatment to the other, depending on the pharmacokinetic profiles of the training and test drugs. Inferences are made regarding the neuropharmacological mechanisms that mediate the discriminative-stimulus effects of the training drug based on the mechanism of action of the test drugs that shift the training-drug dose–response curve.

Advantages and Limitations of Human Drug-Discrimination Procedures Human drug-discrimination procedures offer a number of advantages relative to other assays commonly used in behavioral pharmacology. As mentioned previously, three strengths of human drug discrimination are that it produces data that are orderly and dose-dependent, is pharmacologically selective, and that subjects have virtually identical training and recent drug-exposure histories prior to testing novel drugs and/or drug doses. In addition to these strengths, the relationship between the subjective- and discriminative-effects of drugs may be directly evaluated in human drug-discrimination studies.

Despite these notable strengths, human drug-discrimination procedures also have several potential limitations that warrant consideration. First, drug-discrimination procedures require extensive training before testing can begin and require a considerable investment of time and resources on the part of both volunteers and investigators. An offsetting strength is that fewer subjects are required to achieve adequate statistical power in drug-discrimination studies relative to other procedures that rely more heavily on subjective-effects measures. Second, drug-discrimination tasks specifically provide a relatively limited amount of information (i.e., typically a single outcome measure such as discrimination accuracy) as compared to other behavioral measures that provide information across an array of dimensions (e.g., subjective-effects measures; [9]). However, the interpretation of drug-discrimination data is somewhat less complicated because conclusions may be drawn directly from performance on the discrimination task. The likelihood of Type I errors is also decreased because drug-discrimination procedures rely on a single primary-outcome measure. Third, drug-discrimination performance is relatively insensitive to changes in circulating levels of drug across the time-course of drug effects in that the allocation of responses to the drug-appropriate option does not typically decrease as blood levels decrease (e.g., [18]). Fourth, the investigation of the specific role of various molecular sites of action (e.g., transporters, receptor systems, and specific receptor subtypes) to the discriminative-stimulus effects of drugs in humans are relatively limited because medications that are approved for use with humans by the US Food and Drug Administration are typically used in human drug-discrimination studies. Fifth, as noted above, a significant challenge relative to animal models is that humans vary in their behavioral and pharmacological histories, which can affect study results and complicate the interpretation of the findings. Finally, in the context of the study of substance-use disorders, drug-discrimination procedures lack the face validity of other experimental approaches such as drug self-administration (e.g., [19]). Although the drug-discrimination paradigm may lack a certain degree of face validity relative to other experimental approaches, it has predictive validity with respect to the underlying neurobiological and neuropharmacological mechanisms of drugs and determination of the abuse potential of novel compounds (e.g., [9, 15, 20–22]).

2   Underlying Neuropharmacology of Commonly Abused Illicit Drugs

As indicated in previous chapters, drug-discrimination procedures are pharmacologically selective and, as a result, have been used to assess the underlying neuropharmacology of centrally acting drugs. In addition, findings from human drug-discrimination studies are, in many cases, consistent with the hypothesized neuropharmacological mechanisms of actions of those drugs. According to the most recent epidemiological findings, the three most-used substances in 2013 among persons age 12 years or older were cannabis (19.8 million), psychotherapeutics (including prescription stimulants and opioid pain relievers; 6.5 million), and cocaine (1.5 million; [23]). Therefore, in this section of the chapter, we have chosen to review a portion of the human drug-discrimination literature that demonstrates the utility of this behavioral assay to elucidate the underlying neuropharmacology of stimulants, opioids, and cannabis.

2.1 Stimulants

Basic Neuropharmacology and Mechanism of Action Abused stimulants exert their pharmacodynamic effects via interactions with monoamine transporters (e.g., dopamine [DA], serotonin [5-HT], and norepinephrine [NE]; reviewed in [24–27]). Prior ex vivo studies suggest that stimulants can be classified into two groups based on their differential regulation of these transporters. Amphetamines (e.g., D-amphetamine and methamphetamine) act as substrates for monoamine transporters and are transported into the nerve terminal where they prevent accumulation of neurotransmitter in storage vesicles, inhibit metabolic degradation by monoamine oxidase, and promote neurotransmitter release via carrier-mediated exchange [27]. Although amphetamines can also function as reuptake inhibitors, these effects are more moderate compared to their actions as transporter substrates [28]. By contrast, cocaine is a reuptake inhibitor and may cause firing-dependent reversal of the transporter thereby promoting the accumulation of neurotransmitter in the synapse (for a review, see [24, 29]). Central monoamine systems (e.g., DA, 5-HT and NE) are implicated in the discriminative-stimulus effects of abused stimulants [30–38]. The evidence for the involvement of central monoamine systems, namely DA, in the interoceptive effects of abused stimulants is reviewed below.

Substitution Profile Substitution tests in prior human drug-discrimination studies suggest a prominent role for central monoamine systems in the interoceptive effects of stimulants. For example, in participants discriminating D-amphetamine (i.e., 10 mg) from placebo [39], the D_2 receptor partial agonist phenylpropanolamine (i.e., 25 and 75 mg) and monoamine reuptake inhibitor mazindol (i.e., 0.5 and 2.0 mg) substituted for D-amphetamine, suggesting that central monoamine systems are critically involved in the discriminative-stimulus effects of D-amphetamine. Other studies have shown that drugs that directly modulate monoaminergic tone (e.g., caffeine and methylphenidate; [40, 41]) engender D-amphetamine-appropriate responding; whereas, drugs that do not (e.g., diazepam, hydromorphone, and diazepam) produce partial to minimal drug-appropriate responding [39, 42–50]. These studies demonstrate that D-amphetamine functions as a discriminative stimulus via complex interactions at central monoamine systems.

Central monoamine systems also play a prominent role in the discriminative-stimulus effects of methamphetamine and cocaine. In one study, participants learned to discriminate oral methamphetamine (i.e., 10 mg) from placebo [17]. A range of oral doses of methamphetamine (i.e., 2.5–15 mg), D-amphetamine (i.e., 2.5–15 mg), methylphenidate (i.e., 5–30 mg), and γ-aminobutyric acid-A (GABA$_A$) modulator triazolam (i.e., 0.0625–0.375 mg) was then tested. Figure 2 shows that D-amphetamine and methylphenidate dose-dependently increased methamphetamine-appropriate responding; whereas, triazolam failed to engender methamphetamine-appropriate responding. Similarly, Fig. 3 shows that cocaine and methylphenidate produced similar discriminative-stimulus effects in participants who had learned to discriminate oral cocaine (i.e., 150 mg) from placebo [16]. In contrast, neither

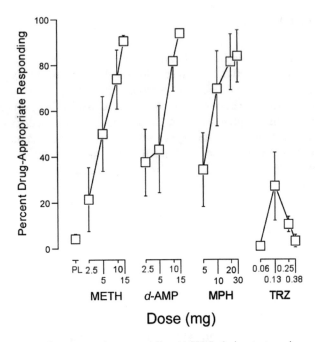

Fig. 2 Mean percent drug-appropriate responding (±SEM) during test sessions with metham-phetamine (METH), D-amphetamine (D-AMP), methylphenidate (MPH), and triazolam (TRZ; negative control) in participants discriminating methamphetamine. D-Amphetamine and methyl-phenidate share discriminative-stimulus effects with methamphetamine but triazolam does not. X-axes: Test doses (mg) of methamphetamine, D-amphetamine, methylphenidate, and triazolam. Data points above PL represent values from test sessions following placebo administration. Y-axis: Percent drug-appropriate responding for methamphetamine. Data points represent the means of seven participants. Reprinted from Sevak et al. [17], with permission

modafinil, a NE releaser with weak affinity for the DA transporter [51, 52], nor the sedative hypnotic drug triazolam fully substituted for cocaine in this study. These findings collectively suggest that drugs that preferentially increase synaptic DA substitute for commonly abused stimulants across a range of doses; whereas, drugs that exert their primary effects through other neurotransmitter systems (e.g., triazolam and modafinil) do not produce discriminative-stimulus effects similar to commonly abused stimulants in humans.

Correspondence with Preclinical Findings The results of substitution tests in preclinical drug-discrimination studies are consistent with the notion that central monoamine systems mediate the discriminative effects of abused stimulants. For example, a range of doses of methamphetamine, cocaine, methylphenidate, D-amphetamine, and GBR 12909 were tested to determine if they shared discriminative-stimulus effects with methamphetamine in rats trained to discrimi-nate 0.3 mg/kg methamphetamine from saline [53]. GBR 12909 is a high-affinity DA transport blocker that is considered to be selective for DA transporters [54, 55]. Each test drug substituted for methamphetamine in a dose-dependent

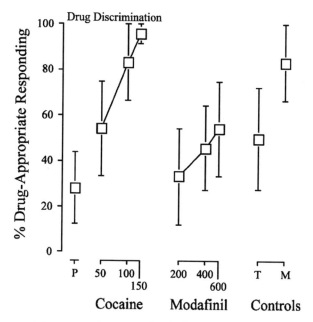

Fig. 3 Mean percent drug-appropriate responding (±SEM) for cocaine, modafinil, triazolam (T; 0.5 mg, negative control), and methylphenidate (M; 60 mg, positive control) in participants discriminating oral cocaine. Modafinil and triazolam did not substitute for cocaine suggesting they may exert their discriminative-stimulus effects through distinct neuropharmacological mechanisms. X-axes: Test doses (mg) of cocaine, modafinil, triazolam, and methylphenidate. Data point above P represents values from test sessions following placebo administration. Y-axis: Percent drug-appropriate responding for cocaine. Data points represent the means of six participants. Reprinted from Rush et al. [16], with permission

manner suggesting that DA neurotransmission contributes to the discriminative-stimulus effects of methamphetamine. Other studies have shown that DA reuptake inhibitors (e.g., bupropion, GBR 12909, and mazindol) fully substitute for cocaine whereas 5-HT and NE reuptake inhibitors do not [36, 38, 56–58]. In addition, D_1- and D_2-receptor agonists (e.g., SKF 38393 and quinpirole, respectively) engender cocaine-appropriate responding [31, 59], suggesting a prominent role for DA signaling in the discriminative-stimulus effects of abused stimulants that are concordant with the results of substitution tests in human drug-discrimination studies.

Pretreatment Studies and Underlying Neuropharmacology Although the lack of selective compounds available for use with humans limits the conclusions that may be made about the specific roles of particular monoamine systems, the results of pretreatment tests in human drug-discrimination studies also suggest that central monoamine systems mediate the discriminative-stimulus effects of commonly abused stimulants. The effects of a range of doses of d-amphetamine (i.e., 0, 2.5, 5, 10, and 15 mg), alone and following pretreatment with the D_2 receptor antagonist fluphenazine (i.e., 0, 3, and 6 mg) were assessed in participants who learned to discriminate 15 mg oral d-amphetamine from placebo [60]. Lower doses of fluphenazine (i.e., 3 mg) did not significantly alter the discriminative-stimulus effects of D-amphetamine in this study, but a higher dose (i.e., 6 mg) produced a marked rightward shift in the D-amphetamine dose–response curve in the one participant that completed the study (Fig. 4). These findings suggest that central DA systems mediate the discriminative-stimulus effects of D-amphetamine in humans. However, these results should be interpreted cautiously because only a single subject completed the study due to the negative side-effect profile of fluphenazine.

Aripiprazole is an atypical antipsychotic that functions as a partial D_2 receptor agonist [61] and is also known to exert effects at $5\text{-}HT_{1A}$, $5\text{-}HT_{2A}$, $5\text{-}HT_{2B}$, and $5\text{-}HT_7$ receptors [62]. Partial agonists can either activate receptors with decreased efficacy relative to full agonists, or conversely function as an antagonist, depending on synaptic neurotransmitter levels. To determine the effects of aripiprazole on the discriminative-stimulus effects of D-amphetamine, a range of doses of D-amphetamine (i.e., 0, 2.5, 5, 10, and 15 mg) were assessed, alone and in combination with aripiprazole (0 and 20 mg), in participants who learned to discriminate oral D-amphetamine (i.e., 15 mg) from placebo [63]. D-Amphetamine functioned as a discriminative stimulus, but aripiprazole did not engender D-amphetamine-appropriate responding when tested alone. Aripiprazole pretreatment significantly attenuated the discriminative-stimulus effects of D-amphetamine, suggesting a role for DA and 5-HT in the interoceptive effects of D-amphetamine. These results are consistent with the ability of a D_2-receptor partial agonist to function as an antagonist in the presence of a drug that elevates synaptic monoamine levels [64]. Other studies have shown similar effects with other antipsychotics and $GABA_A$ modulators such as risperidone and alprazolam, respectively [50, 65].

Pretreatment tests with agonists and antagonists in humans discriminating methamphetamine and cocaine further suggest that central monoamine systems are involved in the discriminative-stimulus effects of commonly abused stimulants (e.g., [12, 66–68] unpublished data). For example, Sevak and colleagues [66] determined the influence of aripiprazole (0 and 20 mg) on the discriminative-stimulus effects of a range of doses of methamphetamine (0, 2.5, 5, and 10 mg) in participants who had learned to discriminate 10 mg methamphetamine. Methamphetamine functioned as a discriminative stimulus and dose-dependently increased drug-appropriate responding. Aripiprazole pretreatment significantly attenuated methamphetamine-appropriate responding (Fig. 5), suggesting that monoamine systems play a role in the discriminative-stimulus effects of methamphetamine. To assess the role of monoamine systems in the discriminative-stimulus effects of cocaine, Lile and colleagues [12] tested a range of doses of oral cocaine (0, 25, 50, 100, and 200 mg) alone and in combination with aripiprazole (15 mg) in participants who had learned to discriminate 150 mg oral cocaine from placebo [12]. Although few effects of aripiprazole were observed, it appeared to attenuate the discriminative-stimulus effects of cocaine. These data collectively suggest that the discriminative-stimulus effects of commonly abused stimulants in humans are mediated by monoamine systems, namely DA and 5-HT.

Correspondence with Preclinical Findings The results of pretreatment tests in preclinical drug-discrimination studies with commonly abused stimulants correspond with those from human drug-discrimination studies and support the

Fig. 4 Mean percent drug-appropriate responding (±SEM) following a range of doses of D-amphetamine alone and in combination with 3 mg (*circles, upper panel*) and 6 mg (*circles, lower panel*) fluphenazine. Squares represent 0 mg of fluphenazine in both panels. The 6 mg dose of fluphenazine shifted the D-amphetamine dose-effect function rightward suggesting that it attenuated the discriminative-stimulus effects of D-amphetamine. *X*-axes: D-Amphetamine dose in mg. Data points above PL represent values from test sessions following placebo administration. *Y*-axes: Percent drug-appropriate responding for D-amphetamine. Data points in the *bottom panel* represent data from one participant. Data from Stoops et al. [60]

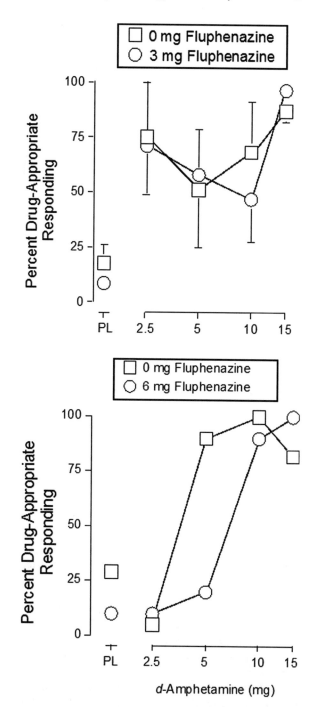

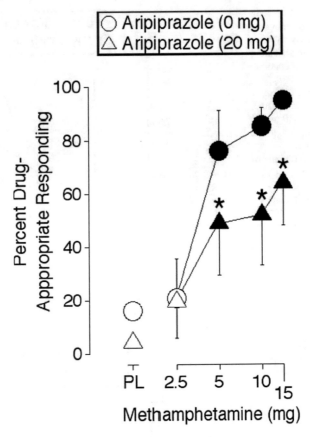

Fig. 5 Mean percent drug-appropriate responding (±SEM) during test sessions with methamphetamine alone and in combination with 0 mg (*circles*) and 20 mg (*triangles*) aripiprazole. Aripiprazole significantly attenuated the discriminative-stimulus effects of methamphetamine. *X*-axes: Methamphetamine dose in mg. Data points above PL represent values from tests with 0 mg methamphetamine (placebo) alone and in combination with 20 mg aripiprazole. *Y*-axis: Percent drug-appropriate responding. Data points represent the means of six participants. *Filled symbols* indicate a significant difference from the placebo–placebo control condition (i.e., circle above PL). An *asterisk* (*) indicate a significant difference between aripiprazole conditions at a given methamphetamine dose. Reprinted from Sevak et al. [66], with permission

hypothesis that central monoamine systems underlie the interoceptive effects of abused stimulants. For example, Mechanic and colleagues [69] determined whether the D_2 and $5\text{-}HT_2$ antagonist olanzapine would attenuate the interoceptive cues elicited by D-amphetamine in rats that were trained to discriminate D-amphetamine (1.0 mg/kg) from saline. Olanzapine (1.5 mg/kg) significantly blunted the discriminative-stimulus effects of D-amphetamine. Similar findings have been obtained with selective D_1 (e.g., SCH39166) and D_2 antagonists (e.g., remoxipride and nemonapride; [70]), as well as the high-affinity dopamine transport blocker

GBR 12909 [71] to suggest a role for DA signaling in the discriminative-stimulus effects of stimulants in laboratory animals. In addition, these DA systems are under the inhibitory control of GABA systems (e.g., [72–75]). For example, Druhan and colleagues [76] showed that pretreatment with the $GABA_A$ receptor modulator midazolam (i.e., 0–0.2 mg/kg) significantly attenuated drug-appropriate responding in rats trained to discriminate D-amphetamine (i.e., 1.0 mg/kg). In sum, the results of drug-discrimination studies with humans and non-human animals suggest that the neuropharmacological mechanisms of the discriminative-stimulus effects of abused stimulants are generally consistent [77].

Summary of Drug-Discrimination Findings with Stimulants In general, data from preclinical and human drug-discrimination studies demonstrate that abused stimulants produce their interoceptive effects via activation of DA and other monoamine systems. Abused stimulants function as discriminative stimuli and readily substitute for one another under a wide range of laboratory conditions and across species. Drugs that share discriminative-stimulus effects with abused drugs might function as effective agonist-replacement therapies to treat stimulant-use disorders [78–81]. Alternatively, drugs that attenuate the discriminative-stimulus effects of abused drugs might function as effective pharmacotherapies for stimulant-use disorder by blunting the interoceptive effects of the drug ([82]; for a review, see [83]).

Collectively, these studies suggest that human drug-discrimination procedures are rigorous behavioral assays that may be used to elucidate the underlying neuropharmacology of the discriminative-stimulus effects of stimulants. Future studies are needed to more fully elucidate the neuropharmacological mechanisms underlying the interoceptive-stimulus effects of abused stimulants in humans. These studies might test blockers of other catecholamines or drug-combinations that may have promise as pharmacotherapies (see [83] for a review). A more comprehensive understanding of the neuropharmacological mechanisms that mediate the interoceptive effects of stimulants in humans will inform the development of putative pharmacotherapies to manage stimulant-use disorders.

2.2   Opioids

Basic Neuropharmacology and Mechanism of Action The basic neuropharmacology of opioid receptors is well known (for reviews see [84, 85]). Briefly, the mu, kappa, and delta opioid receptors belong to the class A (rhodopsin) family of $G_{i/o}$ protein-coupled receptors and are found throughout the central and peripheral nervous systems. These three receptor families mediate the analgesic effects of endogenous opioid peptides and opioid drugs [9, 85, 86]. Opioid drugs are naturally occurring, semi-synthetic, or synthetic formulations (e.g., morphine, hydromorphone, and fentanyl, respectively). They are further classified as full agonists, partial or mixed agonists/antagonists, and full antagonists based on their pharmacological actions, selectivity, affinity and efficacy at the three primary receptor families [9]. The majority of prescribed opioid analgesics are agonists at the mu receptor with relatively limited activity at the other receptor types. The abuse-related behavioral effects of prototypical opioids like morphine, heroin, or hydromorphone have largely been attributed to their interaction with the mu receptor family [87–89]. The mu receptor family, in particular, is known to modulate the neuropharmacological activity of monoamine and GABAergic neurotransmitter systems resulting in increased synaptic dopamine levels [90–92]. The kappa and delta opioid receptor families are structurally and functionally similar to mu opioid receptors [85]. However, the behavioral effects of drugs that activate kappa and delta opioid receptors differ from those that preferentially activate mu receptors. For example, kappa agonists can produce dysphoria and hallucinations and there is evidence that the kappa receptor family is involved in stress responses [93]. Delta receptor agonists are less susceptible to analgesic tolerance compared to mu receptor agonists suggesting that these receptors may produce analgesic effects via different pharmacological mechanisms [94]. This section of the chapter focuses on the mu receptor because most opioids that have been tested affect mu activity and the mu receptor is most clinically relevant with regard to opioid dependence in humans.

Substitution Profile Eleven published clinical studies have examined the discriminative-stimulus effects of opioid drugs [14, 95–104]. A seminal study by Preston and Bigelow [98] illustrates that opioid agonists with similar efficacy and affinity for the mu receptor generalize other mu receptor agonists but do not generalize opioid agonists that differ in these respects. Volunteers with a history of regular opioid use learned to discriminate intramuscular saline, hydromorphone, and butorphanol using the three-choice discrimination procedure (i.e., Drug A, Drug B, or Drug C) to investigate the discriminative-stimulus effects of hydromorphone and other opioid drugs with varying degrees of affinity for mu and kappa opioid receptors. The opioid drugs that were tested included hydromorphone (0.375–3.0 mg), the partial mu and kappa receptor agonist pentazocine (7.5–60 mg), the mu and kappa receptor mixed agonist–antagonist butorphanol (0.75–6 mg), the non-selective opioid agonist nalbuphine (3.0–24 mg), and the partial mu receptor agonist buprenorphine (0.075–0.6 mg). Opioids with greater affinity for the mu receptor fully substituted for hydromorphone regardless of whether the drug was a partial or full agonist. Opioids with lower intrinsic activity at mu receptors did not substitute for the mu agonist hydromorphone. Figure 6 shows that hydromorphone occasioned dose-related increases in hydromorphone-appropriate responding but did not substitute for butorphanol, consistent with their hypothesized neuropharmacological actions at the mu opioid receptor.

Correspondence with Preclinical Findings Preclinical research with pigeons [105, 106], rats [107–111], and non-human primates [112, 113] have consistently shown that the discriminative-stimulus effects of opioids are concordant across species and that these effects follow with their in vitro neuropharmacology. For example, Platt and colleagues [113] investigated the discriminative-stimulus effects

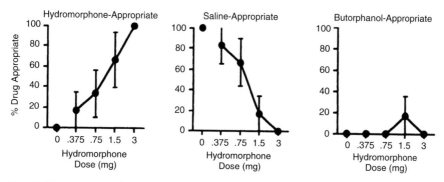

Fig. 6 Mean percent drug-appropriate responding (±SEM) during substitution tests with hydromorphone in participants discriminating hydromorphone, saline, and butorphanol. Hydromorphone significantly increased hydromorphone-appropriate responding but did not substitute for any dose of butorphanol tested. The discriminative-stimulus effects of mu-opioid receptor agonists follow their predicted neuropharmacological actions. X-axes: Hydromorphone dose in mg. Y-axes: Percent drug-appropriate responding for hydromorphone (*left*), saline (*center*), and butorphanol (*right*). Data points represent the means of six participants. Reprinted from Preston and Bigelow [98], with permission

of heroin in non-human primates and showed that the interoceptive effects of heroin were largely attributable to mu opioid receptor activation. Substitution tests with the major metabolites of heroin (i.e., 6-monoacetylmorphine, morphine, morphine-6-glucuronide, and morphine-3-glucuronide) and the mu opioid receptor agonists fentanyl and methadone were conducted with rhesus monkeys trained to discriminate heroin from saline. Each of these drugs occasioned dose-dependent increases in heroin-appropriate responding and, on average, engendered full substitution for heroin.

Pretreatment Tests and Underlying Neuropharmacology We know of two published clinical studies that have used pretreatment strategies to investigate the discriminative-stimulus effects of opioid drugs [104, 114]. For example, Strickland and colleagues [104] utilized antagonist pretreatment in conjunction with substitution strategies to demonstrate that some of the discriminative-stimulus effects of the atypical opioid tramadol are mediated by mu receptor activation. Figure 7 shows representative drug-discrimination data for two subjects following administration of hydromorphone or a range of doses of tramadol alone (circles) or in combination with 50 mg naltrexone (squares). Tramadol occasioned dose-related increases in drug-appropriate responding for tramadol and a test dose of hydromorphone occasioned partial or full substitution for tramadol. Pretreatment with naltrexone (50 mg, p.o.) significantly attenuated the discriminative-stimulus effects of tramadol and hydromorphone. The use of opioid antagonists in human drug-discrimination procedures is an important strategy that provides additional information about the underlying neuropharmacological mechanisms of opioid drugs. Further, the use of this strategy bridges preclinical and clinical research; thereby, strengthening the translational validity of findings from drug-discrimination studies. Unfortunately, there are few clinical studies that have used antagonist

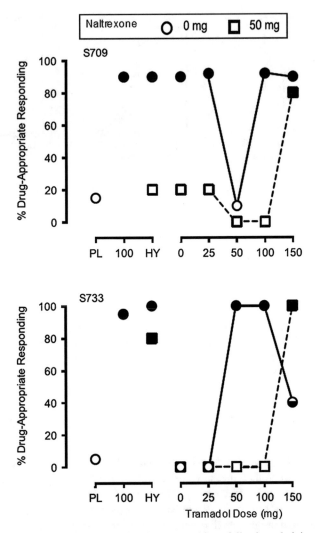

Fig. 7 Percent drug-appropriate responding from two subjects following administration of 4 mg hydromorphone (HY) and a range of doses of oral tramadol alone and in combination with 0 mg (*circles*) and 50 mg (*squares*) naltrexone. Hydromorphone substituted for tramadol in both subjects and naltrexone attenuated the tramadol discriminative stimulus in one subject. Tramadol increased drug-appropriate responding at several doses. Naltrexone attenuated these effects at lower tramadol doses. These findings suggest that mu-opioid receptors are at least partly involved in the discriminative-stimulus effects of tramadol. X-axes: Tramadol dose in mg. Data points above PL and 100 represent values from test of acquisition sessions following administration of placebo and 100 mg tramadol, respectively. *Filled symbols* indicate full tramadol substitution (i.e., ≥80% tramadol-appropriate responding). *Half-filled shapes* indicate partial substitution (i.e., 21–79% tramadol-appropriate responding). Y-axes: Percent drug-appropriate responding for tramadol. Reprinted from Strickland et al. [104], with permission

pretreatment procedures to elucidate the neuropharmacological underpinnings of the discriminative-stimulus effects of opioids.

Correspondence with Preclinical Findings Preclinical work using pretreatment strategies has been crucial for examining the neuropharmacology of the discriminative-stimulus effects of opioid drugs. For example, France and colleagues [115] trained pigeons to discriminate morphine from placebo and then performed substitution tests with morphine and oxymorphazone (a mu opioid receptor agonist). Morphine and oxymorphazone occasioned morphine-appropriate responding in a dose-dependent manner. Pretreatment with naltrexone shifted the dose–response curves to the right, indicating that naltrexone attenuated the discriminative-stimulus effects of these drugs. Antagonism of the discriminative-stimulus effects of opioid drugs by naltrexone pretreatment has also been observed in rhesus monkeys that were trained to discriminate heroin or morphine from vehicle [112, 113, 116].

Summary of Drug-Discrimination Findings with Opioids Opioid drug-discrimination studies in both human and non-human animals using substitution and pretreatment procedures are remarkably consistent with their neuropharmacological binding profiles for the mu receptor. These studies have revealed that although the discriminative-stimulus effects of opioid drugs are not limited to activity at opioid receptors, they are primarily mediated by mu receptor activity. These results are consistent with a primary role for the mu receptor in the ability of repeated opioid administration and dosing cessation to induce dependence and withdrawal, respectively (reviewed in [117]). This neuropharmacological overlap in clinically relevant effects suggests that opioid drug-discrimination procedures could be used for medications development [19]. Opioid drugs with decreased abuse potential that share discriminative-stimulus effects with abused opioids might be effective pharmacotherapies for opioid dependence.

2.3 Δ^9-Tetrahydrocannabinol (Δ^9-THC)

Basic Neuropharmacology and Mechanism of Action Of the more than 60 cannabinoid compounds found in cannabis, Δ^9-tetrahydrocannabinol (Δ^9-THC) is widely considered to be primarily responsible for its psychoactive effects [118]. The behavioral effects of Δ^9-THC are mediated through the endogenous cannabinoid neurotransmitter system, which is composed of two known receptor subtypes: CB_1 and CB_2 [119, 120]. Both cannabinoid receptor subtypes are G-protein-coupled receptors that inhibit adenylate cyclase activity and activate mitogen-activated protein kinase, but they differ to some degree in their interactions with certain ion channels and other G-proteins (e.g., [121–123]). CB_1 and CB_2 receptors also differ in their distribution such that CB_1 receptors are primarily expressed on presynaptic nerve terminals throughout the central and peripheral nervous systems; whereas, CB_2 receptors are expressed on immune cells

[123]. Although Δ^9-THC is a non-selective partial agonist at CB_1 and CB_2 receptors, at least four lines of evidence suggest that the central effects of Δ^9-THC are primarily mediated through CB_1 receptors. First, the in vivo potency of Δ^9-THC correlates with its binding affinity at the CB_1 receptor [124]. Second, the CB_1 receptor subtype is localized in areas of the central nervous system that correspond with Δ^9-THC effects [125]. Third, agonists that are selective for CB_1 receptors produce behavioral effects more similar to Δ^9-THC than selective CB_2 agonists [126–128]. Lastly, the centrally mediated effects of Δ^9-THC are blocked by the administration of CB_1-selective antagonists, but not those selective for CB_2 receptors [129–132]. Given that another principal function of cannabinoid receptors is the modulation of non-cannabinoid neurotransmitter release via retrograde signaling [133], other neurotransmitter systems also likely play a role in the behavioral effects of cannabinoids.

The published literature concerning the discriminative-stimulus effects of Δ^9-THC in humans is much smaller in comparison to the other drug classes discussed in this chapter. To the best of our knowledge, only 8 studies have been published that evaluated the discriminative-stimulus effects of Δ^9-THC in humans [134–141]. In more recent studies, participants learned to discriminate orally administered Δ^9-THC versus placebo. The use of orally administered Δ^9-THC in lieu of smoked cannabis improves pharmacological selectivity (as cannabis contains other cannabinoids), allows better control of dosing parameters, and eliminates peripheral cues associated with smoked cannabis (e.g., [134]). The available literature on the discriminative-stimulus effects of orally administered Δ^9-THC and its underlying neuropharmacology as determined with human drug-discrimination procedures is reviewed below.

Substitution Profile The substitution of other drugs for the discriminative-stimulus effects of Δ^9-THC in humans has been determined in several studies [135–140]. However, most of these studies determined the effects of a test drug alone (i.e., substitution) and in combination (i.e., pretreatment) with Δ^9-THC (i.e., [136–139]). The results of pretreatment tests are discussed below in a separate section for ease of comparison. In the first study by Lile and colleagues [140], eight cannabis users learned to discriminate 25 mg oral Δ^9-THC versus placebo. After learning the discrimination, a range of oral doses of Δ^9-THC (5–25 mg), triazolam (0.0675–0.375 mg), hydromorphone (0.75–4.5 mg), and methylphenidate (5–30 mg) was substituted for the training dose. Figure 8 shows that oral Δ^9-THC engendered dose-related increases in drug-appropriate responding, whereas none of the other drugs occasioned significant Δ^9-THC-like responding. Worth mentioning is that each of the drugs tested produced measurable effects on other study outcomes, confirming that biologically relevant doses were tested. Lile and colleagues [140] determined the substitution profile of the mixed CB receptor agonist nabilone in six human cannabis users who learned to discriminate 25 mg Δ^9-THC from placebo. As shown in Fig. 9, nabilone dose-dependently substituted for the interoceptive-stimulus effects of Δ^9-THC with the highest doses of nabilone (3 and 5 mg) fully substituting for the training dose. In contrast, methylphenidate

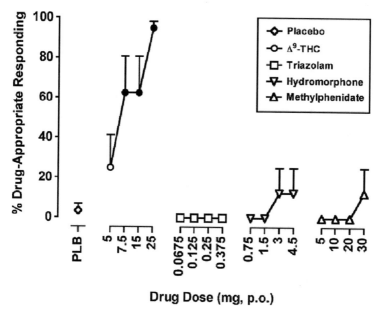

Fig. 8 Mean percent drug-appropriate responding (±SEM) during test sessions with Δ^9-THC (*circles*), triazolam (*squares*), hydromorphone (*inverted triangles*), and methylphenidate (*triangles*) in humans discriminating oral Δ^9-THC. Δ^9-THC functioned as a discriminative stimulus and its discriminative-stimulus effects are not directly mediated by other central neurotransmitter systems. *X*-axes: Oral drug dose in mg per os. The data point (*diamond*) above PLB represents data following placebo administration. All data points represent the means of eight participants. *Filled symbols* indicate a significant difference from placebo (PLB). *Y*-axis: Percent drug-appropriate responding for Δ^9-THC. Reprinted from Lile et al. [140], with permission

did not significantly increase drug-appropriate responding, similar to a previous study [140]. These findings demonstrate the pharmacological selectivity of the discriminative-stimulus effects of Δ^9-THC and suggest that cannabinoid receptors are central to the Δ^9-THC discriminative stimulus but other receptor systems (e.g., GABA) are not.

Correspondence with Preclinical Findings The results of substitution tests with human subjects discriminating Δ^9-THC are relatively consistent with the results of non-human animal studies. Specifically, cannabinoid agonists occasion drug-appropriate responding in animals discriminating Δ^9-THC (e.g., [126, 131, 142–144]), but mu-opioid agonists (e.g., heroin and morphine) generally do not share discriminative-stimulus effects with Δ^9-THC in animals [127, 131, 145–149]. Preclinical studies have also shown that dopaminergic drugs generally do not substitute for the discriminative-stimulus effects of Δ^9-THC [127, 131, 150]. However, the results with triazolam and diazepam in humans [139, 140] do not agree with the preclinical findings that positive modulators of the GABA$_A$ receptor partially substitute for the discriminative-stimulus effects of Δ^9-THC [145, 146, 149, 151–153].

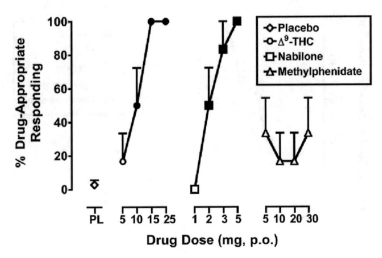

Fig. 9 Mean percent drug-appropriate responding (±SEM) during test sessions with Δ^9-THC (*circles*), nabilone (*squares*), and methylphenidate (*triangles*; negative control) in humans discriminating oral Δ^9-THC. Nabilone dose-dependently substituted for Δ^9-THC suggesting that the discriminative-stimulus effects of Δ^9-THC and nabilone are primarily mediated by activation of the cannabinoid receptor system. X-axes: Oral drug dose in mg per os. The data point (*diamond*) above PL represents data following placebo administration. Y-axis: Percent drug-appropriate responding for Δ^9-THC. Data points represent the means of six participants. Error bars were omitted on certain data points for clarity. Reprinted from Lile et al. [135], with permission

Pretreatment Tests and Underlying Neuropharmacology Five studies have used drug-discrimination procedures to investigate the underlying neuropharmacology of the Δ^9-THC discriminative stimulus in humans [136–139, 141]. These studies used similar procedures to determine the role of the cannabinoid and GABA neurotransmitter systems in the discriminative-stimulus effects of Δ^9-THC. Briefly, participants in these studies learned to discriminate 30 mg of oral Δ^9-THC versus placebo in a two-choice (i.e., Drug vs. Not Drug) procedure. During testing, participants received three doses of nabilone (0, 1, and 3 mg p.o.), tiagabine (0, 6, and 12 mg p.o.), diazepam (0, 5, and 10 mg p.o.), and baclofen (0, 25, and 50 mg p.o.) alone and in combination with oral Δ^9-THC (5, 15, and 30 mg). Figure 10 shows that nabilone occasioned Δ^9-THC-appropriate responding when administered alone and shifted the Δ^9-THC dose-effect function upward and leftward when co-administered with Δ^9-THC [136]. Similarly, the GABA reuptake inhibitor tiagabine fully substituted for the Δ^9-THC discriminative stimulus at the highest dose tested (12 mg) when administered alone and shifted the Δ^9-THC dose–response curve upward and leftward in a dose-related manner [137]. In subsequent studies, the GABA$_A$ positive modulator diazepam did not occasion Δ^9-THC-like responding when administered alone, in agreement with earlier triazolam results [140], and did not systematically affect the discriminative-stimulus effects of Δ^9-THC when administered in combination [139]. In contrast, a high dose of the GABA$_B$ agonist baclofen (50 mg) partially substituted for the Δ^9-THC discriminative stimulus and both doses

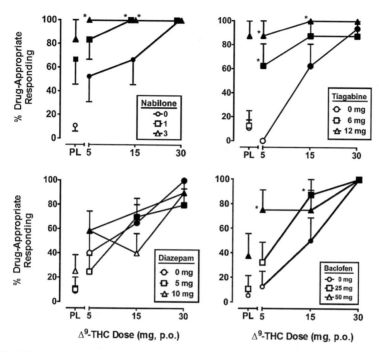

Fig. 10 Mean percent drug-appropriate responding ($\pm$SEM) following Δ^9-THC (5, 15, and 30 mg), alone and in combination with three doses of nabilone (*upper left*), tiagabine (*upper right*), diazepam (*bottom left*) and baclofen (*bottom right*) in humans discriminating oral Δ^9-THC. These findings suggest the involvement of cannabinoid and GABA$_B$ neurotransmitter systems in the discriminative-stimulus effects of Δ^9-THC in humans. X-axes: Oral drug dose in mg per os. Data points above PL represent data for each test drug dose following 0 mg Δ^9-THC. Y-axes: Percent drug-appropriate responding for Δ^9-THC. For nabilone, tiagabine, diazepam, and baclofen, data points represent the means of six, eight, eight, and ten participants, respectively. In all panels, *filled symbols* indicate a significant difference from placebo–placebo (circle above PL). Asterisks (*) indicate a significant difference from a given dose of oral Δ^9-THC alone (circles in each panel). Error bars were omitted in certain instances for clarity. Reprinted from Lile et al. [136–139], with permission

of baclofen significantly enhanced Δ^9-THC-appropriate responding when co-administered [138]. These findings collectively demonstrate the involvement of GABA$_B$ receptor subtype, in the discriminative-stimulus effects of Δ^9-THC in humans.

Correspondence with Preclinical Findings Procedural differences preclude the direct comparison of preclinical and human laboratory studies because most preclinical studies have determined the effects of pretreatment with cannabinoid antagonists on the discriminative-stimulus effects of Δ^9-THC instead of cannabinoid agonists or GABA ligands. For example, pretreatment with the cannabinoid receptor antagonist rimonabant attenuates the discriminative-stimulus effects of Δ^9-THC in laboratory animals (e.g., [143, 154–156]). Despite these differences, some consistent findings emerge. First, drugs that activate the cannabinoid receptor system engender Δ^9-THC-appropriate responding in humans and animals supporting the assertion that the cannabinoid receptor system is critically involved in the discriminative-stimulus effects of Δ^9-THC (e.g., [126, 131, 135, 136, 142–144]). Second, stimulation of GABA neurotransmission appears to play a role in the discriminative-stimulus effects of Δ^9-THC in both humans and preclinical animal models but the mechanisms that mediate these effects may differ between species [137–139, 145, 146, 149, 151–153].

Summary of Drug-Discrimination Findings with Δ^9-THC Although the body of research that has examined the underlying neuropharmacology of Δ^9-THC in human subjects is relatively small, the extant literature demonstrates that cannabinoid and GABA neurotransmitter systems are important contributors to the discriminative-stimulus effects of Δ^9-THC in humans. However, there appear to be species differences in the GABA-specific receptor mechanisms between humans and non-human animals. Lastly, the activation of monoamine (e.g., DA) and mu-opioid receptors does not appear to be involved in the interoceptive effects of Δ^9-THC in humans. These studies also provide insight into potential therapeutic targets for the treatment of cannabis-use disorders. More specifically, these findings suggest that GABA could be targeted in the development of medications for cannabis dependence. In fact, gabapentin, a GABA analog that is approved for treating neuropathic pain and seizures, has recently emerged as a promising candidate pharmacotherapy for cannabis-use disorder [157] and, to date, is the only medication that has demonstrated initial pharmacotherapeutic efficacy in clinical trials in adults. Future research is needed to disentangle the mechanism by which gabapentin reduces cannabis use and also to determine whether a GABA reuptake inhibitor or $GABA_B$ agonist would be useful for managing cannabis dependence. In sum, drug-discrimination studies have greatly enhanced our understanding of the underlying neuropharmacology of Δ^9-THC in humans and have helped to identify potential neuropharmacological targets for the treatment of cannabis dependence.

2.4   General Summary

This section reviewed a number of studies that used human drug-discrimination techniques to investigate the underlying neuropharmacology of stimulants, opioids, and the primary psychoactive constituent in cannabis, Δ^9-THC. At least four overarching conclusions can be drawn from the drug-discrimination literature reviewed above: (1) drugs in each of these classes function as discriminative stimuli in humans, (2) the discriminative-stimulus effects of these drugs are generally consistent with their underlying neuropharmacology, (3) the discriminative-stimulus effects of drugs in these classes are conserved across species, and (4) drug-discrimination techniques allow the determination of the underlying neuropharmacology of commonly abused illicit drugs to identify potential therapeutic targets that may guide the development and evaluation of putative pharmacotherapies for substance-use disorders.

3 Current State and Future of Human Drug-Discrimination Research

The primary objective of this chapter was to provide a basic procedural overview of human drug-discrimination procedures and summarize the extant literature regarding the underlying neuropharmacology of commonly abused drugs (i.e., stimulants, opioids, and cannabis) as determined via human drug-discrimination studies. Although the extant literature firmly establishes human drug discrimination as a highly versatile and useful behavioral assay of in vivo neuropharmacology, interest in human drug-discrimination research and drug-discrimination research in general, has waned somewhat since its peak in the late 1990s. One factor that has potentially led to the decrease in enthusiasm for drug-discrimination studies in substance-abuse research is that the role of discriminative-stimulus effects in substance abuse may be less apparent relative to behavioral processes that are the focus of other experimental approaches. McMahon [19] articulates a particularly poignant example when addressing the downward trend in the publication of drug-discrimination compared with the continued increase in the publication of drug self-administration research. Specifically, he cites that drug discrimination lacks the strong face validity of drug self-administration with regard to substance abuse because operant behavior maintained by a drug reinforcer more closely resembles the behavioral phenomenon of substance abuse [19]. Although behavioral models that have high face validity are intuitively appealing, whether or not they effectively predict the outcome of a manipulation on the phenomenon that they are intended to model is more important. The validity of the drug-discrimination paradigm for identifying the underlying neuropharmacology of centrally acting drugs in whole organisms is virtually unparalleled. However, less research has centered on the role that the discriminative-stimulus effects of drugs play in substance abuse but they may play a particularly important role in relapse and the resumption of problematic drug use.

Although the use of human drug-discrimination procedures in the future is uncertain, the emergence and growing popularity of designer drugs (i.e., bath salts), synthetic marijuana (i.e., spice), and devices that are used to vaporize nicotine (e.g., e-cigarettes) and cannabis will create new opportunities for additional drug-discrimination research. Furthermore, creative thinking about the application of human and laboratory animal drug-discrimination procedures to the investigation of interoceptive events that may contribute to substance abuse (e.g., drug withdrawal, anxiety, stress, etc.) may also provide opportunities for the use of these procedures to investigate the abuse-related behavioral effects of drugs in addition to underlying neuropharmacology.

References

1. Conger JJ (1951) The effects of alcohol on conflict behavior in the albino rat. Q J Stud Alcohol 12(1):1–29. Retrieved from http://www.ncbi.nlm.nih.gov/pubmed/14828044
2. Porter JH, Prus AJ (2009) Drug discrimination: 30 years of progress. Psychopharmacology (Berl) 203(2):189–191. doi:10.1007/s00213-009-1478-7
3. Preston KL (1991) Drug discrimination methods in human drug abuse liability evaluation. Br J Addict 86(12):1587–1594. Retrieved from http://www.ncbi.nlm.nih.gov/pubmed/1786491
4. Jellinek EM (1946) Role of the placebo in tests for drug discrimination. Fed Proc 5(1 Pt 2):184. Retrieved from http://www.ncbi.nlm.nih.gov/pubmed/21064408
5. Preston KL, Bigelow GE (1991) Subjective and discriminative effects of drugs. Behav Pharmacol 2(4 and 5):293–313. Retrieved from http://www.ncbi.nlm.nih.gov/pubmed/11224073
6. Schuster CR, Fischman MW, Johanson CE (1981) Internal stimulus control and subjective effects of drugs. NIDA Res Monogr 37:116–129. Retrieved from http://www.ncbi.nlm.nih.gov/pubmed/6798454
7. Schuster CR, Johanson CE (1988) Relationship between the discriminative stimulus properties and subjective effects of drugs. Psychopharmacol Ser 4:161–175. Retrieved from http://www.ncbi.nlm.nih.gov/pubmed/3293041
8. Kamien JB, Bickel WK, Hughes JR, Higgins ST, Smith BJ (1993) Drug discrimination by humans compared to nonhumans: current status and future directions. Psychopharmacology (Berl) 111(3):259–270. Retrieved from http://www.ncbi.nlm.nih.gov/pubmed/7870962
9. Kelly TH, Stoops WW, Perry AS, Prendergast MA, Rush CR (2003) Clinical neuropharmacology of drugs of abuse: a comparison of drug-discrimination and subject-report measures. Behav Cogn Neurosci Rev 2(4):227–260. doi:10.1177/1534582303262095
10. Rush CR, Critchfield TS, Troisi JR, Griffiths RR (1995) Discriminative stimulus effects of diazepam and buspirone in normal volunteers. J Exp Anal Behav 63(3):277–294. doi:10.1901/jeab.1995.63-277
11. Silverman K, Griffiths RR (1992) Low-dose caffeine discrimination and self-reported mood effects in normal volunteers. J Exp Anal Behav 57(1):91–107. doi:10.1901/jeab.1992.57-91
12. Lile JA, Stoops WW, Glaser PE, Hays LR, Rush CR (2011) Discriminative stimulus, subject-rated and cardiovascular effects of cocaine alone and in combination with aripiprazole in humans. J Psychopharmacol 25(11):1469–1479. doi:10.1177/0269881110385597
13. Oliveto A, Mancino M, Sanders N, Cargile C, Benjamin Guise J, Bickel W, Brooks Gentry W (2013) Effects of prototypic calcium channel blockers in methadone-maintained humans responding under a naloxone discrimination procedure. Eur J Pharmacol 715(1–3):424–435. doi:10.1016/j.ejphar.2013.03.007
14. Preston KL, Bigelow GE, Bickel WK, Liebson IA (1989) Drug discrimination in human postaddicts: agonist–antagonist opioids. J Pharmacol Exp Ther 250(1):184–196. Retrieved from http://www.ncbi.nlm.nih.gov/pubmed/2473187
15. Brauer LH, Goudie AJ, de Wit H (1997) Dopamine ligands and the stimulus effects of amphetamine: animal models versus human laboratory data. Psychopharmacology (Berl) 130 (1):2–13. Retrieved from http://www.ncbi.nlm.nih.gov/pubmed/9089844
16. Rush CR, Kelly TH, Hays LR, Wooten AF (2002) Discriminative-stimulus effects of modafinil in cocaine-trained humans. Drug Alcohol Depend 67(3):311–322. Retrieved from http://www.ncbi.nlm.nih.gov/pubmed/12127202
17. Sevak RJ, Stoops WW, Hays LR, Rush CR (2009) Discriminative stimulus and subject-rated effects of methamphetamine, d-amphetamine, methylphenidate, and triazolam in methamphetamine-trained humans. J Pharmacol Exp Ther 328(3):1007–1018. doi:10.1124/jpet.108.147124
18. Kelly TH, Emurian CS, Baseheart BJ, Martin CA (1997) Discriminative stimulus effects of alcohol in humans. Drug Alcohol Depend 48(3):199–207. Retrieved from http://www.ncbi.nlm.nih.gov/pubmed/9449019

19. McMahon LR (2015) The rise (and fall?) of drug discrimination research. Drug Alcohol Depend 151:284–288. Retrieved from http://www.ncbi.nlm.nih.gov/pubmed/26207268
20. Colpaert FC (1999) Drug discrimination in neurobiology. Pharmacol Biochem Behav 64 (2):337–345. Retrieved from http://www.ncbi.nlm.nih.gov/pubmed/10515310
21. Holtzman SG, Locke KW (1988) Neural mechanisms of drug stimuli: experimental approaches. Psychopharmacol Ser 4:138–153. Retrieved from http://www.ncbi.nlm.nih.gov/pubmed/3293038
22. Huskinson SL, Naylor JE, Rowlett JK, Freeman KB (2014) Predicting abuse potential of stimulants and other dopaminergic drugs: overview and recommendations. Neuropharmacology 87:66–80. doi:10.1016/j.neuropharm.2014.03.009
23. Substance Abuse and Mental Health Services Administration (2014) Results from the 2013 National Survey on Drug Use and Health: Summary of National Findings (NSDUH Series H-48, HHS Publication No. (SMA) 14-4863). Substance Abuse and Mental Health Services Administration, Rockville
24. Fleckenstein AE, Gibb JW, Hanson GR (2000) Differential effects of stimulants on monoaminergic transporters: pharmacological consequences and implications for neurotoxicity. Eur J Pharmacol 406(1):1–13. Retrieved from http://www.ncbi.nlm.nih.gov/pubmed/11011026
25. Johanson CE, Fischman MW (1989) The pharmacology of cocaine related to its abuse. Pharmacol Rev 41(1):3–52. Retrieved from http://www.ncbi.nlm.nih.gov/pubmed/2682679
26. Rothman RB, Glowa JR (1995) A review of the effects of dopaminergic agents on humans, animals, and drug-seeking behavior, and its implications for medication development. Focus on GBR 12909. Mol Neurobiol 11(1–3):1–19. doi:10.1007/BF02740680
27. Seiden LS, Sabol KE, Ricaurte GA (1993) Amphetamine: effects on catecholamine systems and behavior. Annu Rev Pharmacol Toxicol 33:639–677. doi:10.1146/annurev.pa.33.040193.003231
28. Rothman RB, Baumann MH, Dersch CM, Romero DV, Rice KC, Carroll FI, Partilla JS (2001) Amphetamine-type central nervous system stimulants release norepinephrine more potently than they release dopamine and serotonin. Synapse 39(1):32–41. doi:10.1002/1098-2396(20010101)39:1<32::AID-SYN5>3.0.CO;2-3
29. Heal DJ, Gosden J, Smith SL (2014) Dopamine reuptake transporter (DAT) "inverse agonism" – a novel hypothesis to explain the enigmatic pharmacology of cocaine. Neuropharmacology 87:19–40. doi:10.1016/j.neuropharm.2014.06.012
30. Barrett RL, Appel JB (1989) Effects of stimulation and blockade of dopamine receptor subtypes on the discriminative stimulus properties of cocaine. Psychopharmacology (Berl) 99(1):13–16. Retrieved from http://www.ncbi.nlm.nih.gov/pubmed/2528777
31. Callahan PM, Appel JB, Cunningham KA (1991) Dopamine D1 and D2 mediation of the discriminative stimulus properties of d-amphetamine and cocaine. Psychopharmacology (Berl) 103(1):50–55. Retrieved from http://www.ncbi.nlm.nih.gov/pubmed/2006243
32. Callahan PM, Bryan SK, Cunningham KA (1995) Discriminative stimulus effects of cocaine: antagonism by dopamine D1 receptor blockade in the amygdala. Pharmacol Biochem Behav 51(4):759–766. Retrieved from http://www.ncbi.nlm.nih.gov/pubmed/7675856
33. Callahan PM, Cunningham KA (1995) Modulation of the discriminative stimulus properties of cocaine by 5-HT1B and 5-HT2C receptors. J Pharmacol Exp Ther 274(3):1414–1424. Retrieved from http://www.ncbi.nlm.nih.gov/pubmed/7562516
34. Colpaert FC, Niemegeers CJ, Janssen PA (1979) Discriminative stimulus properties of cocaine: neuropharmacological characteristics as derived from stimulus generalization experiments. Pharmacol Biochem Behav 10(4):535–546. Retrieved from http://www.ncbi.nlm.nih.gov/pubmed/37526
35. Johanson CE, Barrett JE (1993) The discriminative stimulus effects of cocaine in pigeons. J Pharmacol Exp Ther 267(1):1–8. Retrieved from http://www.ncbi.nlm.nih.gov/pubmed/8229735

36. Spealman RD (1995) Noradrenergic involvement in the discriminative stimulus effects of cocaine in squirrel monkeys. J Pharmacol Exp Ther 275(1):53–62. Retrieved from http://www.ncbi.nlm.nih.gov/pubmed/7562595

37. Spealman RD, Bergman J, Madras BK, Melia KF (1991) Discriminative stimulus effects of cocaine in squirrel monkeys: involvement of dopamine receptor subtypes. J Pharmacol Exp Ther 258(3):945–953. Retrieved from http://www.ncbi.nlm.nih.gov/pubmed/1679852

38. Terry P, Witkin JM, Katz JL (1994) Pharmacological characterization of the novel discriminative stimulus effects of a low dose of cocaine. J Pharmacol Exp Ther 270(3):1041–1048. Retrieved from http://www.ncbi.nlm.nih.gov/pubmed/7932151

39. Chait LD, Uhlenhuth EH, Johanson CE (1986) The discriminative stimulus and subjective effects of d-amphetamine, phenmetrazine and fenfluramine in humans. Psychopharmacology (Berl) 89(3):301–306. Retrieved from http://www.ncbi.nlm.nih.gov/pubmed/3088654

40. Cauli O, Pinna A, Valentini V, Morelli M (2003) Subchronic caffeine exposure induces sensitization to caffeine and cross-sensitization to amphetamine ipsilateral turning behavior independent from dopamine release. Neuropsychopharmacology 28(10):1752–1759. doi:10.1038/sj.npp.1300240

41. Garrett BE, Griffiths RR (1997) The role of dopamine in the behavioral effects of caffeine in animals and humans. Pharmacol Biochem Behav 57(3):533–541. Retrieved from http://www.ncbi.nlm.nih.gov/pubmed/9218278

42. Chait LD, Johanson CE (1988) Discriminative stimulus effects of caffeine and benzphetamine in amphetamine-trained volunteers. Psychopharmacology (Berl) 96(3):302–308. Retrieved from http://www.ncbi.nlm.nih.gov/pubmed/3146764

43. Chait LD, Uhlenhuth EH, Johanson CE (1984) An experimental paradigm for studying the discriminative stimulus properties of drugs in humans. Psychopharmacology (Berl) 82(3):272–274. Retrieved from http://www.ncbi.nlm.nih.gov/pubmed/6425913

44. Chait LD, Uhlenhuth EH, Johanson CE (1985) The discriminative stimulus and subjective effects of d-amphetamine in humans. Psychopharmacology (Berl) 86(3):307–312. Retrieved from http://www.ncbi.nlm.nih.gov/pubmed/3929301

45. Chait LD, Uhlenhuth EH, Johanson CE (1986) The discriminative stimulus and subjective effects of phenylpropanolamine, mazindol and d-amphetamine in humans. Pharmacol Biochem Behav 24(6):1665–1672. Retrieved from http://www.ncbi.nlm.nih.gov/pubmed/3737634

46. Heishman SJ, Henningfield JE (1991) Discriminative stimulus effects of d-amphetamine, methylphenidate, and diazepam in humans. Psychopharmacology (Berl) 103(4):436–442. Retrieved from http://www.ncbi.nlm.nih.gov/pubmed/2062984

47. Kollins SH, Rush CR (1999). Effects of training dose on the relationship between discriminative-stimulus and self-reported drug effects of d-amphetamine in humans. Pharmacol Biochem Behav 64(2):319–326. Retrieved from http://www.ncbi.nlm.nih.gov/pubmed/10515308

48. Lamb RJ, Henningfield JE (1994) Human d-amphetamine drug discrimination: methamphetamine and hydromorphone. J Exp Anal Behav 61(2):169–180. doi:10.1901/jeab.1994.61-169

49. Rush CR, Kollins SH, Pazzaglia PJ (1998) Discriminative-stimulus and participant-rated effects of methylphenidate, bupropion, and triazolam in d-amphetamine-trained humans. Exp Clin Psychopharmacol 6(1):32–44. Retrieved from http://www.ncbi.nlm.nih.gov/pubmed/9526144

50. Rush CR, Stoops WW, Hays LR, Glaser PE, Hays LS (2003) Risperidone attenuates the discriminative-stimulus effects of d-amphetamine in humans. J Pharmacol Exp Ther 306(1):195–204. doi:10.1124/jpet.102.048439

51. Akaoka H, Roussel B, Lin JS, Chouvet G, Jouvet M (1991) Effect of modafinil and amphetamine on the rat catecholaminergic neuron activity. Neurosci Lett 123(1):20–22. Retrieved from http://www.ncbi.nlm.nih.gov/pubmed/1676498

52. Ferraro L, Antonelli T, O'Connor WT, Tanganelli S, Rambert FA, Fuxe K (1997) Modafinil: an antinarcoleptic drug with a different neurochemical profile to d-amphetamine and

dopamine uptake blockers. Biol Psychiatry 42(12):1181–1183. Retrieved from http://www.ncbi.nlm.nih.gov/pubmed/9426889

53. Desai RI, Paronis CA, Martin J, Desai R, Bergman J (2010) Monoaminergic psychomotor stimulants: discriminative stimulus effects and dopamine efflux. J Pharmacol Exp Ther 333 (3):834–843. doi:10.1124/jpet.110.165746

54. Baumann MH, Ayestas MA, Sharpe LG, Lewis DB, Rice KC, Rothman RB (2002) Persistent antagonism of methamphetamine-induced dopamine release in rats pretreated with GBR12909 decanoate. J Pharmacol Exp Ther 301(3):1190–1197. Retrieved from http://www.ncbi.nlm.nih.gov/pubmed/12023554

55. Howell LL, Kimmel HL (2008) Monoamine transporters and psychostimulant addiction. Biochem Pharmacol 75(1):196–217. doi:10.1016/j.bcp.2007.08.003

56. Baker LE, Riddle EE, Saunders RB, Appel JB (1993) The role of monoamine uptake in the discriminative stimulus effects of cocaine and related compounds. Behav Pharmacol 4 (1):69–79. Retrieved from http://www.ncbi.nlm.nih.gov/pubmed/11224173

57. Broadbent J, Michael EK, Riddle EE, Apple JB (1991) Involvement of dopamine uptake in the discriminative stimulus effects of cocaine. Behav Pharmacol 2(3):187–197. Retrieved from http://www.ncbi.nlm.nih.gov/pubmed/11224062

58. Cunningham KA, Callahan PM (1991) Monoamine reuptake inhibitors enhance the discriminative state induced by cocaine in the rat. Psychopharmacology (Berl) 104(2):177–180. Retrieved from http://www.ncbi.nlm.nih.gov/pubmed/1831559

59. Callahan PM, Cunningham KA (1993) Discriminative stimulus properties of cocaine in relation to dopamine D2 receptor function in rats. J Pharmacol Exp Ther 266(2):585–592. Retrieved from http://www.ncbi.nlm.nih.gov/pubmed/8355192

60. Stoops WW, Glaser PE, Rush CR (2009) Discriminative-stimulus effects of d-amphetamine following pretreatment with fluphenazine. Unpublished data

61. Burris KD, Molski TF, Xu C, Ryan E, Tottori K, Kikuchi T, Yocca FD, Molinoff PB (2002) Aripiprazole, a novel antipsychotic, is a high-affinity partial agonist at human dopamine D2 receptors. J Pharmacol Exp Ther 302(1):381–389. Retrieved from http://www.ncbi.nlm.nih.gov/pubmed/12065741

62. Shapiro DA, Renock S, Arrington E, Chiodo LA, Liu LX, Sibley DR, Roth BL, Mailman R (2003) Aripiprazole, a novel atypical antipsychotic drug with a unique and robust pharmacology. Neuropsychopharmacology 28(8):1400–1411. doi:10.1038/sj.npp.1300203

63. Lile JA, Stoops WW, Vansickel AR, Glaser PE, Hays LR, Rush CR (2005) Aripiprazole attenuates the discriminative-stimulus and subject-rated effects of D-amphetamine in humans. Neuropsychopharmacology 30(11):2103–2114. doi:10.1038/sj.npp.1300803

64. Exner M, Clark D (1992) Agonist and antagonist activity of low efficacy D2 dopamine receptor agonists in rats discriminating d-amphetamine from saline. Behav Pharmacol 3 (6):609–619. Retrieved from http://www.ncbi.nlm.nih.gov/pubmed/11224162

65. Rush CR, Stoops WW, Wagner FP, Hays LR, Glaser PE (2004) Alprazolam attenuates the behavioral effects of d-amphetamine in humans. J Clin Psychopharmacol 24(4):410–420. Retrieved from http://www.ncbi.nlm.nih.gov/pubmed/15232333

66. Sevak RJ, Vansickel AR, Stoops WW, Glaser PE, Hays LR, Rush CR (2011) Discriminative-stimulus, subject-rated, and physiological effects of methamphetamine in humans pretreated with aripiprazole. J Clin Psychopharmacol 31(4):470–480. doi:10.1097/JCP.0b013e318221b2db

67. Vansickel AR, Stoops WW, Glaser PE, Rush CR (2009) Discriminative-stimulus effects of methamphetamine following pretreatment with d-amphetamine. Unpublished data

68. Vansickel AR, Stoops WW, Glaser PE, Rush CR (2009) Discriminative-stimulus effects of methamphetamine following pretreatment with bupropion. Unpublished data

69. Mechanic JA, Wasielewski JA, Carl KL, Holloway FA (2002) Attenuation of the amphetamine discriminative cue in rats with the atypical antipsychotic olanzapine. Pharmacol Biochem Behav 72(4):767–777. Retrieved from http://www.ncbi.nlm.nih.gov/pubmed/12062565

70. Tidey JW, Bergman J (1998) Drug discrimination in methamphetamine-trained monkeys: agonist and antagonist effects of dopaminergic drugs. J Pharmacol Exp Ther 285 (3):1163–1174. Retrieved from http://www.ncbi.nlm.nih.gov/pubmed/9618419
71. Czoty PW, Ramanathan CR, Mutschler NH, Makriyannis A, Bergman J (2004) Drug discrimination in methamphetamine-trained monkeys: effects of monoamine transporter inhibitors. J Pharmacol Exp Ther 311(2):720–727. doi:10.1124/jpet.104.071035
72. Dewey SL, Chaurasia CS, Chen CE, Volkow ND, Clarkson FA, Porter SP, Straughter-Moore RM, Alexoff DL, Tedeschi D, Russo NB, Fowler JS, Brodie JD (1997) GABAergic attenuation of cocaine-induced dopamine release and locomotor activity. Synapse 25(4):393–398. doi:10.1002/(SICI)1098-2396(199704)25:4<393::AID-SYN11>3.0.CO;2-W
73. Kalivas PW, Duffy P, Eberhardt H (1990) Modulation of A10 dopamine neurons by gamma-aminobutyric acid agonists. J Pharmacol Exp Ther 253(2):858–866. Retrieved from http://www.ncbi.nlm.nih.gov/pubmed/2160011
74. Kita H, Kitai ST (1988) Glutamate decarboxylase immunoreactive neurons in rat neostriatum: their morphological types and populations. Brain Res 447(2):346–352. Retrieved from http://www.ncbi.nlm.nih.gov/pubmed/3390703
75. Zetterstrom T, Fillenz M (1990) Local administration of flurazepam has different effects on dopamine release in striatum and nucleus accumbens: a microdialysis study. Neuropharmacology 29(2):129–134. Retrieved from http://www.ncbi.nlm.nih.gov/pubmed/2109839
76. Druhan JP, Fibiger HC, Phillips AG (1991) Influence of some drugs of abuse on the discriminative stimulus properties of amphetamine. Behav Pharmacol 2(4 and 5):391–403. Retrieved from http://www.ncbi.nlm.nih.gov/pubmed/11224082
77. Rush CR, Vansickel AR, Stoops WW (2011) Human drug discrimination: methodological considerations and application to elucidating the neuropharmacology of amphetamines. In: Glennon RA, Young R (eds) Drug discrimination: applications to medicinal chemistry and drug studies. Wiley, Hoboken
78. Klee H, Wright S, Carnwath T, Merrill J (2001) The role of substitute therapy in the treatment of problem amphetamine use. Drug Alcohol Rev 20:417–429
79. Shearer J, Sherman J, Wodak A, van Beek I (2002) Substitution therapy for amphetamine users. Drug Alcohol Rev 21(2):179–185. doi:10.1080/09595230220139082
80. Shearer J, Wodak A, Mattick RP, Van Beek I, Lewis J, Hall W, Dolan K (2001) Pilot randomized controlled study of dexamphetamine substitution for amphetamine dependence. Addiction 96(9):1289–1296. doi:10.1080/09652140120070346
81. Tiihonen J, Kuoppasalmi K, Fohr J, Tuomola P, Kuikanmaki O, Vorma H, Sokero P, Haukka J, Meririnne E (2007) A comparison of aripiprazole, methylphenidate, and placebo for amphetamine dependence. Am J Psychiatry 164(1):160–162. doi:10.1176/ajp.2007.164.1.160
82. de Wit H, Stewart J (1981) Reinstatement of cocaine-reinforced responding in the rat. Psychopharmacology (Berl) 75(2):134–143. Retrieved from http://www.ncbi.nlm.nih.gov/pubmed/6798603
83. Stoops WW, Rush CR (2014) Combination pharmacotherapies for stimulant use disorder: a review of clinical findings and recommendations for future research. Expert Rev Clin Pharmacol 7(3):363–374. doi:10.1586/17512433.2014.909283
84. Janecka A, Fichna J, Janecki T (2004) Opioid receptors and their ligands. Curr Top Med Chem 4(1):1–17. Retrieved from http://www.ncbi.nlm.nih.gov/pubmed/14754373
85. Waldhoer M, Bartlett SE, Whistler JL (2004) Opioid receptors. Annu Rev Biochem 73:953–990. doi:10.1146/annurev.biochem.73.011303.073940
86. Borg L, Kreek MJ (1998) Pharmacology of opiates. In: Tarter RE, Ammerman RO, Peggy J (eds) Handbook of substance abuse. Springer, New York, pp 331–341
87. Mello NK, Mendelson JH, Bree MP (1981) Naltrexone effects on morphine and food self-administration in morphine-dependent rhesus monkeys. J Pharmacol Exp Ther 218 (2):550–557. Retrieved from http://www.ncbi.nlm.nih.gov/pubmed/7195937

88. Sullivan MA, Vosburg SK, Comer SD (2006) Depot naltrexone: antagonism of the reinforcing, subjective, and physiological effects of heroin. Psychopharmacology (Berl) 189(1):37–46. doi:10.1007/s00213-006-0509-x

89. Walsh SL, Sullivan JT, Preston KL, Garner JE, Bigelow GE (1996) Effects of naltrexone on response to intravenous cocaine, hydromorphone and their combination in humans. J Pharmacol Exp Ther 279(2):524–538. Retrieved from http://www.ncbi.nlm.nih.gov/pubmed/8930154

90. Baldauf K, Braun K, Gruss M (2005) Opiate modulation of monoamines in the chick forebrain: possible role in emotional regulation? J Neurobiol 62(2):149–163. doi:10.1002/neu.20076

91. Chefer VI, Denoroy L, Zapata A, Shippenberg TS (2009) Mu opioid receptor modulation of somatodendritic dopamine overflow: GABAergic and glutamatergic mechanisms. Eur J Neurosci 30(2):272–278. doi:10.1111/j.1460-9568.2009.06827.x

92. Vaughan CW, Ingram SL, Connor MA, Christie MJ (1997) How opioids inhibit GABA-mediated neurotransmission. Nature 390(6660):611–614. doi:10.1038/37610

93. Land BB, Bruchas MR, Lemos JC, Xu M, Melief EJ, Chavkin C (2008) The dysphoric component of stress is encoded by activation of the dynorphin kappa-opioid system. J Neurosci 28(2):407–414. doi:10.1523/JNEUROSCI.4458-07.2008

94. Varga EV, Navratilova E, Stropova D, Jambrosic J, Roeske WR, Yamamura HI (2004) Agonist-specific regulation of the delta-opioid receptor. Life Sci 76(6):599–612. doi:10.1016/j.lfs.2004.07.020

95. Bickel WK, Bigelow GE, Preston KL, Liebson IA (1989) Opioid drug discrimination in humans: stability, specificity and relation to self-reported drug effect. J Pharmacol Exp Ther 251(3):1053–1063. Retrieved from http://www.ncbi.nlm.nih.gov/pubmed/2481029

96. Duke AN, Bigelow GE, Lanier RK, Strain EC (2011) Discriminative stimulus effects of tramadol in humans. J Pharmacol Exp Ther 338(1):255–262. doi:10.1124/jpet.111.181131

97. Jones HE, Bigelow GE, Preston KL (1999) Assessment of opioid partial agonist activity with a three-choice hydromorphone dose-discrimination procedure. J Pharmacol Exp Ther 289(3):1350–1361. Retrieved from http://www.ncbi.nlm.nih.gov/pubmed/10336526

98. Preston KL, Bigelow GE (1994) Drug discrimination assessment of agonist–antagonist opioids in humans: a three-choice saline-hydromorphone-butorphanol procedure. J Pharmacol Exp Ther 271(1):48–60. Retrieved from http://www.ncbi.nlm.nih.gov/pubmed/7525929

99. Preston KL, Bigelow GE (1998) Opioid discrimination in humans: discriminative and subjective effects of progressively lower training dose. Behav Pharmacol 9(7):533–543. Retrieved from http://www.ncbi.nlm.nih.gov/pubmed/9862079

100. Preston KL, Bigelow GE (2000) Effects of agonist–antagonist opioids in humans trained in a hydromorphone/not hydromorphone discrimination. J Pharmacol Exp Ther 295(1):114–124. Retrieved from http://www.ncbi.nlm.nih.gov/pubmed/10991968

101. Preston KL, Bigelow GE, Bickel W, Liebson IA (1987) Three-choice drug discrimination in opioid-dependent humans: hydromorphone, naloxone and saline. J Pharmacol Exp Ther 243(3):1002–1009. Retrieved from http://www.ncbi.nlm.nih.gov/pubmed/2447262

102. Preston KL, Bigelow GE, Liebson IA (1990) Discrimination of butorphanol and nalbuphine in opioid-dependent humans. Pharmacol Biochem Behav 37(3):511–522. Retrieved from http://www.ncbi.nlm.nih.gov/pubmed/1708145

103. Preston KL, Liebson IA, Bigelow GE (1992) Discrimination of agonist–antagonist opioids in humans trained on a two-choice saline-hydromorphone discrimination. J Pharmacol Exp Ther 261(1):62–71. Retrieved from http://www.ncbi.nlm.nih.gov/pubmed/1373189

104. Strickland JC, Rush CR, Stoops WW (2015) Mu opioid mediated discriminative-stimulus effects of tramadol: an individual subjects analysis. J Exp Anal Behav 103(2):361–374. doi:10.1002/jeab.137

105. Morgan D, Picker MJ (1998) The mu opioid irreversible antagonist beta-funaltrexamine differentiates the discriminative stimulus effects of opioids with high and low efficacy at the

mu opioid receptor. Psychopharmacology (Berl) 140(1):20–28. Retrieved from http://www.ncbi.nlm.nih.gov/pubmed/9862398

106. Picker MJ, Yarbrough J, Hughes CE, Smith MA, Morgan D, Dykstra LA (1993) Agonist and antagonist effects of mixed action opioids in the pigeon drug discrimination procedure: influence of training dose, intrinsic efficacy and interanimal differences. J Pharmacol Exp Ther 266(2):756–767. Retrieved from http://www.ncbi.nlm.nih.gov/pubmed/8394915

107. Beardsley PM, Aceto MD, Cook CD, Bowman ER, Newman JL, Harris LS (2004) Discriminative stimulus, reinforcing, physical dependence, and antinociceptive effects of oxycodone in mice, rats, and rhesus monkeys. Exp Clin Psychopharmacol 12(3):163–172. doi:10.1037/1064-1297.12.3.163

108. Morgan D, Cook CD, Picker MJ (1999) Sensitivity to the discriminative stimulus and antinociceptive effects of mu opioids: role of strain of rat, stimulus intensity, and intrinsic efficacy at the mu opioid receptor. J Pharmacol Exp Ther 289(2):965–975. Retrieved from http://www.ncbi.nlm.nih.gov/pubmed/10215676

109. Shannon HE, Holtzman SG (1977) Further evaluation of the discriminative effects of morphine in the rat. J Pharmacol Exp Ther 201(1):55–66. Retrieved from http://www.ncbi.nlm.nih.gov/pubmed/15104

110. Shannon HE, Holtzman SG (1977) Discriminative effects of morphine administered intracerebrally in the rat. Life Sci 21(4):585–594. Retrieved from http://www.ncbi.nlm.nih.gov/pubmed/904437

111. Shannon HE, Holtzman SG (1979) Morphine training dose: a determinant of stimulus generalization to narcotic antagonists in the rat. Psychopharmacology (Berl) 61 (3):239–244. Retrieved from http://www.ncbi.nlm.nih.gov/pubmed/156379

112. Platt DM, Rowlett JK, Izenwasser S, Spealman RD (2004) Opioid partial agonist effects of 3-O-methylnaltrexone in rhesus monkeys. J Pharmacol Exp Ther 308(3):1030–1039. doi:10.1124/jpet.103.060962

113. Platt DM, Rowlett JK, Spealman RD (2001) Discriminative stimulus effects of intravenous heroin and its metabolites in rhesus monkeys: opioid and dopaminergic mechanisms. J Pharmacol Exp Ther 299(2):760–767. Retrieved from http://www.ncbi.nlm.nih.gov/pubmed/11602692

114. Oliveto AH, Rosen MI, Kosten TA, Hameedi FA, Woods SW, Kosten TR (1998) Hydromorphone-naloxone combinations in opioid-dependent humans under a naloxone novel-response discrimination procedure. Exp Clin Psychopharmacol 6(2):169–178. Retrieved from http://www.ncbi.nlm.nih.gov/pubmed/9608349

115. France CP, Jacobson AE, Woods JH (1984) Discriminative stimulus effects of reversible and irreversible opiate agonists: morphine, oxymorphazone and buprenorphine. J Pharmacol Exp Ther 230(3):652–657. Retrieved from http://www.ncbi.nlm.nih.gov/pubmed/6206224

116. Bowen CA, Fischer BD, Mello NK, Negus SS (2002) Antagonism of the antinociceptive and discriminative stimulus effects of heroin and morphine by 3-methoxynaltrexone and naltrexone in rhesus monkeys. J Pharmacol Exp Ther 302(1):264–273. Retrieved from http://www.ncbi.nlm.nih.gov/pubmed/12065726

117. Bailey CP, Connor M (2005) Opioids: cellular mechanisms of tolerance and physical dependence. Curr Opin Pharmacol 5(1):60–68. doi:10.1016/j.coph.2004.08.012

118. Ashton CH (2001) Pharmacology and effects of cannabis: a brief review. Br J Psychiatry 178:101–106. Retrieved from http://www.ncbi.nlm.nih.gov/pubmed/11157422

119. Matsuda LA, Lolait SJ, Brownstein MJ, Young AC, Bonner TI (1990) Structure of a cannabinoid receptor and functional expression of the cloned cDNA. Nature 346 (6284):561–564. doi:10.1038/346561a0

120. Munro S, Thomas KL, Abu-Shaar M (1993) Molecular characterization of a peripheral receptor for cannabinoids. Nature 365(6441):61–65. doi:10.1038/365061a0

121. Onaivi ES (2006) Neuropsychobiological evidence for the functional presence and expression of cannabinoid CB2 receptors in the brain. Neuropsychobiology 54(4):231–246. doi:10.1159/000100778

122. Pertwee RG (1997) Pharmacology of cannabinoid CB1 and CB2 receptors. Pharmacol Ther 74(2):129–180. Retrieved from http://www.ncbi.nlm.nih.gov/pubmed/9336020
123. Pertwee RG (2006) The pharmacology of cannabinoid receptors and their ligands: an overview. Int J Obes (Lond) 30(Suppl 1): S13–S18. doi:10.1038/sj.ijo.0803272
124. Compton DR, Rice KC, De Costa BR, Razdan RK, Melvin LS, Johnson MR, Martin BR (1993) Cannabinoid structure-activity relationships: correlation of receptor binding and in vivo activities. J Pharmacol Exp Ther 265(1):218–226. Retrieved from http://www.ncbi. nlm.nih.gov/pubmed/8474008
125. Breivogel CS, Childers SR (1998) The functional neuroanatomy of brain cannabinoid receptors. Neurobiol Dis 5(6 Pt B):417–431. doi:10.1006/nbdi.1998.0229
126. Järbe TU, Liu Q, Makriyannis A (2006) Antagonism of discriminative stimulus effects of delta(9)-THC and (R)-methanandamide in rats. Psychopharmacology (Berl) 184(1):36–45. doi:10.1007/s00213-005-0225-y
127. McMahon LR (2006) Discriminative stimulus effects of the cannabinoid CB1 antagonist SR 141716A in rhesus monkeys pretreated with Delta9-tetrahydrocannabinol. Psychopharmacology (Berl) 188(3):306–314. doi:10.1007/s00213-006-0500-6
128. Valenzano KJ, Tafesse L, Lee G, Harrison JE, Boulet JM, Gottshall SL, Mark L, Pearson MS, Miller W, Shan S, Rabadi L, Rotshteyn Y, Chaffer SM, Turchin PI, Elsemore DA, Toth M, Koetzner L, Whiteside GT (2005) Pharmacological and pharmacokinetic characterization of the cannabinoid receptor 2 agonist, GW405833, utilizing rodent models of acute and chronic pain, anxiety, ataxia and catalepsy. Neuropharmacology 48(5):658–672. doi:10.1016/j. neuropharm.2004.12.008
129. Compton DR, Aceto MD, Lowe J, Martin BR (1996) In vivo characterization of a specific cannabinoid receptor antagonist (SR141716A): inhibition of delta 9-tetrahydrocannabinol-induced responses and apparent agonist activity. J Pharmacol Exp Ther 277(2):586–594. Retrieved from http://www.ncbi.nlm.nih.gov/pubmed/8627535
130. Huestis MA, Gorelick DA, Heishman SJ, Preston KL, Nelson RA, Moolchan ET, Frank RA (2001) Blockade of effects of smoked marijuana by the CB1-selective cannabinoid receptor antagonist SR141716. Arch Gen Psychiatry 58(4):322–328. Retrieved from http://www.ncbi. nlm.nih.gov/pubmed/11296091
131. Järbe TU, Lamb RJ, Liu Q, Makriyannis A (2006) Discriminative stimulus functions of AM-1346, a CB1R selective anandamide analog in rats trained with Delta9-THC or (R)-methanandamide (AM-356). Psychopharmacology (Berl) 188(3):315–323. doi:10.1007/ s00213-006-0517-x
132. Zuurman L, Roy C, Schoemaker RC, Amatsaleh A, Guimaeres L, Pinquier JL, Cohen AF, van Gerven JM (2010) Inhibition of THC-induced effects on the central nervous system and heart rate by a novel CB1 receptor antagonist AVE1625. J Psychopharmacol 24(3):363–371. doi:10.1177/0269881108096509
133. Szabo B, Schlicker E (2005) Effects of cannabinoids on neurotransmission. Handb Exp Pharmacol (168):327–365. Retrieved from http://www.ncbi.nlm.nih.gov/pubmed/16596780
134. Chait LD, Evans SM, Grant KA, Kamien JB, Johanson CE, Schuster CR (1988) Discriminative stimulus and subjective effects of smoked marijuana in humans. Psychopharmacology (Berl) 94(2) 206–212. Retrieved from http://www.ncbi.nlm.nih.gov/pubmed/3127846
135. Lile JA, Kelly TH, Hays LR (2010) Substitution profile of the cannabinoid agonist nabilone in human subjects discriminating delta9-tetrahydrocannabinol. Clin Neuropharmacol 33 (5):235–242. doi:10.1097/WNF.0b013e3181e77428
136. Lile JA, Kelly TH, Hays LR (2011) Separate and combined effects of the cannabinoid agonists nabilone and Delta(9)-THC in humans discriminating Delta(9)-THC. Drug Alcohol Depend 116(1–3):86–92. doi:10.1016/j.drugalcdep.2010.11.019
137. Lile JA, Kelly TH, Hays LR (2012) Separate and combined effects of the GABA reuptake inhibitor tiagabine and Delta9-THC in humans discriminating Delta9-THC. Drug Alcohol Depend 122(1–2):61–69. doi:10.1016/j.drugalcdep.2011.09.010

138. Lile JA, Kelly TH, Hays LR (2012) Separate and combined effects of the GABA(B) agonist baclofen and Delta9-THC in humans discriminating Delta9-THC. Drug Alcohol Depend 126 (1–2):216–223. doi:10.1016/j.drugalcdep.2012.05.023

139. Lile JA, Kelly TH, Hays LR (2014) Separate and combined effects of the GABAA positive allosteric modulator diazepam and Delta(9)-THC in humans discriminating Delta(9)-THC. Drug Alcohol Depend 143:141–148. doi:10.1016/j.drugalcdep.2014.07.016

140. Lile JA, Kelly TH, Pinsky DJ, Hays LR (2009) Substitution profile of Delta9-tetrahydrocannabinol, triazolam, hydromorphone, and methylphenidate in humans discriminating Delta9-tetrahydrocannabinol. Psychopharmacology (Berl) 203(2):241–250. doi:10.1007/s00213-008-1393-3

141. Lile JA, Wesley MJ, Kelly TH, Hays LR (2015) Separate and combined effects of gabapentin and Delta(9)-tetrahydrocananbinol in humans discriminating Delta(9)-tetrahydrocananbinol. Behav Pharmacol. doi:10.1097/FBP.0000000000000187

142. De Vry J, Jentzsch KR (2003) Intrinsic activity estimation of cannabinoid CB1 receptor ligands in a drug discrimination paradigm. Behav Pharmacol 14(5–6):471–476. doi:10.1097/01.fbp.0000087739.21047.d8

143. Järbe TU, Li C, Vadivel SK, Makriyannis A (2010) Discriminative stimulus functions of methanandamide and delta(9)-THC in rats: tests with aminoalkylindoles (WIN55,212-2 and AM678) and ethanol. Psychopharmacology (Berl) 208(1):87–98. doi:10.1007/s00213-009-1708-z

144. Järbe TU, Tai S, LeMay BJ, Nikas SP, Shukla VG, Zvonok A, Makriyannis A (2012) AM2389, a high-affinity, in vivo potent CB1-receptor-selective cannabinergic ligand as evidenced by drug discrimination in rats and hypothermia testing in mice. Psychopharmacology (Berl) 220(2):417–426. doi:10.1007/s00213-011-2491-1

145. Browne RG, Weissman A (1981) Discriminative stimulus properties of delta 9-tetrahydrocannabinol: mechanistic studies. J Clin Pharmacol 21(8–9 Suppl):227S–234S. Retrieved from http://www.ncbi.nlm.nih.gov/pubmed/6271828

146. Järbe TU, Hiltunen AJ (1988) Limited stimulus generalization between delta 9-THC and diazepam in pigeons and gerbils. Psychopharmacology (Berl) 94(3):328–331. Retrieved from http://www.ncbi.nlm.nih.gov/pubmed/2833760

147. Solinas M, Goldberg SR (2005) Involvement of mu-, delta- and kappa-opioid receptor subtypes in the discriminative-stimulus effects of delta-9-tetrahydrocannabinol (THC) in rats. Psychopharmacology (Berl) 179(4):804–812. doi:10.1007/s00213-004-2118-x

148. Solinas M, Zangen A, Thiriet N, Goldberg SR (2004) Beta-endorphin elevations in the ventral tegmental area regulate the discriminative effects of Delta-9-tetrahydrocannabinol. Eur J Neurosci 19(12):3183–3192. doi:10.1111/j.0953-816X.2004.03420.x

149. Wiley JL, Huffman JW, Balster RL, Martin BR (1995) Pharmacological specificity of the discriminative stimulus effects of delta 9-tetrahydrocannabinol in rhesus monkeys. Drug Alcohol Depend 40(1):81–86. Retrieved from http://www.ncbi.nlm.nih.gov/pubmed/8746928

150. Bueno OF, Carlini EA, Finkelfarb E, Suzuki JS (1976) Delta 9-Tetrahydrocannabinol, ethanol, and amphetamine as discriminative stimuli-generalization tests with other drugs. Psychopharmacologia 46(3):235–243. Retrieved from http://www.ncbi.nlm.nih.gov/pubmed/951459

151. Barrett RL, Wiley JL, Balster RL, Martin BR (1995) Pharmacological specificity of delta 9-tetrahydrocannabinol discrimination in rats. Psychopharmacology (Berl) 118(4), 419–424. Retrieved from http://www.ncbi.nlm.nih.gov/pubmed/7568628

152. Mokler DJ, Nelson BD, Harris LS, Rosecrans JA (1986) The role of benzodiazepine receptors in the discriminative stimulus properties of delta-9-tetrahydrocannabinol. Life Sci 38 (17):1581–1589. Retrieved from http://www.ncbi.nlm.nih.gov/pubmed/3010019

153. Wiley JL, Martin BR (1999) Effects of SR141716A on diazepam substitution for delta9-tetrahydrocannabinol in rat drug discrimination. Pharmacol Biochem Behav 64(3):519–522. Retrieved from http://www.ncbi.nlm.nih.gov/pubmed/10548265

154. Järbe TU, Gifford RS, Makriyannis A (2010) Antagonism of (9)-THC induced behavioral effects by rimonabant: time course studies in rats. Eur J Pharmacol 648(1–3):133–138. doi:10.1016/j.ejphar.2010.09.006

155. Järbe TU, LeMay BJ, Halikhedkar A, Wood J, Vadivel SK, Zvonok A, Makriyannis A (2014) Differentiation between low- and high-efficacy CB1 receptor agonists using a drug discrimination protocol for rats. Psychopharmacology (Berl) 231(3):489–500. doi:10.1007/s00213-013-3257-8

156. Wiley JL, Breivogel CS, Mahadevan A, Pertwee RG, Cascio MG, Bolognini D, Huffman JW, Walentiny DM, Vann RE, Razdan RK, Martin BR (2011) Structural and pharmacological analysis of O-2050, a putative neutral cannabinoid CB(1) receptor antagonist. Eur J Pharmacol 651(1–3):96–105. doi:10.1016/j.ejphar.2010.10.085

157. Mason BJ, Crean R, Goodell V, Light JM, Quello S, Shadan F, Buffkins K, Kyle M, Adusumalli M, Begovic A, Rao S (2012) A proof-of-concept randomized controlled study of gabapentin: effects on cannabis use, withdrawal and executive function deficits in cannabis-dependent adults. Neuropsychopharmacology 37(7):1689–1698. doi:10.1038/npp.2012.14

Conditioned Taste Avoidance Drug Discrimination Procedure: Assessments and Applications

Anthony L. Riley, Matthew M. Clasen, and Mary A. Friar

Abstract In the present chapter, we summarize much of the work on the taste avoidance drug discrimination procedure, presenting the logic for its initial introduction and the extension of the procedure in the investigation of the discriminative properties of various drugs. Results from these assessments parallel those from more traditional operant and maze designs in classifying and characterizing the discriminative properties of drug. At the same time, this design reveals a procedure that is sensitive in such assessments by indexing these stimulus properties more rapidly and at lower doses than in the more traditional procedures (in some cases for drugs heretofore resistant in their detection). Importantly, much remains to be learned about the taste avoidance procedure in that the nature of such learning remains unknown and the specific parameters under which it can be established and generalized and its neurochemical and neuroanatomical bases are largely unexplored. The application of drug discrimination learning to human drug abuse continues to be an important consideration for this specific design (as well as that of drug discrimination procedures in general), and recent parallels between drug use and food intake in terms of its regulation by interoceptive stimuli suggests a possible role of the loss of stimulus control in drug escalation and addiction (with possible therapeutic implications via the modulation of these interoceptive cues).

Keywords Conditioned taste avoidance • Drug discrimination learning • Drug classification and characterization • Drug use and abuse

A.L. Riley (✉), M.M. Clasen, and M.A. Friar
Psychopharmacology Laboratory, Center for Behavioral Neuroscience, American University, Washington, DC 20016, USA
e-mail: alriley@american.edu

© Springer International Publishing Switzerland 2016
Curr Topics Behav Neurosci (2018) 39: 297–318
DOI 10.1007/7854_2016_8
Published Online: 23 May 2016

Contents

1   Introduction

Although operant and maze designs have been used primarily in assessing the stimulus properties of drugs (see Overton and Porter, this volume), Riley and his colleagues (see [1]; see also [2–4]) described a relatively unique preparation that utilized classical conditioning in the assessment of a drug's subjective effects, specifically, conditioned taste avoidance learning (see below). The logic for the use of this procedure was based on early work by John Garcia and his colleagues on the tendency of rats to avoid consuming various foods that had been paired or associated with toxicosis. Specifically, Garcia et al. [5] reported that rats that had been irradiated following consumption of a novel saccharin solution decreased consumption of the solution on subsequent exposures. The avoidance was presumably based on the rat's association of the sweet saccharin solution with the aversive effects of the radiation. This avoidance was rapidly acquired (often in a single pairing; [5]), very robust (to the point of complete suppression of consumption; [6]), occurred despite long delays between consumption of the solution and the onset of radiation [7] and was relatively selective to taste, i.e., environmental cues paired with the radiation were not readily associated with its effects [7]. Garcia and his colleagues argued that these characteristics of taste avoidance learning reflected evolutionary pressures on mammals that contributed to their avoidance of poisoned foods (for reviews, see [8, 9]).

Although taste avoidance learning was initially discussed and assessed for its relatively unique characteristics (all of which challenged traditional learning theory; see [9, 10]), subsequent work with the preparation focused on the various conditions under which it was established and expressed, e.g., other drugs, species, tastes, temporal parameters (see [10]), its physiological, biochemical and molecular underpinnings (see [11]) and relatively recently its translational application to a variety of clinical and behavioral issues (for reviews, see [10, 12]). In relation to the latter point, taste avoidance learning was applied to the suppression of predation in wild animals, the control and treatment of alcohol abuse, the elimination or reduction of cancer-induced anorexia, the nature and control of immunosuppression, and the assessment of the biochemical mediation of learning and memory.

2 Taste Avoidance Drug Discrimination Procedure

Given the relatively unique ability of animals to rapidly and robustly associate taste with illness, using a drug to signal this pairing might allow the development of a rapid assay of the stimulus effects of drugs. Specifically, if an animal is given a drug prior to the pairing of a taste and some aversive agent and the drug vehicle prior to a presentation of saccharin alone, it might learn the signaling function of the drug. Under such conditions, the animal would come to avoid the taste when it was preceded by the drug (that signaled the taste-toxin pairing) and consume the same taste when it was preceded by an injection of the drug vehicle. That is, the animal would learn the drug discrimination within the taste avoidance preparation. The function of the drug in this preparation is functionally identical to that seen in the more traditional drug discrimination procedures, i.e., the drug signals some programmed contingency. Again, the difference is that the general taste avoidance design is so rapidly acquired and robust that the taste avoidance procedure may provide an efficient and effective assay of drug discrimination learning.

In this vein, my laboratory tested this prediction with the glutamate channel blocker phencyclidine (PCP; see [1]). In this assessment, adult female Long-Evans rats were adapted to a restricted water schedule (20-min day) until consumption stabilized. They were then given a novel saccharin solution for 3 habituation days during which a vehicle injection was given 10 min prior to each saccharin access. On the 1st drug day, subjects in Group PL were given an injection of PCP (1.8 mg/kg) 10 min prior to access to the saccharin solution (for 20 min). This saccharin access was then followed immediately by an injection of the toxin LiCl (1.8 mEq). For this group, P stood for the PCP pretreatment; L stood for administration of LiCl following saccharin consumption. Subjects in a second group (Group PW) were also injected with PCP prior to the 20-min saccharin access but they were injected with the LiCl vehicle immediately following its consumption. For this group, P again stood for the PCP pretreatment; W stood for administration of the LiCl vehicle following saccharin consumption. On the following 3 days, subjects in both groups were given the PCP vehicle prior to access to saccharin which was followed by the LiCl vehicle, i.e., 3 safe days. This alternating procedure was repeated until all subjects had received five complete cycles. Following the fifth cycle, the above-mentioned procedure was repeated except that on the 2nd recovery day of each cycle various doses of PCP, ketamine, and D-amphetamine were given prior to saccharin access to assess their ability to substitute for PCP. On these probe sessions, saccharin access was followed by the LiCl vehicle (see Fig. 1 for a schematic of the general procedures used in this design).

As expected from work with PCP in more traditional assessments of drug discrimination learning (see [13, 14]), the animals learned the PCP/vehicle discrimination. Specifically, subjects in Group PL avoided the saccharin solution when it was preceded by the injection of PCP and consumed the same saccharin solution when it was preceded by an injection of the PCP vehicle. Importantly, and consistent with our earlier predictions, the discrimination was acquired very rapidly. In

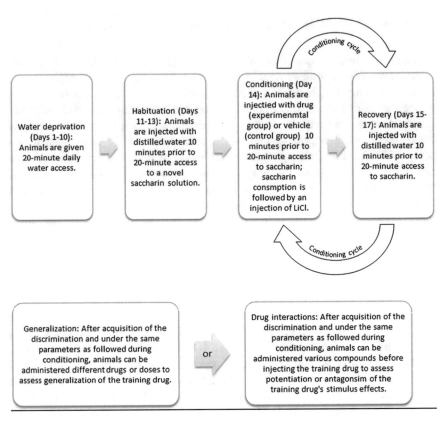

Fig. 1 Schematic diagram depicting the stages and timeline of the conditioned taste avoidance drug discrimination procedure (*top panel*). Once the discrimination has been acquired, slight modifications in the procedure allow for the assessment of drugs to substitute, potentiate or block the stimulus effects of the training drug by administering other compounds on test days given on the second recovery session of the conditioning cycle (*bottom panel*)

fact, after only two conditioning cycles (8 days total) subjects in Group PL drank significantly less saccharin on the PCP treatment days than on the vehicle treatment days (see Fig. 2). That this difference did not reflect any unconditioned effects of PCP on fluid consumption was evident in the fact that subjects in Group PW, those injected with PCP prior to a saccharin-vehicle pairing, drank saccharin at high levels following both PCP and vehicle pretreatment days, and saccharin consumption on these days did not differ. Clearly, PCP came to serve a discriminative function, in this case that either saccharin was aversive or that saccharin (or drinking) was paired with LiCl (see below). On subsequent probe sessions, different doses of PCP (1, 1.8, and 3.2 mg/kg) substituted for the training dose (1.8 m/kg), with the greatest substitution occurring at the highest dose. The PCP-like compound ketamine dose-dependently substituted for PCP, while the psychostimulant D-amphetamine did not (see Fig. 3).

In this same study, a different group of animals (Group WL) was given the PCP vehicle prior to the saccharin-LiCl pairing and PCP prior to a pairing of saccharin

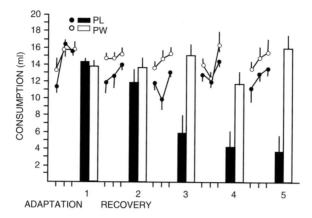

Fig. 2 Mean absolute saccharin consumption for subjects in Groups PL and PW during adaptation and throughout the repeated conditioning and recovery cycles (see text for details). From Mastropaolo et al. [1]

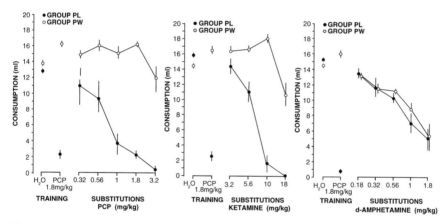

Fig. 3 Mean absolute saccharin consumption for subjects in Groups PL and PW during training and when given various does of PCP (*left panel*), ketamine (*middle panel*) and D-amphetamine (*right panel*) during multiple generalization tests following conditioning (see text for details). Figure redrawn from Mastropaolo et al. [1]

with the LiCl vehicle (for a fourth group, Group WW, both the PCP vehicle and PCP signaled a pairing of saccharin with the LiCl vehicle). This procedure was utilized to test whether the drug had to signal a taste-toxin pairing to be effective as a discriminative cue. Under these conditions, the discrimination was again rapidly learned (after two conditioning cycles), except here animals in Group WL avoided consumption of saccharin when access was preceded by the vehicle injection and consumed the saccharin when it was preceded by PCP. Control subjects in Group WW drank saccharin at high levels following both the vehicle and PCP (see Fig. 4). During substitution probes, consumption was suppressed at low doses of PCP and gradually increased as the probe dose approached the training dose. Similar

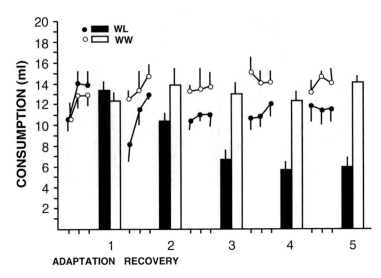

Fig. 4 Mean absolute saccharin consumption for subjects in Groups WL and WW during adaptation and throughout the repeated conditioning and recovery cycles (see text for details). From Mastropaolo et al. [1]

dose–response functions were evident with ketamine. Amphetamine suppressed consumption at every dose, suggesting that it was discriminated from PCP and treated as the vehicle.

It was clear from our initial assessment that a drug effective as a discriminative stimulus in more traditional drug discrimination designs served effectively as a discriminative stimulus in the conditioned taste avoidance procedure we use in our laboratory. Interestingly, discriminative control under the conditioned taste avoidance procedure was achieved with a lower dose (1.8 mg/kg compared to 4.0 mg/kg) than the previously published literature exploring the stimulus properties of PCP [14]. Further, the rate of acquisition of the discrimination was more rapid in the taste avoidance procedure than typically reported in more traditional assessments. Importantly, the rate and patterns of substitution did not appear to be dependent upon whether the drug signaled a taste-toxin or a taste-vehicle pairing (though see below). What was important was that the drug signaled some behavioral contingency.

Concurrent with our work with PCP in this preparation, others reported similar drug discrimination learning using the taste avoidance procedure. Lucki [3], for example, demonstrated the procedure's ability to characterize the specific receptor systems mediating the discriminative effects of the serotonergic agonists 8-OH-DPAT (5-HT$_{1A}$) and TFMPT (5HT$_{1B/1C}$) (see also [15]; [16–19]). Martin and his colleagues ([4]; see also [20]) attempted to characterize the nature of the learning occurring in the taste avoidance procedure and reported that the animals undergoing discrimination training with morphine had not learned anything specific about the taste-toxin pairing but instead had learned that morphine signaled something about a class of responses, e.g., drinking, and its consequences (see Mastropaolo et al. for a similar suggestion with PCP; see also [21–23]). Jaeger and Mucha [2] added one final twist to the initial findings with the taste avoidance drug discrimination procedure by demonstrating that while the procedure was a sensitive index of discriminative control (with the acquisition of the discrimination rapid under training conditions for both pentobarbital and fentanyl), the control established, the generalization produced, and the drug substitutions seen were a function of the training drug and the training procedures, i.e., drug danger vs. drug safe; see above. (For a description of an alternative model using the taste avoidance procedure, specifically the cross-familiarization design, see [24–27]; [28].)

3 Sensitivity of the Taste Avoidance Procedure

These first four studies established the taste avoidance procedure to assess the discriminative stimulus effects of drugs (and displayed the procedure's ability to classify and characterize the drug stimulus), initiated assessments into the specific receptor mechanisms mediating the stimulus effects of a number of drugs, addressed the nature of such learning, and illustrated differences in discriminative control as a function of the training drug and training conditions. What followed from these initial reports was a series of studies that extended these initial investigations to other drugs and to other conditions under which such learning could occur. Several studies also addressed the neurochemical and neuroanatomical substrates of such learning via specific agonist probes and antagonist challenges [19, 29] and selective chemical placement and lesions (see [20, 30, 31]). In relation to the specific training drug, a wide variety of compounds, e.g., acetaldehyde [32], alprazolam [33], amphetamine [34, 35], buprenorphine [36, 37], cocaine [38], chlordiazepoxide [39–41], D-amphetamine [42, 43], diprenorphine [44], ethanol [24, 45, 46], fluvoxamine [27], indorenate [16–19], morphine [4, 20, 21, 30, 31, 47–54], nalorphine [37], naloxone [55, 56], tetrahydrocannabinol (THC) [57], and rimonabant [58], have now been assessed for their ability to establish discriminative control. Although discriminative control can be established to each of these drugs, there was another feature of these assessments that characterized the taste avoidance procedure, i.e., control was rapidly acquired and to a degree often significantly faster than the control seen under the more traditional operant procedures, arguing that the taste avoidance design is a sensitive index of such learning. Further, for several drugs, e.g., cholecystokinin [29, 59–61], estradiol [62], naloxone [63], and testosterone ([64]; for a comprehensive table describing each study which utilizes the CTA procedure of DDL contact, alriley@american.edu), discriminative control was established at lower doses than required in operant assessments. That is, the taste avoidance procedure was not only more sensitive in the speed with which discriminations were acquired, but also in terms of the dose needed to acquire the discrimination.

One example of such a drug is the opioid antagonist naloxone. Although naloxone had been reported to serve as a discriminative cue in opiate-dependent animals (via precipitated withdrawal; [65]), it was generally ineffective as such a cue unless high doses were used or animals were subjected to extended training. For example, in one of the first attempts at establishing naloxone stimulus control in morphine-naïve rats, Colpaert et al. [66] reported that naloxone (at a dose range of 10–160 mg/kg) failed to serve as a discriminative cue in an operant procedure in which food served as the reinforcer (see also [65, 67]). Similarly, Overton and Batta [68] reported that the majority of rats in a shock-escape T-maze procedure failed to reach criterion performance at a dose of 25 mg/kg naloxone, even after 60 training sessions (see [69]). Interestingly, although Carter and Leander [70] reported that pigeons could acquire a discrimination based on naloxone in a food-reinforced design, these effects were only at 30 mg/kg and after an average of 79 sessions, again documenting the relatively weak stimulus effects of naloxone. In this context, Kautz and her colleagues [63] attempted to establish naloxone discriminative control using the taste avoidance procedure. The logic for this attempt paralleled that used above with other assessments within this procedure, i.e., taste avoidance learning is robust and rapidly acquired and thus may provide a more sensitive index of drug discrimination learning. Using this procedure, Kautz et al. injected rats with 1 or 3 mg/kg naloxone 10 min prior to a pairing of a saccharin solution with the emetic LiCl (1.8 mEq, 0.15 M; 76 mg/kg). On subsequent recovery days, the animals were given an injection of the naloxone vehicle prior to a pairing of saccharin with the LiCl vehicle. The conditioning cycle was repeated until the discrimination was acquired. Control subjects were injected with naloxone prior to saccharin consumption as well, but saccharin was never followed by LiCl for this group. As with PCP (and other compounds previously tested in this procedure), the naloxone discrimination (at both doses) was rapidly acquired (in this case by the third conditioning cycle – 12 consecutive days) with subjects injected with naloxone drinking significantly less saccharin than controls (see Fig. 5). The fact that controls did not show the same suppression indicated that the suppression of consumption by the conditioned subjects was not a function of any unconditioned effects of naloxone. Subsequent generalization tests revealed that the relative selective mu opioid antagonist naltrexone substituted completely for naloxone (and at lower doses than the training drug); whereas, the mu agonist morphine did not. This work revealed that opioid antagonists (like opioid agonists) could serve as discriminative stimuli in rats and that this discriminative control could be established at low doses and after only a few training trials. Although the basis for the discriminative control was not assessed in the Kautz et al. work, it was likely that naloxone's discriminative effects are a function of antagonism of endogenous opioid tone. Independent of the specific basis of these effects, the discriminative effects of naloxone were likely mediated at the mu receptor subtypes of the opioid receptor given that animals trained to discriminate naloxone (1 mg/kg) from its vehicle in the taste avoidance procedure generalize control to the relatively selective mu antagonist naltrexone, but not to the selective delta antagonist, naltrindole, nor the selective kappa antagonist, MR2266 (Fig. 6; see [55]; for other assessments

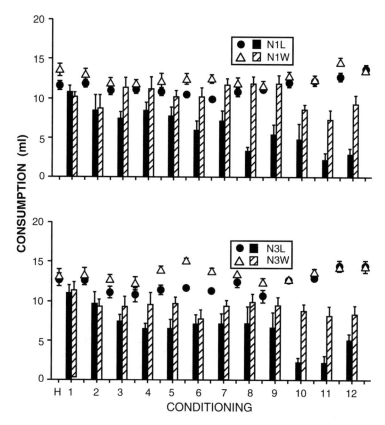

Fig. 5 Mean absolute saccharin consumption for subjects in Groups NL and NW during adaptation and throughout the repeated conditioning and recovery cycles (see text for details). Subjects were trained with either 1 (N1; *upper panel*) or 3 (N3; *lower panel*) mg/kg of naloxone. From Kautz et al. [63]

of opiate antagonist and mixed agonist/antagonist discriminative control, see [36, 37, 44, 56, 71, 72]).

The speed with which drug discrimination is acquired and the dose required for such acquisition relative to the traditional operant design have been used by us and others to argue that taste avoidance may be a more sensitive index of such learning [2–4, 15, 73]. While suggestive, it is important to note several caveats on this position. First, although the criteria used to index discriminative control is generally well defined for operant procedures, e.g., 80% drug-appropriate responding following the drug [74], there are no established criteria used in the taste avoidance procedure. Individual researchers have indicated stimulus control when consumption following the drug is significantly different than that following the vehicle (for a discussion of this issue, see [45]) or when consumption in the conditioned group following the drug is significantly different than that in the control group following the drug. While each of these comparisons in the taste avoidance procedure indexes

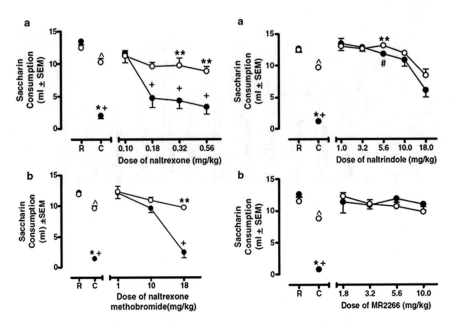

Fig. 6 Mean absolute saccharin consumption for subjects in Groups NL (*closed circles*) and NW (*open circles*) during conditioning (C) and recovery (R) and when given various does of naltrexone (*upper left panel*), naltrindole (*upper right panel*), naltrexone methobromide (*lower left panel*), and MR2266 (*lower right panel*) during multiple generalization tests following conditioning. Figure redrawn from Davis et al. [55]

the fact that discrimination has been acquired, it is not clear at all that this is a comparable measure to that used in the operant designs. As such, comparative statements about relative sensitivity must be cautiously made.

Secondly, the parameters used in the taste avoidance and operant procedures are not always comparable. In fact, most comparisons between the two designs are generally made across studies and across laboratories (for an exception, see [26]). It is crucial that direct comparisons be made in the speed of acquisition and the dose sufficient to establish control. Only then can relative statements about sensitivity and additional comparisons in terms of the drugs that can substitute for or block discriminative control, species, sex and age differences, and neurochemical and neuroanatomical substrates be confidently made. It is interesting to note that what initially appeared to be significant differences in the processing of the drug stimulus in the operant and taste avoidance designs (i.e., graded or quantal; for a discussion, see [75–80]) eventually was reported to be similar when the two procedures were assessed under comparable parametric procedures (see [72]).

Finally, although stimulus control within the taste avoidance procedure is generally discussed as being consistent across all parametric analysis, there are differences dependent upon the training drug and training conditions, indicating that its relative sensitivity may be dependent upon these factors (see [2]). It is interesting in this context that although discriminative control with estradiol is difficult to demonstrate in traditional operant procedures (see above), Gorzalka et al. [81] were able to establish control with estradiol in a Y-maze shock-escape task. Such comparisons suggest that relative sensitivity may be evident in other designs as well and is likely a function of the specific nature and conditions of the procedures used to train and assess stimulus control.

4 Caveats

As described above, the conditioned taste avoidance drug discrimination procedure appears to be able to classify and characterize the discriminative properties of drugs in a manner similar to that seen in more traditional operant and maze assessments (although few comparisons have been directly made between the various designs). Importantly, such discriminative effects are more rapidly acquired and are evident at lower doses in the taste avoidance procedure. There are several caveats with the procedure, however, which need to be addressed (for a discussion, see [40, 82]). One issue raised about this procedure is the necessity of multiple groups when assessing the stimulus properties of any drug. As described above, in any general assessment of the discriminative effects of a given drug in the avoidance procedure, the animal is injected with a drug prior to a pairing of a fluid with an injection of an aversion-inducing agent and injected with the drug vehicle prior to consumption of the same fluid but not followed by the aversive agent. If the drug used as the discriminative cue unconditionally suppresses consumption, e.g., has unconditional adispsogenic effects, the suppression seen following the drug may be a result not of its cueing function but instead its unconditioned effects on fluid consumption. Because of this, control subjects are generally tested that receive the drug prior to fluid access alone (but not paired with an emetic-like LiCl) on both conditioning and recovery days. These subjects serve as a control for the effects of the drug on fluid intake and provide a comparison with the conditioned subjects to substantiate the cueing function of the drug in this group. In operant designs, the animal serves as its own control, injected with the drug prior to the responding on one lever for food and the drug vehicle prior to responding on another lever for the same reinforcer. Given that the measured response is an active lever or key press, one can immediately see if the drug unconditionally suppresses response rate. In the taste avoidance design, suppression of consumption is the measured response and has to be controlled by this second group. Consequently, the number of animals used in the taste avoidance procedure is doubled.

Although this increase in the number of subjects is somewhat offset by the fact that multiple animals can be concurrently run with little increase in time (or equipment), others have addressed the potential confounding issue of unconditioned drug-induced suppression of fluid consumption by other procedures that do not necessarily require additional subjects. For example, following discrimination training in which an animal is injected with a stimulus drug prior to the pairing of a taste with a toxin (and on recovery days, injected with vehicle prior to the taste alone), one can give the animal access to both the taste and water in a two-bottle test of the ability of the stimulus drug to suppress consumption (see [3, 4, 15–17, 19, 34, 35, 42, 83]). If the drug is simply suppressing consumption in general, both saccharin and water intake would be affected; if the drug suppresses consumption by virtue of its signaling function, only that taste would be affected and the taste would be selectively avoided relative to water which is what generally is reported under such assessments.

Others have addressed this need for a control for the unconditionally suppressive effects of the drug by injecting the animal with the drug prior to a taste paired with the toxin vehicle, i.e., as a safety cue. In this design (see above), animals are injected with the drug vehicle prior to the pairing of the taste with the toxin and injected with the drug prior to the presentation of the taste alone [1, 2, 4, 19, 52]. Animals readily acquire the drug discrimination, avoiding saccharin when it is preceded by vehicle and consuming the same saccharin when preceded by the drug (see above, Fig. 4). If the drug has unconditioned suppressive effects on fluid consumption, the discrimination would be impaired (as animals decrease consumption of the taste preceded by the drug). As noted, under these conditions, animals readily acquire the discrimination with a variety of drugs, e.g., pentobarbital, morphine, PCP, and indorenate.

Although several caveats have been discussed in relation to the taste avoidance drug discrimination procedure, it remains highly effective as a design in detecting the stimulus effects of a variety of drugs and to and does so at low doses and relatively rapidly. The extent to which the procedure is effective in behavioral pharmacology and generally used will be weighed against its possible limitations (and adaptations to circumvent potential confounds).

5 Drugs as Interoceptive Stimuli for Regulated Drug Intake

Drug discrimination learning in general is an effective tool in the classification and characterization of drugs (see [84, 85]). In so doing, the procedure can identify and class a drug according to the stimulus properties it shares with other drugs, e.g., opioids, stimulants, and depressants. It can also be used to identify the biological and neurochemical substrates of drug action as other pharmacological probes potentiate or antagonize the drug's stimulus effects, yielding information on the specific receptor systems mediating the drug's effects and the efficacy of the drug at a specific receptor (e.g., as a partial or full agonist). The model has gone beyond these basic pharmacological assessments to its use in clinical pharmacology as a procedure to identify drugs that induce or abate anxiety or drugs that induce dependence and produce withdrawal (see [86]). One area in which the basic drug discrimination procedure has been extended (and often with controversy; see [87]) is in its assessment of the abuse liability of various drugs. The use of the design in this vein stems from the fact that drugs of abuse are used by humans for their subjective effects, and if the drug discrimination procedure is a reliable and robust index of such effects it may be useful in corroborating or predicting use, abuse and dependence. Although often described in this context, there is little evidence of its utility in this area, and several arguments and demonstrations have been made against its general use as such a predictor. For example, although most drugs of abuse can serve as discriminative stimuli, other drugs, many with no abuse liability, serve such a function as well [84]. Further, when direct assessments have been made between abuse liability and degree of discriminability, there was no relationship (see [68]).

Importantly, most investigators in drug discrimination research recognize the limitation of this design in indexing abuse vulnerability and argue instead that the procedure is simply one additional tool in assessments of drug abuse and one that should be used in conjunction with other assays (ones often with more face validity, e.g., self-administration) to index the rewarding and addicting properties of drugs (see [74]). Although drug discrimination learning may be limited in its capacity to identify abuse vulnerability (at least in the manner by which it is typically used; see above), Panlilio et al. [88] have presented an interesting model that may, nonetheless, have importance to drug use and abuse. Specifically, Panlilio et al. argue that drug taking (as indexed in animal self-administration models) displays highly regular patterns of responding and pauses. They argue that the initiation and maintenance of responding is a function of the drug's rewarding effects. The pauses are indicative of the fact that the drug is either not rewarding or aversive. That is, animals initiate responding when the drug level is low (or absent) and responding at these times is highly rewarding to the animal. Immediate responding after this initial bout is not rewarding given that the system mediating the drug's rewarding effect is saturated. With normal metabolic processes, the drug levels are reduced below some trigger or set point such that an infusion at this point is again rewarding. The issue is how does the animal know what the blood level is to induce or suppress responding? Given the well-established phenomenon that drugs do have subjective effects in animals (as assessed by the drug discrimination procedure), Panlilio et al. argue that these effects serve as cues to when a behavioral response (e.g., bar press) will or will not be reinforced. In this view, regulated drug intake is monitored and controlled by the discriminative effects of the drug itself (see also [89–92]).

This view on the patterning of drug self-administration parallels work by Davidson and his colleagues examining regulatory food intake (for a review, see [93]). They have argued that the initiation of food intake is based heavily on food cues that drive consumption. Like regulated drug intake, animals also regulate the amount consumed. That is, following a bout of eating, animals stop food intake, although the very same cues present at its initiation may still be available. Regulation of feeding, according to Davidson comes from satiety cues that inhibit further eating. These cues work to inform the animal that continued eating is either no longer reinforcing or is aversive. This hypothesis necessitates the ability of the animal to detect such cues, and in direct assessments of this hypothesis (using a drug discrimination learning design), he and his colleagues have shown that animals can use deprivation (satiety) cues to control behavior. For example, in one such experiment (see [94]) animals were given a 1 ma foot shock for 0.5 s when 24 h deprived but not when 0 h deprived (another group had these contingencies reversed such that they were shocked when 0 h deprived and not shocked under the 24 h deprivation condition). The amount of time spent freezing was used to index learning for both groups. Although both groups displayed more freezing than a non-shocked control group, the level of freezing was a function of the deprivation level, i.e., those shocked under the 24 h condition froze more when 24 h deprived than when 0 h deprived, whereas those shocked under the 0 h deprivation condition froze more when 0 h deprived than when 24 h deprived, indicative of the ability of the deprivation states to serve as discriminative cues for responding (see also [95, 96]; see [97] for similar assessments in a food-motivated task).

The parallels between eating and drug taking are interesting for other reasons beyond a potentially common process, i.e., discriminative control by interoceptive stimuli. Another parallel is that both types of behaviors can be dysregulated. In the case of eating, such dysregulated behavior can lead to excessive eating and obesity. In the case of drug use, dysregulated behavior can lead to escalated drug intake and addiction. Interestingly, a model proposed by Davidson and his group suggests the basis for this dysregulation may be similar between the two conditions (see [98, 99]). Specifically, Davidson and his colleagues have shown that animals with hippocampal damage have difficulty in behavioral tasks dependent upon inhibitory control, e.g., serial feature negative discriminations in which one cue signals some outcome but not when that cue is preceded by another stimulus [100]. In this case, the second stimulus informs the animal that the original stimulus-outcome contingency is not available and the animal inhibits responding in the presence of this feature negative stimulus (negative occasion setter). Animals with hippocampal lesions have deficits in displaying this inhibition and consequently have deficits in serial feature negative tasks [101]. Davidson and his group has further demonstrated that animals exposed chronically to high fat, western diets have loss of integrity of the blood–brain-barrier relatively selective to the hippocampus and also display deficits in hippocampal mediated task, e.g., serial feature negative discriminations [93, 102, 103]. The relevance to obesity comes from the fact that such animals (those exposed to high fat diets) may be displaying overeating and obesity as a function of the hippocampal damage that reduces the animal's ability to inhibit responding to food-related cues, i.e., they can't use these interoceptive cues to inform them that food is no longer rewarding and to inhibit feeding. Thus, in this model, excessive food intake and obesity are a consequence of high fat diet exposure that impacts discrimination learning or its expression.

Although little has been done in the context of drug abuse in this model, it is interesting and important to note that both acute and chronic cocaine (as well as methamphetamine and morphine) impairs blood–brain-barrier integrity (see [104–106]) and impairs performance on serial feature negative tasks (see [107]). Chronic drug use (as use goes from impulsive to compulsive that is characteristic of addiction; see [108, 109]) has been suggested to be a function of a host of neuroplastic changes including downregulation of brain reward pathways and upregulation or recruitment of brain stress systems that shift basic drug use from its positively to its negatively reinforcing effects (see [110]). If there are also drug-induced changes in BBB permeability that allow for the movement of cytokines and glia into the brain (and selectively into the hippocampus) that result in hippocampal damage and the loss of inhibitory control, changes in drug intake (as already demonstrated for food intake) may also be mediated by these neuroplastic changes. Traditionally, this latter position of reduced inhibitory control has been thought to be regulated by the cortex (dorsolateral, orbitofrontal, and anterior cingulate cortex), but as the parallels between dysregulated food intake and drug intake show, this deficit may also be a function of the loss of discriminative control of regulated drug intake due to selective hippocampal damage (which may be mediated via its interconnections with the prefrontal cortex; see [111, 112]). More research will be needed to assess the role of the hippocampus and discriminative control in drug abuse and addiction, as well as the ability of the taste avoidance drug discrimination procedure to detect differences in drug-naïve and drug-exposed animals. It is interesting in this context that the specific taste avoidance drug discrimination procedure in which the drug signals the safe presentation of a taste (and the drug vehicle signals the taste-toxin pairing; see above) is an example of negative occasion setting or feature negative learning. The taste avoidance drug discrimination procedure should provide an additional model to assess the effects of drugs on BBB integrity, hippocampal function, and behavioral deficits, expanding the use of this model to understanding the potentially important role that drug discrimination learning may play in both regulated and dysregulated behavior in general and human drug abuse and addiction, specifically.

6 Conclusions

Work with the taste avoidance drug discrimination procedure has revealed a preparation that is both rapid and sensitive in its ability to index the discriminative stimulus properties of drugs. The design further allows for assessments of the ability of other drugs to substitute for the training drug (providing an ability to classify compounds with shared stimulus effects) and for manipulations that block and/or potentiate its stimulus control (providing a procedure to characterize the neurochemical basis and neuroanatomical locus of the training drug's stimulus effects). Work with this procedure parallels the effects reported in other designs, although the speed of acquiring stimulus control and the sensitivity in its detection may make this a useful behavioral tool in such assessments. Further, although a host of drugs have been examined in the taste avoidance drug discrimination procedure, much remains to be done to determine the conditions under which stimulus control is established, the degree to which this control is dependent upon specific parameters, the nature of the learning within this design, its neurochemical and neuroanatomical mediation and its limitations. The investigation of the role of drug discrimination learning in both normal and dysregulated behavior in humans continues to be an important consideration, and recent work with cognitive deficits (and obesity) as well as drug use and abuse illustrates the potential utility of drug discrimination learning in understanding the basis for these behaviors. As such, the experimental assessment and clinical applications of drug discrimination learning (within both traditional operant and the taste avoidance drug discrimination procedures) may continue to provide insights to basic pharmacology and translational opportunities for the understanding and treatment of human pathology.

Acknowledgements Preparation of this chapter and the research described from the authors' laboratory were supported in part from grants from the Mellon Foundation to ALR. Requests for reprints should be sent to alriley@american.edu.

References

1. Mastropaolo JP, Moskowitz KH, Dacanay RJ, Riley AL (1989) Conditioned taste aversions as a behavioral baseline for drug discrimination learning: an assessment with phencyclidine. Pharmacol Biochem Behav 32(1):1–8
2. Jaeger TV, Mucha RF (1990) A taste aversion model of drug discrimination learning: training drug and condition influence rate of learning, sensitivity and drug specificity. Psychopharmacology (Berl) 100(2):145–150
3. Lucki I (1988) Rapid discrimination of the stimulus properties of 5-hydroxytryptamine agonists using conditioned taste aversion. J Pharmacol Exp Ther 247(3):1120–1127
4. Martin GM, Gans M, van der Kooy D (1990) Discriminative properties of morphine that modulate associations between tastes and lithium chloride. J Exp Psychol Anim Behav Process 16(1):56
5. Garcia J, Kimeldorf DJ, Koelling RA (1955) Conditioned aversion to saccharin resulting from exposure to gamma radiation. Science 122:157–158
6. Garcia J, Kimeldorf DJ (1957) Temporal relationship within the conditioning of a saccharine aversion through radiation exposure. J Comp Physiol Psychol 50(2):180–183
7. Garcia J, Koelling RA (1966) Relation of cue to consequence in avoidance learning. Psychon Sci 4(1):123–124
8. Garcia J, Ervin F (1968) Gustatory-visceral and telereceptor-cutaneous conditioning: adaptation in internal and external milieus. Commun Behav Biol 1(5):389–415
9. Rozin P, Kalat JW (1971) Specific hungers and poison avoidance as adaptive specializations of learning. Psychol Rev 78(6):459–486
10. Freeman KB, Riley AL (2009) The origins of conditioned taste aversion learning: a historical analysis. In: Reilly S, Schachtman TR (eds) Conditioned taste aversion: behavioral and neural processes. Oxford University Press, Oxford, pp 9–33
11. Reilly S, Schachtman T (eds) (2009) Conditioned taste aversion: behavioral and neural processes. Oxford University Press, Oxford
12. Davidson TL, Riley AL (2015) Taste, sickness, and learning. Am Sci 103(3):204–211
13. Brady KT, Balster RL (1981) Discriminative stimulus properties of phencyclidine and five analogues in the squirrel monkey. Pharmacol Biochem Behav 14(2):213–218
14. Poling A, White F, Appel J (1979) Discriminative stimulus properties of phencyclidine. Neuropharmacology 18(5):459–463

15. Lucki I, Marcoccia JM (1991) Discriminated taste aversion with 5-HT1A agonist measured using saccharin preference. Behav Pharmacol 2:335–44

16. Miranda F, Hong E, Velázquez-Martínez D (2001) Discriminative stimulus properties of indorenate in a conditioned taste aversion paradigm. Pharmacol Biochem Behav 68 (3):427–433

17. Miranda F, Orozco G, Velazquez-Martinez D (2002) Full substitution of the discriminative cue of a 5-HT1A/1B/2C agonist with the combined administration of a 5-HT1B/2C and a 5-HT1A agonist. Behav Pharmacol 13(4):303–311

18. Miranda F, Hong E, López CM, Velázquez MD (1998) Modulation of the discriminative stimulus properties of indorenate by 5-HT receptors. Proc West Pharmacol Soc 41:59–60

19. Miranda F, Hong E, Sánchez H, Velázquez-Martínez D (2003) Further evidence that the discriminative stimulus properties of indorenate are mediated by 5-HT 1A/1B/2C receptors. Pharmacol Biochem Behav 74(2):371–380

20. Martin GM, Bechara A, van der Kooy D (1991) The perception of emotion: parallel neural processing of the affective and discriminative properties of opiates. Psychobiology 19 (2):147–152

21. Järbe T, Lamb R (1999) Effects of lithium dose (UCS) on the acquisition and extinction of a discriminated morphine aversion: tests with morphine and [DELTA] 9-THC. Behav Pharmacol 10(4):349–358

22. Skinner DM (2000) Modulation of taste aversions by a pentobarbital drug state: an assessment of its transfer properties. Learn Motiv 31(4):381–401

23. Skinner DM, Martin GM, Pridgar A, Van Der Kooy D (1994) Conditional control of fluid consumption in an occasion setting paradigm is independent of Pavlovian associations. Learn Motiv 25(4):368–400

24. Bienkowski P, Piasecki J, Koros E, Stefanski R, Kostowskia W (1998) Studies on the role of nicotinic acetylcholine receptors in the discriminative and aversive stimulus properties of ethanol in the rat. Eur Neuropsychopharmacol 8(2):79–87

25. De Boer T (1996) The pharmacologic profile of mirtazepine. J Clin Psychiatry 57(Suppl 4):19–25

26. De Beun R, Lohmann A, Schneider R, De Vry J (1996) Ethanol intake-reducing effects of ipsapirone in rats are not due to simple stimulus substitution. Pharmacol Biochem Behav 53 (4):891–898

27. Olivier B, Gommans J, Van der Gugten J, Bouwknecht J, Herremans A, Patty T, Hijzen T (1999) Stimulus properties of the selective 5-HT reuptake inhibitor fluvoxamine in conditioned taste aversion procedures. Pharmacol Biochem Behav 64(2):213–220

28. Serafine KM, Riley AL (2012) Cocaine-induced conditioned taste aversions: role of monoamine reuptake inhibition. In: Hall FS (ed) Serotonin: biosynthesis, regulation and health implications. NOVA Science Publishers, Hauppauge, pp 257–291

29. Melton PM, Riley AL (1994) Receptor mediation of the stimulus properties of cholecystokinin. Pharmacol Biochem Behav 48(1):275–279

30. Jaeger TV, van der Kooy D (1993) Morphine acts in the parabrachial nucleus, a pontine viscerosensory relay, to produce discriminative stimulus effects. Psychopharmacology (Berl) 110(1–2):76–84

31. Jaeger TV, van der Kooy D (1996) Separate neural substrates mediate the motivating and discriminative properties of morphine. Behav Neurosci 110(1):181–201

32. Redila VA, Aliatas E, Smith BR, Amit Z (2002) Effects of ethanol on an acetaldehyde drug discrimination with a conditioned taste aversion procedure. Alcohol 28(2):103–109

33. Glowa JR, Jeffreys RD, Riley AL (1991) Drug discrimination using a conditioned taste-aversions paradigm in rhesus monkeys. J Exp Anal Behav 56(2):303–312

34. Herrera F, Martinez DV (1997) Discriminative stimulus properties of amphetamine in a conditioned taste aversion paradigm. Behav Pharmacol 8(5):458–464

35. Miranda F, Sandoval-Sánchez A, Cedillo LN, Jiménez JC, Millán-Mejía P, Velázquez-Martínez DN (2007) Modulatory role of 5-HT1B receptors in the discriminative signal of amphetamine in the conditioned taste aversion paradigm. Pharmacol Rep 59(5):517–524

36. Pournaghash S, Riley AL (1993) Buprenorphine as a stimulus in drug discrimination learning: an assessment of mu and kappa receptor activity. Pharmacol Biochem Behav 46 (3):593–604

37. Smurthwaite ST, Riley AL (1994) Nalorphine as a stimulus in drug discrimination learning: assessment of the role of μ- and κ-receptor subtypes. Pharmacol Biochem Behav 48 (3):635–642

38. Awasaki Y, Nojima H, Nishida N (2011) Application of the conditioned taste aversion paradigm to assess discriminative stimulus properties of psychostimulants in rats. Drug Alcohol Depend 118(2):288–294

39. Fox MA, Levine ES, Riley AL (2001) The inability of CCK to block (or CCK antagonists to substitute for) the stimulus effects of chlordiazepoxide. Pharmacol Biochem Behav 69 (1):77–84

40. van Hest A, Hijzen T, Slangen J, Olivier B (1992) Assessment of the stimulus properties of anxiolytic drugs by means of the conditioned taste aversion procedure. Pharmacol Biochem Behav 42(3):487–495

41. Woudenberg F, Hijzen TH (1991) Discriminated taste aversion with chlordiazepoxide. Pharmacol Biochem Behav 39(4):859–863

42. Miranda F, Jiménez JC, Cedillo LN, Sandoval-Sánchez A, Millán-Mejía P, Sánchez-Castillo H, Velázquez-Martínez DN (2009) The GABA-B antagonist 2-hydroxysaclofen reverses the effects of baclofen on the discriminative stimulus effects of D-amphetamine in the conditioned taste aversion procedure. Pharmacol Biochem Behav 93(1):25–30

43. Revusky S, Coombes S, Pohl RW (1982) Drug states as discriminative stimuli in a flavor-aversion learning experiment. J Comp Physiol Psychol 96(2):200–211

44. Smurthwaite ST, Riley AL (1992) Diprenorphine as a stimulus in drug discrimination learning. Pharmacol Biochem Behav 43(3):839–846

45. Quertemont E (2003) Discriminative stimulus effects of ethanol with a conditioned taste aversion procedure: lack of acetaldehyde substitution. Behav Pharmacol 14(4):343–350

46. Redila VA, Smith BR, Amit Z (2000) The effects of aminotriazole and acetaldehyde on an ethanol drug discrimination with a conditioned taste aversion procedure. Alcohol 21 (3):279–285

47. Grabus SD, Smurthwaite ST, Riley AL (1999) Nalorphine's ability to substitute for morphine in a drug discrimination procedure is a function of training dose. Pharmacol Biochem Behav 63(3):481–488

48. Järbe T, Lamb R (1995) Discriminated conditioned taste aversion for studying multi-element stimulus control. Behav Pharmacol 6(2):149–155

49. Järbe T, Lamb R (1999) Discriminated taste aversion and context: a progress report. Pharmacol Biochem Behav 64(2):403–407

50. Skinner DM, Martin GM (1992) Conditioned taste aversions support drug discrimination learning at low dosages of morphine. Behav Neural Biol 58(3):236–241

51. Skinner DM, Martin GM, Harley C, Kolb B, Pridgar A, Bechara A, van der Kooy D (1994) Acquisition of conditional discriminations in hippocampal lesioned and decorticated rats: evidence for learning that is separate from both simple classical conditioning and configural learning. Behav Neurosci 108(5):911–926

52. Skinner DM, Martin GM, Howe RD, Pridgar A, van der Kooy D (1995) Drug discrimination learning using a taste aversion paradigm: an assessment of the role of safety cues. Learn Motiv 26(4):343–369

53. Stevenson GW, Pournaghash S, Riley AL (1992) Antagonism of drug discrimination learning within the conditioned taste aversion procedure. Pharmacol Biochem Behav 41(1):245–249

54. Stevenson GW, Cañadas F, Zhang X, Rice KC, Riley AL (2000) Morphine discriminative control is mediated by the mu opioid receptor: assessment of delta opioid substitution and antagonism. Pharmacol Biochem Behav 66(4):851–856

55. Davis CM, Stevenson GW, Cañadas F, Ullrich T, Rice KC, Riley AL (2009) Discriminative stimulus properties of naloxone in Long–Evans rats: assessment with the conditioned taste aversion baseline of drug discrimination learning. Psychopharmacology (Berl) 203 (2):421–429

56. Smurthwaite ST, Kautz MA, Geter B, Riley AL (1992) Naloxone as a stimulus in drug discrimination learning: generalization to other opiate antagonists. Pharmacol Biochem Behav 41(1):43–47

57. Järbe TU, Harris MY, Li C, Liu Q, Makriyannis A (2004) Discriminative stimulus effects in rats of SR-141716 (rimonabant), a cannabinoid CB1 receptor antagonist. Psychopharmacology (Berl) 177(1–2):35–45

58. Järbe TU, Li C, Vadivel SK, Makriyannis A (2008) Discriminative stimulus effects of the cannabinoid CB1 receptor antagonist rimonabant in rats. Psychopharmacology (Berl) 198 (4):467–478

59. Melton PM, Riley AL (1993) An assessment of the interaction between cholecystokinin and the opiates within a drug discrimination procedure. Pharmacol Biochem Behav 46 (1):237–242

60. Melton PM, Kopman JA, Riley AL (1993) Cholecystokinin as a stimulus in drug discrimination learning. Pharmacol Biochem Behav 44(2):249–252

61. Riley AL, Melton PM (1997) Effects of μ- and δ-opioid–receptor antagonists on the stimulus properties of cholecystokinin. Pharmacol Biochem Behav 57(1):57–62

62. De Beun R, Heinsbroek R, Slangen J, van de Poll N (1991) Discriminative stimulus properties of estradiol in male and female rats revealed by a taste-aversion procedure. Behav Pharmacol 2(6):439–445

63. Kautz MA, Geter B, McBride SA, Mastropaolo JP, Riley AL (1989) Naloxone as a stimulus for drug discrimination learning. Drug Dev Res 16(2–4):317–326

64. De Beun R, Jansen E, Slangen JL, van de Poll NE (1992) Testosterone as appetitive and discriminative stimulus in rats: sex- and dose-dependent effects. Physiol Behav 52 (4):629–634

65. Weissman A (1978) Discriminability of naloxone in rats depends on concomitant morphine treatment. Psychopharmacology (Berl) 58(2):2–12

66. Colpaert FC, Niemegeers CJ, Janssen PA (1976) On the ability of narcotic antagonists to produce the narcotic cue. J Pharmacol Exp Ther 197(1):180–187

67. Lal H, Miksic S, McCarten M (1978) A comparison of discriminative stimuli produced by naloxone, cyclazocine and morphine in the rat. In: Colpaert FC, Rosecrans JA (eds) Stimulus properties of drugs: ten years of progress. Elsevier, Amsterdam, pp 177–180

68. Overton DA, Batta SK (1977) Relationship between abuse liability of drugs and their degree of discriminability in the rat. In: Thompson T, Unna KR (eds) Predicting dependence liability of stimulant and depressant drugs. University Park Press, Baltimore, pp 125–135

69. Overton DA (1982) Comparison of the degree of discriminability of various drugs using the T-maze drug discrimination paradigm. Psychopharmacology (Berl) 76(4):385–395

70. Carter RB, Leander JD (1982) Discriminative stimulus properties of naloxone. Psychopharmacology (Berl) 77(4):305–308

71. Riley AL, Pournaghash S (1995) The effects of chronic morphine on the generalization of buprenorphine stimulus control: an assessment of kappa antagonist activity. Pharmacol Biochem Behav 52(4):779–787

72. Sobel B-F, Wetherington C, Riley A (1995) The contribution of within-session averaging of drug- and vehicle-appropriate responding to the graded dose–response function in drug discrimination learning. Behav Pharmacol 6(4):348–358

73. Rowan GA, Lucki I (1992) Discriminative stimulus properties of the benzodiazepine receptor antagonist flumazenil. Psychopharmacology (Berl) 107(1):103–112

74. Solinas M, Panlilio LV, Justinova Z, Yasar S, Goldberg SR (2006) Using drug-discrimination techniques to study the abuse-related effects of psychoactive drugs in rats. Nat Protoc 1 (3):1194–1206
75. Barrett RJ, Caul WF, Huffman EM, Smith RL (1994) Drug discrimination is a continuous rather than a quantal process following training on a VI-TO schedule of reinforcement. Psychopharmacology (Berl) 113(3–4):289–296
76. Barrett RJ, Caul WF, Huffman EM, Smith RL (1994) Reply to comments by Overton, Emmett-Oglesby and Gauvin and Holloway. Psychopharmacology (Berl) 113(3):302–303
77. Emmett-Oglesby M (1994) Commentary on "Drug discrimination is a continuous rather than a quantal process following training on a VI-TO schedule of reinforcement" by Barrett et al. Psychopharmacology (Berl) 113(3):300–301
78. Mathis D, Emmett-Oglesby M (1990) Quantal vs. graded generalization in drug discrimination: measuring a graded response. J Neurosci Methods 31(1):23–33
79. Overton DA (1994) Disadvantages of quantal drug discrimination procedures. Psychopharmacology (Berl) 113(3):298–299
80. Riley A, Kautz M, Geter B, Pournaghash S, Melton P, Ferrari C (1991) A demonstration of the graded nature of the generalization function of drug discrimination learning within the conditioned taste aversion procedure. Behav Pharmacol 2(4–5):323–334
81. Gorzalka BB, Wilkie DM, Hanson LA (1995) Discrimination of ovarian steroids by rats. Physiol Behav 58(5):1003–1011
82. van Hest A, Slangen JL, Olivier B (1991) Is the conditioned taste aversion procedure a useful tool in drug discrimination research? In: Oliver B, Mos J, Slangen JL (eds) Animal models in psychopharmacology. Birkhauser, Basel, pp 399–405
83. Lucki I, Singh A, Kreiss DS (1994) Antidepressant-like behavioral effects of serotonin receptor agonists. Neurosci Biobehav Rev 18(1):85–95
84. Järbe T (1989) Discrimination learning with drug stimuli: methods and applications. In: Boulten AA, Baker GB, Greenshaw AJ (eds) Neuromethods, vol 13. Psychopharmacology. Humana Press, Clifton, pp 513–563
85. Slangen J (1991) Drug discrimination and animal models. In: Oliver B, Mos J, Slangen JL (eds) Animal models in psychopharmacology. Birkhauser, Basel, pp 359–373
86. Emmett-Oglesby M, Mathis D, Moon R, Lal H (1990) Animal models of drug withdrawal symptoms. Psychopharmacology (Berl) 101(3):292–309
87. McMahon LR (2015) The rise (and fall?) of drug discrimination research. Drug Alcohol Depend 151:284–288
88. Panlilio LV, Thorndike EB, Schindler CW (2008) A stimulus-control account of regulated drug intake in rats. Psychopharmacology (Berl) 196(3):441–450
89. Panlilio LV, Thorndike EB, Schindler CW (2009) A stimulus-control account of dysregulated drug intake. Pharmacol Biochem Behav 92(3):439–447
90. Suto N, Wise RA (2011) Satiating effects of cocaine are controlled by dopamine actions in the nucleus accumbens core. J Neurosci 31(49):17917–17922
91. Tsibulsky VL, Norman AB (1999) Satiety threshold: a quantitative model of maintained cocaine self-administration. Brain Res 839(1):85–93
92. Tsibulsky VL, Norman AB (2001) Satiety threshold during maintained cocaine self-administration in outbred mice. Neuroreport 12(2):325–328
93. Davidson T, Hargrave S, Swithers S, Sample C, Fu X, Kinzig K, Zheng W (2013) Inter-relationships among diet, obesity and hippocampal-dependent cognitive function. Neuroscience 253C:110–122
94. Davidson TL (1987) Learning about deprivation intensity stimuli. Behav Neurosci 101 (2):198–208
95. Davidson TL, Carretta JC (1993) Cholecystokinin, but not bombesin, has interoceptive sensory consequences like 1-h food deprivation. Physiol Behav 53(4):737–745
96. Davidson T, Flynn FW, Grill HJ (1988) Comparison of the interoceptive sensory consequences of CCK, LiCl, and satiety in rats. Behav Neurosci 102(1):134–140

97. Davidson T, Kanoski SE, Tracy AL, Walls EK, Clegg D, Benoit SC (2005) The interoceptive cue properties of ghrelin generalize to cues produced by food deprivation. Peptides 26 (9):1602–1610

98. Davidson T, Sample C, Swithers S (2014) An application of Pavlovian principles to the problems of obesity and cognitive decline. Neurobiol Learn Mem 108C:172–184

99. Davidson TL, Tracy AL, Schier LA, Swithers SE (2014) A view of obesity as a learning and memory disorder. J Exp Psychol Anim Learn Cogn 40(3):261–279

100. Davidson TL, Jarrard LE (1993) A role for hippocampus in the utilization of hunger signals. Behav Neural Biol 59(2):167–171

101. Holland PC, Lamoureux JA, Han JS, Gallagher M (1999) Hippocampal lesions interfere with Pavlovian negative occasion setting. Hippocampus 9(2):143–157

102. Davidson TL, Monnot A, Neal AU, Martin AA, Horton JJ, Zheng W (2012) The effects of a high-energy diet on hippocampal-dependent discrimination performance and blood–brain barrier integrity differ for diet-induced obese and diet-resistant rats. Physiol Behav 107 (1):26–33

103. Sample CH, Martin AA, Jones S, Hargrave SL, Davidson TL (2015) Western-style diet impairs stimulus control by food deprivation state cues: implications for obesogenic environments. Appetite 93:13–23

104. Martins T, Baptista S, Gonçalves J, Leal E, Milhazes N, Borges F, Ribeiro C, Quintela O, Lendoiro E, López-Rivadulla M (2011) Methamphetamine transiently increases the blood–brain barrier permeability in the hippocampus: role of tight junction proteins and matrix metalloproteinase-9. Brain Res 1411:28–40

105. Sharma HS, Ali SF (2006) Alterations in blood–brain barrier function by morphine and methamphetamine. Ann N Y Acad Sci 1074(1):198–224

106. Yao H, Duan M, Buch S (2011) Cocaine-mediated induction of platelet-derived growth factor: implication for increased vascular permeability. Blood 117(8):2538–2547

107. Riley A, Kearns D, Hargrave S, Davidson T (2015) Cocaine impairs serial feature negative learning: implications for cocaine abuse. Drug Alcohol Depend 156, e190

108. Koob GF (2015) The dark side of emotion: the addiction perspective. Eur J Pharmacol 753:73–87

109. Volkow ND, Morales M (2015) The brain on drugs: from reward to addiction. Cell 162 (4):712–725

110. Koob GF, Arends MA, Le Moal M (2014) Drugs, addiction, and the brain. Academic, Oxford

111. Anderson MC, Bunce JG, Barbas H (2015) Prefrontal-hippocampal pathways underlying inhibitory control over memory. Neurobiol Learn Mem (in press)

112. Brincat SL, Miller EK (2015) Frequency-specific hippocampal-prefrontal interactions during associative learning. Nat Neurosci 18(4):576–581

A Prospective Evaluation of Drug Discrimination in Pharmacology

Ellen A. Walker

Abstract As investigators, we use many methodologies to answer both practical and theoretical questions in our field. Occasionally, we must stop and collect the latest findings or trends and then look forward to where our ideas, findings, and hypotheses may take us. Similar to volumes that were published in previous years on drug discrimination (Glennon and Young, Drug discrimination applications to medicinal chemistry and drug studies. Wiley, Hoboken, 2011; Ho et al., Drug discrimination and state dependent learning. Academic Press, New York, 1978), this collection in Current Topics in Behavioral Neurosciences serves as a current analysis of the continued value of the drug discrimination procedure to the fields of pharmacology, neuroscience, and psychology and as a stepping stone to where drug discrimination methodology can be applied next, in both a practical and theoretical sense. This final chapter represents one investigator's perspective on the utility and possibilities for a methodology that she fell in love with over 30 years ago.

Keywords Abuse liability testing · Complex cues · Drug discrimination · Interoceptive states · Receptor theory

Contents

For several decades, drug discrimination has been used as a tool to understand the pharmacology of different drug classes or has been involved in the discovery of new drug targets or receptors (Porter et al. 2018). This trend continues today. In a practical sense, drug discrimination is an excellent procedure to understand the

E. A. Walker (✉)
Department of Pharmaceutical Sciences, Temple University School of Pharmacy, Philadelphia, PA, USA
e-mail: ellen.walker@temple.edu

© Springer International Publishing AG, part of Springer Nature 2018 319
Curr Topics Behav Neurosci (2018) 39: 319–328
DOI 10.1007/7854_2018_59
Published Online: 29 July 2018

underlying pharmacology, mechanisms, and functional outcomes for drug-receptor interactions. Bar none, the pharmacological selectivity, orderly adherence to biological principles, and sensitivity to antagonism made drug discrimination a key tool in neuropharmacology. As stated by the late Francis Colpaert "... the DD [*drug discrimination*] paradigm offers an exquisitely specific, selective, and sensitive approach to the in vivo analysis of drug-receptor interactions ..." (Colpaert 2011).

1 Drug Discrimination as a Tool to Define Receptor Pharmacology

In the current volume, a number of excellent chapters have reviewed the history and our current understanding of drug-receptor interactions as defined by drug discrimination methodology for a range of drug classes in various species, including humans. For example, Mori and Suzuki (2016) nicely outlined the necessity for 5-HT_2 receptor activation as the critical component with a clear role for 5-HT_{1A} modulatory function for the discriminative stimulus effects of hallucinogens such as MDMA and LSD. Furthermore, through drug discrimination, investigators were able to differentiate the contributions of 5-HT to the effects of MDMA and distinguish substitution patterns for different psychostimulants such as N,N-DMT, 5-MeO-DMT, and methamphetamine. These patterns could then be compared and contrasted to cocaine and opioid discriminative stimuli (Mori and Suzuki 2016). In the opioid field, the high selectivity of opioid drug discrimination is readily demonstrated as only MOP, KOP, or DOP receptor ligands substitute for morphine, U50,488, SNC80, and BW373U86 discriminative stimuli and only receptor selective antagonists such as CTAP, nor-BNI, or naltrindole will block these cues, respectively (Butelman and Kreek 2016). More recently, drug discrimination has been extended to the selectivity of NOP or nociceptin receptor ligands. For example, when the NOP receptor agonist Ro 64-6198 was trained as a discriminative stimulus in rats, morphine, U50,488, and SNC80 failed to substitute for Ro 64-6198 and Ro 64-6198 failed to substitute for morphine in rats trained to discriminate morphine suggesting this NOP receptor agonist is selective for NOP and no other opioid receptors (Recker and Higgins 2004). Finally, a classic collection of studies on drug discrimination in receptor classification was reviewed by Rosecrans and Young (2017). In these studies, investigators demonstrated that the (−)-nicotine discriminative stimulus was blocked by antagonists such as mecamylamine and DHβE (dihydro-β-erythroidine), which demonstrated the roles of α4β2 nicotinic acetylcholine receptors in the brain and for underlying the discriminative stimulus effects of (−)-nicotine.

An additional requirement in the classification of drug-receptor interactions and the understanding of pharmacological action is the demonstration of stereoselectivity, sensitivity to time course, and pharmacokinetics to substitution patterns. For example, the stereoselectivity or time course for opioids (Butelman and Kreek 2016) and stimulants (Berquist and Fantegrossi 2017; Rosecrans and Young 2017) has long played an important role in determining patterns of stimulus substitution and discriminability for different training drugs. Interestingly, Negus and Banks (2016) actually use the relationship of pharmacokinetics (PK) to pharmacodynamics (PD) to analyze the variable relationship over time for the discriminative stimulus effects of cocaine and various metabolites which influences conclusions of drug action. This interesting PK/PD relationship allows a unique perspective of potential species differences in the discriminative stimulus effects of drugs.

Taken as a whole, the studies reviewed in this volume are just a fraction of the literature demonstrating the high receptor selectivity, stereoselectivity, and susceptibility to competitive antagonism for drugs trained as discriminative stimuli, the classic receptor pharmacology principles required to define a drug class. In the future, drug discrimination will still be needed to characterize new ligands, new enantiomers, and novel antagonists especially those agents with likely CNS activity. Although radioligand binding assays or functional GPCR assays are clearly the first steps to screen new compounds, a functional assay in a whole animal, such as drug discrimination, will always be needed to validate the results of more molecular characterizations.

2 Drug Discrimination as a Tool to Reveal Complex Cues and Pharmacological Actions

The early characterization of fentanyl as a discriminative stimulus and the corresponding receptor neuropharmacology of this direct acting opioid agonist (Colpaert 2011) led to using training drugs with more indirect or unique mechanisms of action as discriminative stimuli. Indeed, drug discrimination studies were key in distinguishing potential underlying neural mechanisms. For example, drug discrimination studies differentiated the stimulus effects of PCP (phencyclidine) and MK-801 (dizocilpine) as noncompetitive NMDA antagonists as opposed to direct acting NMDA (N-methyl-D-aspartate) receptor antagonists revealing a complex or compound cue involving the regulation of dopaminergic and serotoninergic systems with sigma1 receptor function likely involved (Mori and Suzuki 2016). Psychostimulants, such as cocaine, amphetamine, and more recently synthetic cathinones that possess a mix of transporter inhibition, release, or reverse transporters have been trained as discriminative stimulus and reviewed in this volume (Berquist and Fantegrossi 2017). Drug discrimination techniques can be very useful for studying and classifying opioids with complex pharmacology at multiple receptors, as these can vary significantly across species due to likely different receptor proportions or signaling across species (e.g., Zhu et al. 1997). Drug discrimination techniques have been critical for understanding the role of endogenous cannabinoids and their various metabolic activities and for the pharmacological effects of phytocannabinoids and the synthetic cannabinoid agents (Wiley et al. 2016). For example, the complexity that can be revealed by training metabolic enzyme inhibitors as a discriminative stimuli to tap into endocannabinoid function was recently demonstrated by training SA-57, a dual fatty acid amide hydrolase (FAAH) and mono-acylglycerol lipase101 (MAGL) inhibitor and by training selective MAGL inhibitor MJN110. Using the patterns of substitution for other dual FAAH/MAGL, MAGL, or FAAH inhibitors to substitute for MJN110 as well as cannabinoid agonists, these authors suggest that the MJN110 discriminative stimulus through selective MAGL inhibition is mediated through 2-AG-mediated stimulation of CB1 receptors. Furthermore, under normal endogenous conditions, MAGL may reduce endocannabinoid-mediated overstimulation of the CB1 receptor, thereby preventing induction of a cannabimimetic subjective state (Owens et al. 2017). This example and others reviewed in this volume highlight the manner in which investigators can use drug discrimination techniques to enhance our understanding of the roles of endogenous regulators of drug action.

There are numerous examples highlighted in the current volume that reveal certain drugs can have compound pharmacological cues and drug discrimination methodology has been used to dissect the relative contributions of each component. For example, cocaine, scopolamine, and D_1 and D_2 agonists substitute for the bupropion discriminative stimulus, and these effects were either fully or partially blocked by DA receptor antagonists (Prus and Porter 2016). Similarly, the discriminative stimulus effects of competitive and noncompetitive NMDA receptor antagonists tap into dopaminergic and serotonergic systems as well as sigma-1 receptor actions suggesting that training these agents can result in a compound cue (Mori and Suzuki 2016). Inhalants as a class of discriminative stimuli also fall into the category of interacting with multiple receptor systems such as $GABA_A$-positive modulators and NMDA, for example, depending on the particular inhalant trained as the discriminative stimulus (Shelton 2016). Using drug discrimination to characterize the inhalants allows an investigator to meaningfully group these substance inhalants together despite being such a heterogeneous pharmacological group. The most studied complex discriminative stimulus is ethanol in which GABA (*gamma-aminobutyric acid*) and glutamate ionotropic receptors and serotonergic mechanisms all contribute to the discriminative stimulus effects especially dependent on training dose (Allen et al. 2017). Indeed, there have been clever control experiments designed to separate exteroceptive vs. interoceptive cue components such as route of administration studies to eliminate odor as providing a key role in the discriminative stimulus effects of toluene as reviewed by Shelton (2016).

Interestingly, drug discrimination procedures can be modified to further separate out complex cues for drugs with overlapping pharmacological mechanisms by training dose-dose or three-choice discriminations. For example, Berquist and Fantegrossi (2017) nicely review the usefulness of three-choice discriminations, especially for analyzing the effects of enantiomers in the substitution patterns of MDMA (3,4-methylenedioxymethamphetamine, commonly known as ecstasy). Three-choice discriminations for MDMA, saline, and d-amphetamine reveal a likely serotoninergic-dopaminergic continuum for the underlying neuropharmacological mechanisms of MDMA (Harper et al. 2011; Goodwin and Baker 2000) based on substitution patterns of different psychostimulants and doses. Three-choice discriminations can also be established with high and low doses of drugs to parcel out the role of efficacy in discriminative stimulus effects (e.g., Jones et al. 1999; Vanecek and Young 1995). Leveraging different mouse strains to further triangulate on components of a complex discriminative stimulus such as clozapine has been a fruitful strategy (Porter et al. 2017) essentially similar to varying a training dose. Narrowing the conditions under which generalization will occur with each new cue or dose that can be trained is a sophisticated strategy to dissect out pharmacological mechanisms under particular contingencies and may explain individual subject substitution patterns.

The observation that individuals can attend to one component of a complex cue more than others has precedence in the literature. In a classic experiment, Reynolds (1961) demonstrated that when two individual pigeons were trained to respond in the presence of a white triangle on a red key and tested with either the triangle or red background alone, one pigeon exclusively attended to the triangle while the other the red background. Drugs with multiple pharmacological components could certainly serve similar functions in individual subjects so that in a group of subjects, some could attend more to one component of the complex stimulus or the other or perhaps even only the Gestalt of the multiple components together. Possibly, component pharmacology or cues could be a contributing factor to some of the inter-subject variability obtained in drug discrimination experiments and one of the reasons examining a pattern of substitution and antagonism in individual subjects is an important part of data analysis in this field. Indeed, this notion has been well-studied by researchers investigating mixtures of drugs (e.g., Stolerman et al. 1999).

3 Drug Discrimination to Study Internal States

Whereas the use of drug discrimination to understand contributions of complex underlying pharmacological mechanisms to drug effects has been invaluable to researchers, one may argue that the ability of discrimination methodologies to tap into the various interoceptive effects of drug stimuli that control behavior makes it a unique procedure without parallel. As described in the first chapter of this volume, drug discrimination grew out of the interest in the effects of drugs on memory retrieval and state-dependent learning (Porter et al. 2018). Two examples of "states" produced by drugs, or the withdrawal of drugs, are worth mentioning because these examples reveal what is especially novel about the results from drug discrimination studies. Rosecrans and Young (2017) reviewed a study in which rats were trained to discriminate pentylenetetrazol from saline and suggested that the basis for the discrimination was pentylenetetrazol-induced anxiety (Harris et al. 1986). When the pentylenetetrazol-trained rats were administered high doses of nicotine for a 3-week period and then were withdrawn from nicotine dosing, the rats responded partially on the pentylenetetrazol-appropriate lever 24 h after the cessation of dosing. These investigators suggested that rats in nicotine withdrawal may be experiencing "anxiety" as measured by their pentylenetetrazol generalization response. The possibility that pentylenetetrazol as a discriminative stimulus may represent a state akin to anxiety in animals was followed up with additional pharmacological characterization (Jung et al. 2002), and ethologically relevant drug discrimination experiments demonstrating an interoceptive state associated with species-specific defense reactions in rats produced by exposure to cat predators were similar to the discriminative stimulus cues produced by pentylenetetrazol (Gauvin and Holloway 1991).

Other withdrawal states have been modelled in drug discrimination, including those from repeated agonist administration followed up by later discrimination training sessions with antagonists. Excellent examples include experiments where opioid withdrawal substitutes for the discriminative stimulus effects of naltrexone (e.g., Becker et al. 2008) or partial agonist nalbuphine (Walker et al. 2004) and THC withdrawal substitutes for the discriminative stimulus effects of cannabinoid antagonist rimonabant (e.g., Stewart and McMahon 2010). Peptides and drugs with potential anorexic effects have been tested in rats trained to discriminate between 22- and 2-h food deprivations, a methodology of studying the internal state of "hunger" (Jewett et al. 2006, 2009).

Antagonists in general can be difficult to train as discriminative stimuli although there is a long history of training and testing antipsychotic agents (Prus and Porter 2016; Porter 2011) and noncompetitive NMDA antagonists (Balster 1991; Koek 1999). Often many of these antagonists reveal complex, compound cues which may or may not be reversed by agonist administration and the cue may depend on the species studied (Porter 2011). For some drug classes, modifications of procedures are employed such as maintaining the subjects dependent on an agonist as described above. The maintenance of a subject on chronic agonist treatment induces a certain change in homeostasis or an increase in endogenous tone that can be disrupted with antagonists or drug withdrawal. Another modification of the drug discrimination assay to train antagonists such as phencyclidine, diprenorphine, naloxone, naltrexone, and rimonabant as discriminative stimuli without chronic agonist treatment is the conditioned taste aversion methodology reviewed by Riley et al. (2016). One possibility for the establishment of antagonists as discriminative stimuli to control behavior has been suggested to be the disruption of an endogenous tone by the antagonist. In drug-naïve subjects, one might simply suspect basal endogenous tone would be the same after the injection of a given dose of antagonist irrespective of the training procedure. Yet, antagonists can easily serve as discriminative stimuli to control taste aversion learning at lower doses than previously attempted using operant-based training techniques, and these antagonist doses can be trained much more quickly using conditioned taste aversion. These studies demonstrate that the discriminative stimulus properties of a drug are not inviolate properties of the pharmacology but more so intimately tied to the training conditions and predictive consequences of that discriminative stimulus.

In humans, investigators are able to compare subjective effect questionnaires to the results obtained from drug discrimination assays allowing for an assessment of whether drug discrimination is a model of subjective effects. Overall, there is a relatively good correspondence between the discriminative stimulus and subjective effects in humans across the different pharmacological classes; however, there are some interesting exceptions. Bolin et al. (2016) provide an interesting discussion regarding the face validity and some potential limitations of drug discrimination procedures in humans for studying the abuse potential of drugs (see also McMahon 2015). For example, drug discrimination in humans is relatively insensitive to circulating blood levels of drug such that the time course of the discriminative stimulus effects, or the proportion of responses to the drug-appropriate option, does not always follow the measured blood levels (Kelly et al. 1997). Although we believe that humans are able to articulate the stimuli that may be controlling their behavior, this is probably an overstatement. For example, in humans responding to receive i.m. injections of morphine, much lower doses of morphine were self-administered as compared to those doses that occasioned positive reports of subjective drug effects (Lamb et al. 1991). The notion that to be a discriminative stimulus, a drug must produce something akin to a subjective effect leaves out some discriminative stimuli that likely do not possess strong subjective effects. For example, MAO inhibitors such as iproniazid, nialamide, phenelzine, and tranylcypromine can be discriminated using a T-maze procedure (Overton 1982), and Ca++ channel blockers can be discriminative stimuli in traditional operant procedures (Schechter 1995) when these agents are not likely to have what would be considered strong subjective effects. Finally, the observation that antidepressants can be trained in rats and mice that are not depressed suggests that the underlying pharmacology of these agents interacts with underlying basal states to support a salient enough stimulus to control behavior (Prus and Porter 2016) and reveal how the drug discrimination procedure is an exceedingly sensitive methodology.

4 In Praise of Drug Discrimination

As outlined in the many chapters of this volume, there are few experimental models we have available today that are as pharmacologically selective, sensitive, and such an objective measure an interoceptive state in an organism. As Berquist and Fantegrossi (2017) state in the current volume, "Nevertheless, the drug discrimination assay, in its most basic form, reveals pharmacological effects that occur within the central nervous system in species that display little to no verbal communication. We consider this an achievement in scientific research in general, and we submit that the drug discrimination approach is among the most useful in vivo analyses available to behavioral pharmacology." Drug discrimination is essentially unchallenged as a method to characterize drug stimuli and resulting behavior. Even with the advanced technologies available today, the ability to study pharmacologically and disease-relevant doses with such specificity in a preclinical experiment is readily available using drug discrimination. Drug discrimination will likely continue to contribute to our understanding of drug-receptor interactions and basic pharmacological characterization in combination with other technologies such as imaging, optogenetics, gene delivery strategies, RNA interference technology, and designer receptors exclusively activated by designer drugs (DREADD)-based chemogenetic tools. All of these more recent technologies provide exquisite detail on molecular and cellular signaling and brain circuitry; however, to deliver a representation of either drug stimuli or internal states of physiology, a particular cue will have to be specifically trained in an experimental animal. For any question that requires a functional output and a precise, selective pharmacological result, drug discrimination will always be the answer. The only limitation is our creativity.

References

Allen DC, Ford MM, Grant KA (2017) Cross-species translational findings in the discriminative stimulus effects of ethanol. Curr Top Behav Neurosci. https://doi.org/10.1007/7854_2017_2

Balster RL (1991) Discriminative stimulus properties of phencyclidine and other NMDA antagonists. In: Glennon RA, Järbe TUC, Frankenheim J (eds) Drug discrimination: applications to drug abuse research. NIDA Research Monograph 116. US Government Printing Office, Washington, pp 163–180

Becker GL, Gerak LR, Koek W, France CP (2008) Antagonist-precipitated and discontinuation-induced withdrawal in morphine-dependent rhesus monkeys. Psychopharmacology 201(3): 373–382

Berquist MD 2nd, Fantegrossi WE (2017) Discriminative stimulus effects of psychostimulants. Curr Top Behav Neurosci. https://doi.org/10.1007/7854_2017_5

Bolin BL, Alcorn JL 3rd, Reynolds AR, Lile JA, Stoops WW, Rush CR (2016) Human drug discrimination: elucidating the neuropharmacology of commonly abused illicit drugs. Curr Top Behav Neurosci. https://doi.org/10.1007/7854_2016_10

Butelman ER, Kreek MJ (2016) Discriminative stimulus properties of opioid ligands: progress and future directions. Curr Top Behav Neurosci. https://doi.org/10.1007/7854_2016_9

Colpaert FC (2011) Drug discrimination: a perspective. In: Glennon R, Young R (eds) Drug discrimination applications to medicinal chemistry and drug studies. Wiley, Hoboken

Gauvin DV, Holloway FA (1991) Cross-generalization between an ethologically [correction of ecologically] relevant stimulus and a pentylenetetrazole-discriminative cue. Pharmacol Biochem Behav 39(2):521–523

Goodwin AK, Baker LE (2000) A three-choice discrimination procedure dissociates the discriminative stimulus effects of d-amphetamine and (±)-MDMA in rats. Exp Clin Psychopharmacol 8:415–423

Harper DN, Crowther A, Schenk S (2011) A comparison of MDMA and amphetamine in the drug discrimination paradigm. Open Addiction J 4:22–23

Harris CM, Emmett-Oglesby MW, Robinson NG, Lal H (1986) Withdrawal from chronic nicotine substitutes partially for the interoceptive stimulus produced by pentylenetetrazol (PTZ). Psychopharmacology 90(1):85–89

Jewett DC, Lefever TW, Flashinski DP, Koffarnus MN, Cameron CR, Hehli DJ, Grace MK, Levine AS (2006) Intraparaventricular neuropeptide Y and ghrelin induce learned behaviors that report food deprivation in rats. Neuroreport 17(7):733–737

Jewett DC, Hahn TW, Smith TR, Fiksdal BL, Wiebelhaus JM, Dunbar AR, Filtz CR, Novinska NL, Levine AS (2009) Effects of sibutramine and rimonabant in rats trained to discriminate between 22- and 2-h food deprivation. Psychopharmacology 203(2):453–459

Jones HE, Bigelow GE, Preston KL (1999) Assessment of opioid partial agonist activity with a three-choice hydromorphone dose-discrimination procedure. J Pharmacol Exp Ther 289(3): 1350–1361

Jung ME, Lal H, Gatch MB (2002) The discriminative stimulus effects of pentylenetetrazol as a model of anxiety: recent developments. Neurosci Biobehav Rev 26(4):429–439

Kelly TH, Emurian CS, Baseheart BJ, Martin CA (1997) Discriminative stimulus effects of alcohol in humans. Drug Alcohol Depend 48(3):199–207

Koek W (1999) N-methyl-D-aspartate antagonists and drug discrimination. Pharmacol Biochem Behav 64:275–281

Lamb RJ, Preston KL, Schindler CW, Meisch RA, Davis F, Katz JL, Henningfield JE, Goldberg SR (1991) The reinforcing and subjective effects of morphine in post-addicts: a dose-response study. J Pharmacol Exp Ther 259(3):1165–1173

McMahon LR (2015) The rise (and fall?) of drug discrimination research. Drug Alcohol Depend 151:284–288

Mori T, Suzuki T (2016) The discriminative stimulus properties of hallucinogenic and dissociative anesthetic drugs. Curr Top Behav Neurosci. https://doi.org/10.1007/7854_2016_29

Negus SS, Banks ML (2016) Pharmacokinetic-pharmacodynamic (PKPD) analysis with drug discrimination. Curr Top Behav Neurosci. https://doi.org/10.1007/7854_2016_36

Overton DA (1982) Comparison of the degree of discriminability of various drugs using the T-maze drug discrimination paradigm. Psychopharmacology 76:385–395

Owens RA, Mustafa MA, Ignatowska-Jankowska BM, Damaj MI, Beardsley PM, Wiley JL, Niphakis MJ, Cravatt BF, Lichtman AH (2017) Inhibition of the endocannabinoid-regulating enzyme monoacylglycerol lipase elicits a CB1 receptor-mediated discriminative stimulus in mice. Neuropharmacology 125:80–86

Porter JH (2011) Discriminative stimulus properties of receptor antagonists. In: Glennon R, Young R (eds) Drug discrimination applications to medicinal chemistry and drug studies. Wiley, Hoboken, pp 287–322

Porter JH, Webster KA, Prus AJ (2017) Translational value of drug discrimination with typical and atypical antipsychotic drugs. Curr Top Behav Neurosci. https://doi.org/10.1007/7854_2017_4

Porter JH, Prus AJ, Overton DA (2018) Drug discrimination: historical origins, important concepts, and principles. Curr Top Behav Neurosci. https://doi.org/10.1007/7854_2018_40

Prus AJ, Porter JH (2016) The discriminative stimulus properties of drugs used to treat depression and anxiety. Curr Top Behav Neurosci. https://doi.org/10.1007/7854_2016_27

Recker MD, Higgins GA (2004) The opioid receptor like-1 receptor agonist Ro 64-6,198 (1S,3aS-8-2,3,3a,4,5,6-hexahydro-1H-phenalen-1-yl-1-phenyl-1,3,8-triaza-spiro[4.5]decan-4-one) produces a discriminative stimulus in rats distinct from that of a mu, kappa, and delta opioid receptor agonist cue. J Pharmacol Exp Ther 311:652–658

Reynolds GS (1961) Attention in the pigeon. J Exp Anal Behav 4:203–208

Riley AL, Clasen MM, Friar MA (2016) Conditioned taste avoidance drug discrimination procedure: assessments and applications. Curr Top Behav Neurosci. https://doi.org/10.1007/7854_2016_8

Rosecrans JA, Young R (2017) Discriminative stimulus properties of S(−)-nicotine: "a drug for all seasons". Curr Top Behav Neurosci. https://doi.org/10.1007/7854_2017_3

Schechter MD (1995) Discriminative stimulus properties of isradipine: effect of other calcium channel blockers. Pharmacol Biochem Behav 50(4):539–543

Shelton KL (2016) Discriminative stimulus effects of abused inhalants. Curr Top Behav Neurosci. https://doi.org/10.1007/7854_2016_22

Stewart JL, McMahon LR (2010) Rimonabant-induced Delta9-tetrahydrocannabinol withdrawal in rhesus monkeys: discriminative stimulus effects and other withdrawal signs. J Pharmacol Exp Ther 334(1):347–356

Stolerman IP, Mariathasan EA, White JA, Olufsen KS (1999) Drug mixtures and ethanol as compound internal stimuli. Pharmacol Biochem Behav 64(2):221–228

Vanecek SA, Young AM (1995) Pharmacological characterization of an operant discrimination among two doses of morphine and saline in pigeons. Behav Pharmacol 6(7):669–681

Walker EA, Picker MJ, Granger A, Dykstra LA (2004) Effects of opioids in morphine-treated pigeons trained to discriminate among morphine, the low-efficacy agonist nalbuphine, and saline. J Pharmacol Exp Ther 310(1):150–158

Wiley JL, Owens RA, Lichtman AH (2016) Discriminative stimulus properties of phyto-cannabinoids, endocannabinoids, and synthetic cannabinoids. Curr Top Behav Neurosci. https://doi.org/10.1007/7854_2016_24

Zhu J, Luo LY, Li JG, Chen C, Liu-Chen LY (1997) Activation of the cloned human kappa opioid receptor by agonists enhances [35S]GTPgammaS binding to membranes: determination of potencies and efficacies of ligands. J Pharmacol Exp Ther 282(2):676–684